Recent Advances in Engineering Fluid Dynamics

Recent Advances in Engineering Fluid Dynamics

Editor: Henley Fisher

NY Research Press
New York

Published by NY Research Press
118-35 Queens Blvd., Suite 400,
Forest Hills, NY 11375, USA
www.nyresearchpress.com

Recent Advances in Engineering Fluid Dynamics
Edited by Henley Fisher

International Standard Book Number: 978-1-64725-464-3 (Hardback)

Cataloging-in-Publication Data

Recent advances in engineering fluid dynamics / edited by Henley Fisher.
p. cm.
Includes bibliographical references and index.
ISBN 978-1-64725-464-3
1. Fluid dynamics. 2. Fluid mechanics. 3. Mechanics, Applied. I. Fisher, Henley.
TA357 .R43 2023
620.106--dc23

Contents

Preface

Over the recent decade, advancements and applications have progressed exponentially. This has led to the increased interest in this field and projects are being conducted to enhance knowledge. The main objective of this book is to present some of the critical challenges and provide insights into possible solutions. This book will answer the varied questions that arise in the field and also provide an increased scope for furthering studies.

Fluid dynamics is an area of fluid mechanics that deals with the movement of liquids and gases. It has many applications in engineering as well as non-engineering domains such as identifying the mass flow rate of petroleum through pipelines, modeling detonation of fission weapon, calculating moments and forces on aircraft, and predicting weather patterns. It provides a systematic structure and uses empirical and semi-empirical laws obtained from flow measurement for solving various practical problems. Some of the significant technological uses of fluid dynamics include oil pipelines, rocket engines, air conditioning systems, and wind turbines. Experimental fluid dynamics (EFD) and computational fluid dynamics (CFD) techniques are widely used across various branches of engineering. This book provides comprehensive insights into the recent advances in the field of engineering fluid dynamics. It presents researches and studies performed by experts across the globe. Scientists and students actively engaged in this field will find this book full of crucial and unexplored concepts.

I hope that this book, with its visionary approach, will be a valuable addition and will promote interest among readers. Each of the authors has provided their extraordinary competence in their specific fields by providing different perspectives as they come from diverse nations and regions. I thank them for their contributions.

Editor

1

Comparison of Two Single Stage Low-Pressure Rotary Lobe Expander Geometries in Terms of Operation

Michalina Kurkus-Gruszecka * and Piotr Krawczyk

Institute of Heat Engineering, Faculty of Power and Aeronautical Engineering,
Warsaw University of Technology, Nowowiejska 21/25, 00-665 Warsaw, Poland; piotr.krawczyk@pw.edu.pl
* Correspondence: michalina.kurkus-gruszecka.dokt@pw.edu.pl

Abstract: In the article the computational fluid dynamics (CFD) simulation and calculated operational parameters of the single stage low-pressure rotary lobe expander compared with the values obtained from a different geometry simulation are presented. Low-pressure rotary lobe expanders are rotary engines that use a compressed gas to produce mechanical energy, which in turn can be converted into another form, i.e., electric energy. Currently, expanders are used in narrow areas, but have a large potential in the energy production from gases of low thermodynamic parameters. The first geometry model was designed on the basis of an industrial device and validated with the empirical data. Simulation of the second geometry was made based on a validated model in order to estimate the operational parameters of the device. The CFD model included the transient simulation of compressible fluid in the geometry changing over time and the rotors motion around two rotation axes. The numerical model was implemented in ANSYS CFX software. After obtaining simulation results in the form of parameters monitors for each time step, a number of calculations were performed using a written code analysing the CFD program output files. The article presents the calculation results and the geometries comparison in terms of work efficiency. The research indicated that the construction of the device on a small scale could cause a significant decrease in the aforementioned parameter, caused by medium leaks in the expander clearances.

Keywords: rotary lobe expander; low-pressure gas; transient CFD

1. Introduction

In the era of energy recovery, one of the main trends in addition to renewable energy development is the search for solutions that enable energy generation from waste sources. There are many technological solutions enabling such energy production, e.g., direct combustion, pyrolysis, fermentation, conventional, and plasma gasification [1]. All of the previously mentioned technologies lead to thermal energy through the final combustion process of solid, liquid, and gas fuels [2]. In addition to the previously mentioned forms of energy recovery, there is also the possibility of the direct use of waste gases from industrial technological processes. At large industrial plants built, there are gas discharge stations with relatively low thermodynamic parameters. Previously, for economic purposes, the release of waste gases was considered reasonable. However, all effort is being made today to increase the energy efficiency of industrial processes and reduce energy wastage [3,4]. There are several device types that can be applied for aforementioned systems: Dynamic expanders, i.e., axial, cantilever, radial [5,6], or volumetric expanders i.e., piston, screw, vane, rotary lobe, and scroll [7–11]. Each of the devices has a different operating characteristic, which means it can be adapted to the system in terms of performance, operation time, start-up frequency, and the thermodynamic parameters of the working medium. A comprehensive comparison of expanders with an attempt to identify the best, in terms of power generation below 100 kW is presented in [6]. The author points out the pros and cons of selected devices, e.g., a decrease in the efficiency of turbine expanders along with a decrease in power, high unit investment cost of screw expanders, or lubrication and friction losses of vane expanders. Unlike turbines in a higher power range, despite 30 years of research and experience, the best technology of low-power expanders has still not clearly emerged [5]. This fact may partly be caused by the still low demand for low-power generators powered by waste sources, high unit cost, and a combination of pros and cons of each expanders type. Research and experiments of expanders are still being carried out [5,9–17] in order to, for example, optimize the geometry. However, they are associated with relatively high costs of constructing the device often related to the required manufacturing precision [6]. There are also computational fluid dynamics (CFD) calculations on the expanders' operation [5,17,18], but since most of them are performed for time-varying geometry, these calculations are labour-intensive and require a lot of computing power. Furthermore, they should be validated with an operating device.

This work concerns rotary lobe expanders which are volume-type ones, where the torque is generated by two rotors whose power is transferred to the output shaft located in the axis of one of the rotors. The rotors are connected through synchronized gear which enables operation without friction. They can be used to produce mechanical or electrical energy from low-pressure waste gases or as part of a system with thermal energy obtained in the waste incineration process [6]. Rotary lobe expanders show a number of advantages that led to this expander type selection and to a series of CFD simulations afterwards. An additional feature considered by the authors is the presence of this type commercial devices [14,19–21] enabling the validation of the CFD model. To this day, rotary lobe expanders have been used for over 30 years as pneumatic motors in industry related to explosive environments, e.g., in mines or drilling platforms [21]. Since up to now mainly electric motors have been used in the considered power range, expander technology requiring the supply of compressed gas was only mechanically improved within the mentioned industry. However, after years of manufacturer's experience, the declared technological features can be considered reliable and encourage further research such as the optimization of the expander geometry. Rotary lobe expander features [19,21]:

- Both rotors move without physical contact, which eliminates friction and creates torque with mechanical efficiency exceeding 98%. It results in a long service life without maintenance and downtime;
- Minimal maintenance enhanced by hermetic bearings;
- The completely enclosed motor housing without vent holes allows for its usage in humid or polluted environments without the risk of corrosion inside the expander;
- Relatively low-pressure range of two to eight bars;
- Imperceptible vibration during operation, even at high speeds;
- Can be mounted in all directions with the shaft vertical or horizontal;
- Ability to stop under load;
- Work in hazardous condition provided by a sealed housing.

Expander modelling with the computational fluid dynamics methods involves the following aspects: Rotating machine modelling with computational domain geometry changing in time, transient case, and the main flow direction not being compatible with the rotors axes. The above-mentioned aspects combined may be the reason that CFDs of rotary lobe expanders have not been found in the literature. However, articles regarding models of devices with a similar type of operation or modelling methods have been found. The 3D semi-transient screw expander CFD analysis were conducted by Papes et al. [16]. The immersed solid method was introduced in rotary volumetric pumps and compressors by Voorde et al. at [22] and by Schiffer&Klomberg [23]. A single state Wankel expander CFD model was introduced in [17]. The screw rotors mathematical models are developed in articles [12,24,25] and patents [26–30]. The rotary vane expander CFD model with a customized grid generation methodology was developed by Binch et al. [31] and Montenego et al. [32,33].

This document is a continuation of work presented, among others in [18]. In the course of previous works, the CFD rotary lobe expander model was built based on industrial geometry. Then the series of simulations results were compared with empirical data. After a successful validation process, a new modified geometry was designed and CFD analysis was performed. Presented work was performed in order to select the appropriate expander geometrical parameters for the test stand. The new geometry model was subject to certain construction requirements such as expanders inlet pressure or rotational speed. The results comparison and analysis are presented in the graphical and tabular form.

2. Model Description

2.1. Flow Governing Equations

The numerical environment used for CFD calculations is based on continuity, momentum, energy, and state equations. The k-ω based Shear-Stress-Transport (SST) turbulence model was implemented. The k-ω SST model combines the advantages of the k-ε model and the k-ω model by introducing an additional factor limiting the overproduction of turbulence kinetic energy in areas of strong positive pressure gradients (accumulation points, boundary layer detachment areas) [34,35] that occur in the expander chamber near rotating lobes. The k-ε model well reflects turbulence in free flow and shear layers. Additionally, it has low sensitivity to inlet conditions for quantities describing turbulence. This is a desirable feature when creating the expander model because these quantities are not exactly known. The k-ω models turbulent flow in the boundary layer much better, which corresponds to areas near rotating lobes, while it is very sensitive to turbulent quantities in free flow (in the areas of inlet and outlet chambers), so the combination of features of both models is valuable in the expander model. The applied turbulence model is well described in [34,36–39]. The discrete form of the aforementioned equations has been implemented in the ANSYS CFX (18.2, ANSYS, Inc., Canonsburg, PA, USA) software, which was used to perform calculations.

The immersed solid model implemented in Ansys CFX software was also applied, which allows to model a transient state with solids that can move through a fluid domain. During simulation, CFX-Solver calculates which parts of the fluid domain coincide with the immersed solid and applies the momentum source to the fluid inside the immersed solid domain in order to force the flow to move along with the solid [35]. Additional conditions are applied at the boundary to improve flow behaviour near the wall of the immersed solid boundary. To calculate the immersed solid impact on the surrounding fluid the volume of fluid that corresponds to the solid volume is forced to move with the solid by using the source term in the momentum equation:

$$S_i = -\alpha C\left(U_i - U_i^F\right) \tag{1}$$

where $i = 1,2,3$, U_i, U_i^F are components of fluid velocity and components of forcing velocity due to the immersed solid, C, α are momentum source coefficient and momentum force scaling factor. The momentum source coefficient is evaluated as the average of the three diagonal coefficients in the momentum equation [36]. The momentum source scaling factor value is a balance between accuracy and robustness. A higher value leads to a more accurate solution but is less robust and gives convergence difficulties during calculations. The default value introduced by Ansys CFX is 10 and was increased to 25 in presented calculations as the relatively high value was obtained with satisfactory compliance.

2.2. Geometries

The geometers of the rotary lobe expanders consist of two rotors of various shapes rotating in opposite directions, enclosed in a stationary housing. The designed geometries with the rotation direction indicated are presented in Figure 1. The first geometry based on the commercial device was used to carry out the comparison with the geometry designed according to new criteria. As mentioned, the new geometry model was subject to certain construction requirements for the planned test stand, which have a lesser air consumption and speed ratio of 3000:1500 rpm resulting in a pitch diameter ratio of 2. The geometry parameters comparison is presented in Table 1. As easily seen, most of the significant parameters characterizing the compared geometries differ. In the previously modelled and validated geometry, which was reproduced on the basis of a commercial device, the smaller rotor was equipped with two lobes. This type of construction caused a maximum difference in pressure between the two halves of the lobe, which in turn resulted in a high value of the force acting on the smaller rotor. At the same time, a large pressure difference acting on a single lobe causes relatively large leaks on the radial expander clearance. Therefore, in the new geometry a larger number of smaller rotor lobes was implemented to prevent an excessive leakage of the medium. Another modification introduced in the new geometry was the change in the inclination of inlet and outlet channels.

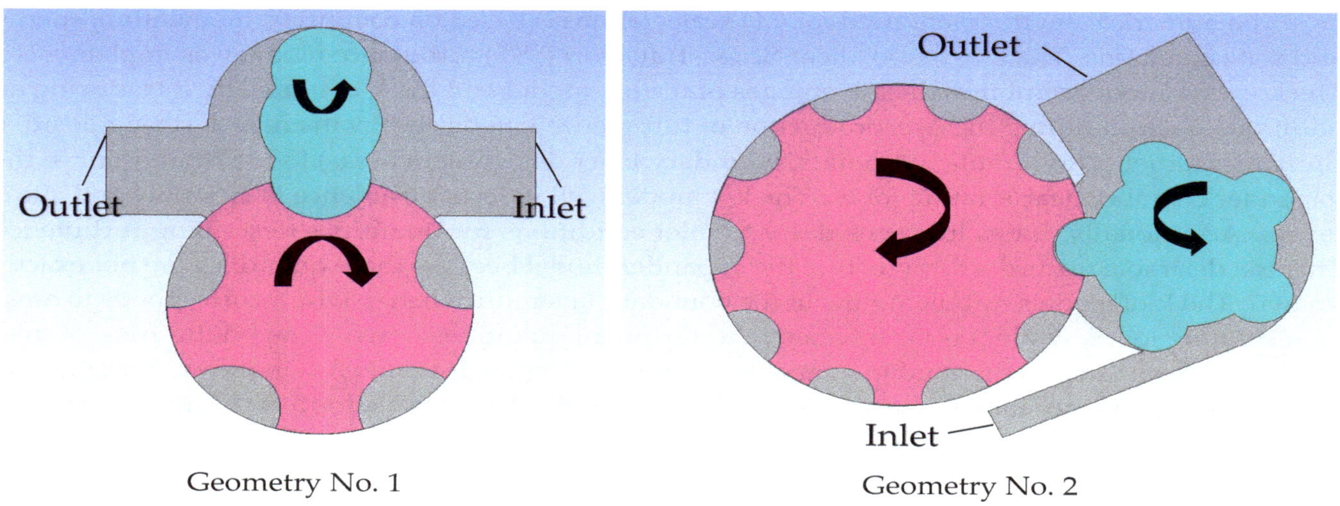

Figure 1. Rotary lobe expander geometries with inlets, outlets, and rotational directions as marked.

Table 1. Geometry parameters comparison.

Parameter	Unit	Geometry No. 1	Geometry No. 2
Number of lobes—smaller rotor	-	2	5
Number of lobes—larger rotor	-	5	10
Height of lobes—smaller rotor	mm	41.62	8.75
Height of lobes—larger rotor	mm	30.00	6.37
Angle between inlet and outlet channel	degrees	180	90
Inlet channel width	mm	60	8
Outlet to inlet channel width ratio	-	1	5
Inlet channel to radial clearances width ratio	-	3000	400
Radial clearance	mm	0.2	0.2

A commercial device that was the basis for building the original geometry is often used as a drive in mine lifts where its main advantage is the possibility of working in two directions. However, this feature meant that the inlet and outlet of the geometry No. 1 had to have the same cross-sections. Since the new expander geometry is designed for purposes other than those used so far, it was decided to design an outlet chamber with a larger cross-section, thus allowing a medium expansion without throttling. The second geometry dimensions are much smaller due to the design associated with the expander test installation construction planned for the future. Due to the computational cost reduction, the thickness of the calculation domain is 1 mm.

2.3. Meshes

As the immersed solid calculation method requires the use of separate grids for the expander chamber and rotors, three independent calculation grids have been created for each geometry—see

Figure 2. The calculation grids were modelled according to the meshes quality criteria (e.g., skewness, element quality, and aspect ratio). The grids density was selected by the gradients of parameters criteria such as speed, density, and pressure. It was generated particularly carefully near the clearance between the rotors and the expander chamber as well as in the area of rotors interlocking (Figures 3 and 4).

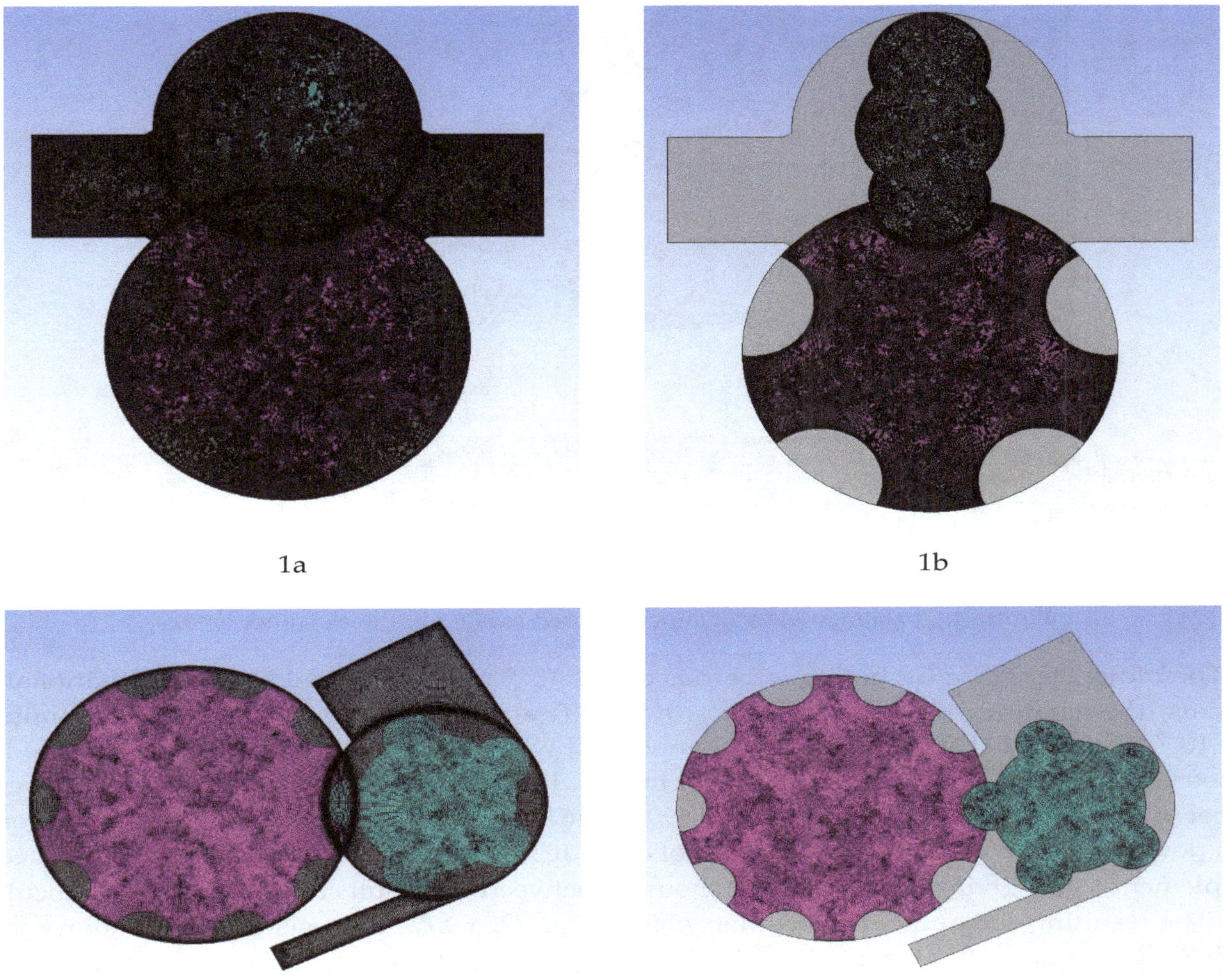

1a 1b

2a 2b

Figure 2. Computational grids of geometry No.1 (**1a**,**1b**) and geometry No. 2 (**2a**,**2b**). Fluid domains (**1a**,**2a**), rotors (**1b**,**2b**).

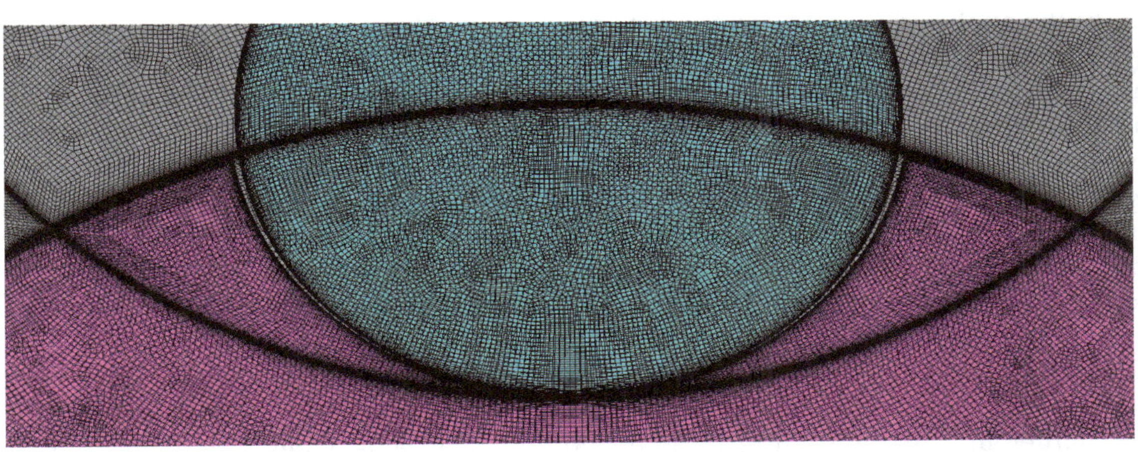

Figure 3. A zoomed view of superimposed meshes in the area of rotors interlocking.

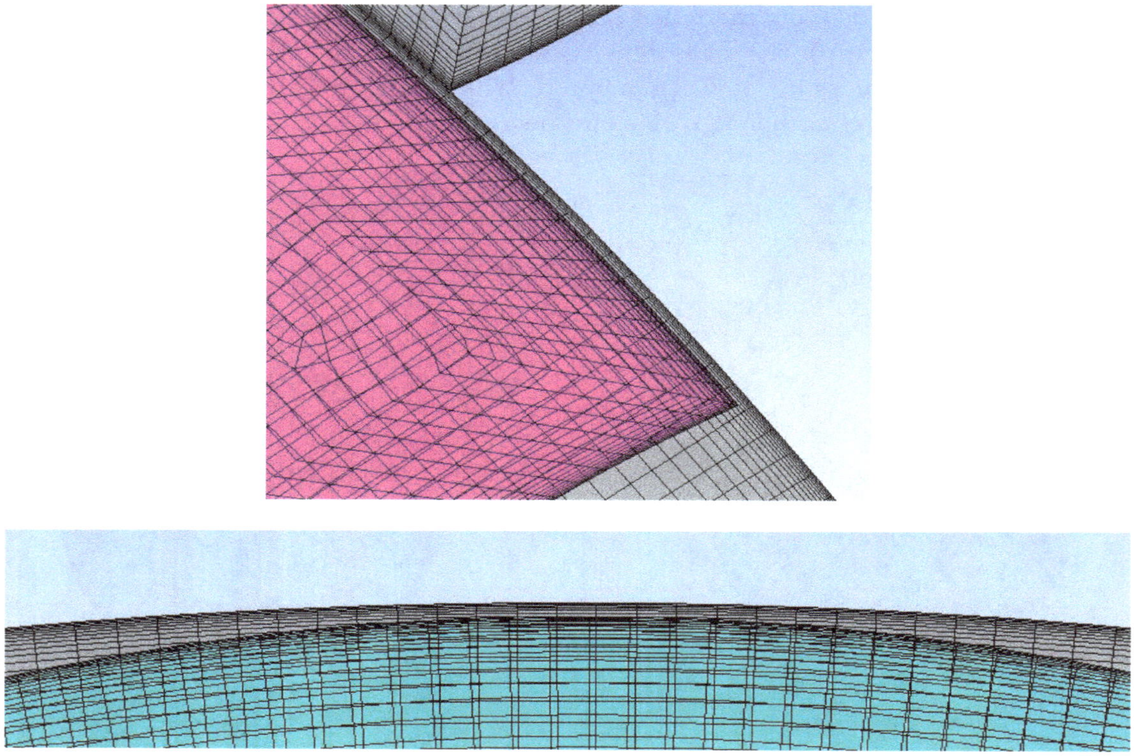

Figure 4. A zoomed view of superimposed meshes in the clearance area.

The radial clearance for both geometries of 2×10^{-4} m was divided by 7 computational grid elements, the smallest of which was 2×10^{-5} m. The maximum size of cell wall as limited to 2.5×10^{-7} m^2. Geometry No. 2 mesh was modelled with the same criteria and sizes as the No. 1 geometry. The sensitivity analysis of geometry the No. 1 mesh was carried out with the following total number of elements: 1,763,339; 1,356,414; 770,690; and 678,207 with the result presented as a percentage of the power obtained at the highest number of cells: 100%, 98%, 97%, and 92%. The authors decided to implement a grid giving a score of 97%, choosing between the results reliability and the calculation time. The resulting cell number of geometry No. 1 was 770,690 and consisted of domain rotor cells (495,548), as well as larger rotor cells (190,582) and smaller rotor cells (84,560). The cell number of geometry No. 2 domain, larger, and smaller rotor equalled respectively 335,735; 121,920, and 61,640.

2.4. Boundary Conditions and Simulations Settings

The immersed solid modelling method requires the definition of a fluid and the rigid solid domains. In the analysed case, the expander chamber was defined as a fluid domain and rotors as solids with a specified rotational speed. At the domain inlet, a boundary condition was defined as a constant pressure. At the outlet, a pressure boundary condition of 1 bar was applied. The outlet type was defined as "opening", which allows the medium to move back into the domain and the boundary condition for the remaining walls of the fluid domain was adopted as adiabatic. The gas model was implemented as air ideal gas with thermodynamic properties set according to the Ansys CFX programme library. In order to validate the numerical model, four simulations were carried out for various operational conditions and then compared with empirical data provided by the expander manufacturer. The important operational parameters that changed in the individual simulations are presented in the Table 2.

After the validation process, two more simulations were carried out, one with geometry No. 1 and the other with geometry No. 2. The rotational speed of the validated geometry was 1500 and 3750 rpm for the larger and smaller rotor respectively. Rotational speeds of new geometry rotors were 1500 and 3000 rpm for larger and smaller rotor. The difference in the selected rotational speeds results from the ratio of the diameters (and the number of lobes) of the larger and smaller rotor.

The key adopted solver settings in the ANSYS CFX are: Advection scheme: High resolution; Transient scheme: Second order backward Euler; and Residual type: Root mean square (RMS); Convergence criteria: 10^{-4}.

Table 2. Validation simulations input parameters.

Parameter	Unit	No. of Simulation			
		1	2	3	4
Speed of the larger rotating wheel	rev/min	1000	1000	1500	1500
Speed of the smaller rotating wheel	rev/min	2500	2500	3750	3750
Expanders inlet pressure	bar	4	8	4	3
Total simulation time	s	0.05	0.05	0.03	0.032
Time step	s	5×10^{-5}	5×10^{-5}	3×10^{-5}	3.2×10^{-5}
Number of time steps for a single simulation	-	1000	1000	1000	1000
Number of revolutions of the smaller rotor during simulation	-	2.083	2.083	1.875	2
Number of revolutions of the larger rotor during simulation	-	0.833	0.833	0.750	0.8

In order to calculate the expander power according to the method presented in the next chapter of this paper, monitors of appropriate values were set in each simulation. Mass flows at the inlet and outlet of the expanders as well as mean pressures on the surface of the halves of each lobes were monitored. Such criteria resulted in 16 monitors for validated geometry simulation and 32 monitors for the new geometry. Separated surfaces for pressure monitoring are shown in Figure 5. Values of crucial parameters were recorded at each timestep: Pressure, temperature, and gas velocity. Simulations were carried out using the Intel®Xeon®CPU E5-2600 2.20 GHz computing server. The time needed to perform a single simulation was ~26 hours with the utilization of 16 partition.

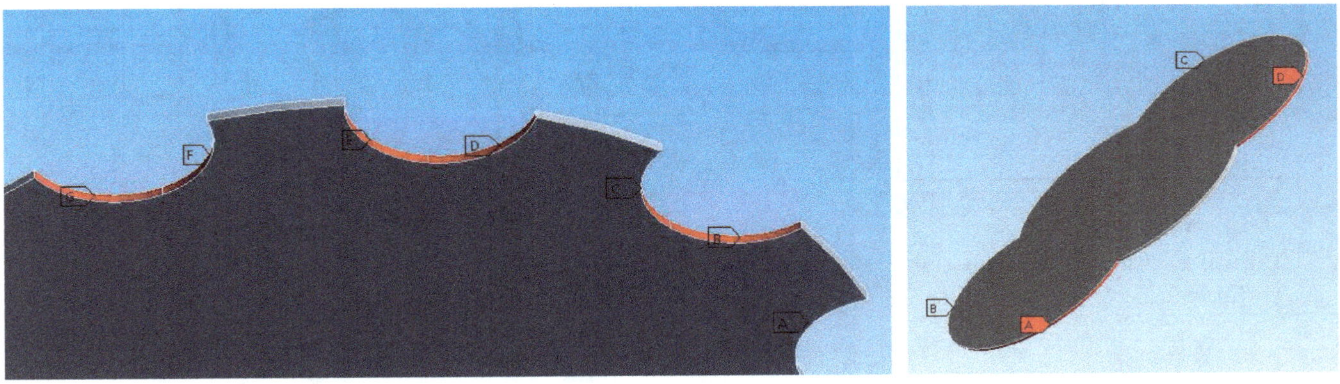

Figure 5. Separated surfaces for pressure monitoring at the halves of each rotor lobes.

3. Results, Validation, and Discussions

3.1. Post-Simulation Data Analysis

Since the simulations were completed, in order to obtain the power values of each of the expander, a series of calculations based on the monitor data were made. The calculations were carried out with a program created in the MATLAB (R2017b, MathWorks, Natick, MA, USA) environment. Firstly, the mean pressure values acting on the lobe halves for each time step from the ANSYS CFX output files were loaded. Examples of average pressure waveforms are presented in Figure 6. Afterwards, the pressure difference Δp acting on each lobe for each time step and both rotors was calculated according to the Equations (2)–(5):

$$\Delta p_{t,r,l} = p_{h1,t,r,l} - p_{h2,t,r,l} \tag{2}$$

$$t = 1, 2, 3 \ldots, T \tag{3}$$

$$l = 1, 2, 3 \ldots, L \tag{4}$$

where p_{h1} and p_{h2} are the average pressures acting on lobes halves and T, r, and L are the number of timesteps, rotor, and lobe. Then, each lobe active surface area A was calculated as a projection of the lobe half area on the surface passing through the rotor axis and through the lobe center of mass (see Figure 7). The force F with direction perpendicular to the rotors axes and the momentum M acting on each lobe were calculated according to these equations:

$$F_{t,r,l} = \Delta p A_r \tag{5}$$

$$M_{t,r,l} = F_{t,r,l} R \tag{6}$$

$$M_{t,r} = \sum_l M_{t,r,l} \tag{7}$$

$$P_{t,r} = M_{t,r} \Omega_r \tag{8}$$

where R is the arm of each lobe calculated as the distance from the rotor axis to the center of lobe weight. For each rotor and each time step, the moments acting on individual lobes were summed up (Equation (7)). The power was obtained by multiplying the momentum acting on each of the rotors with rotor angular velocity (Equation (8)).

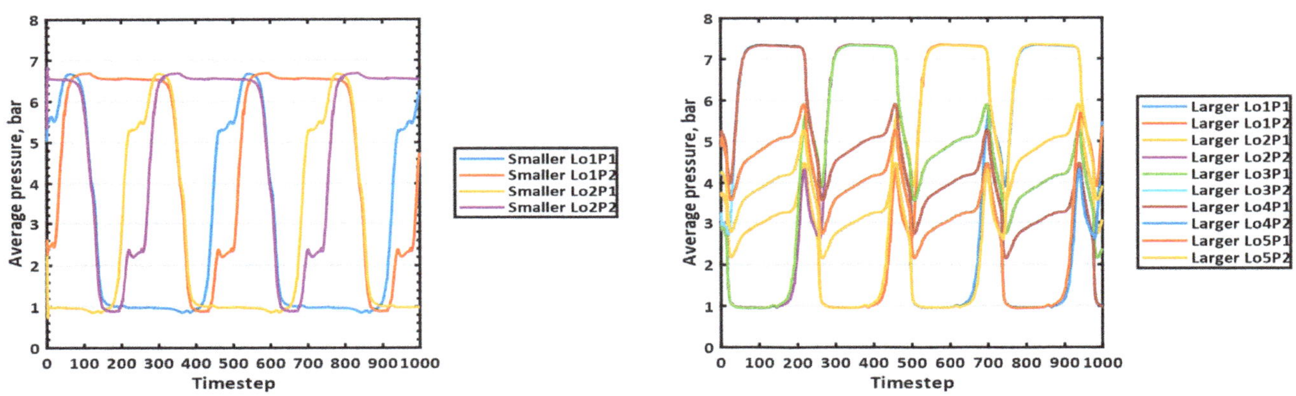

Figure 6. Average pressure waveforms for smaller and larger rotors. Simulation No. 2 according to Table 2.

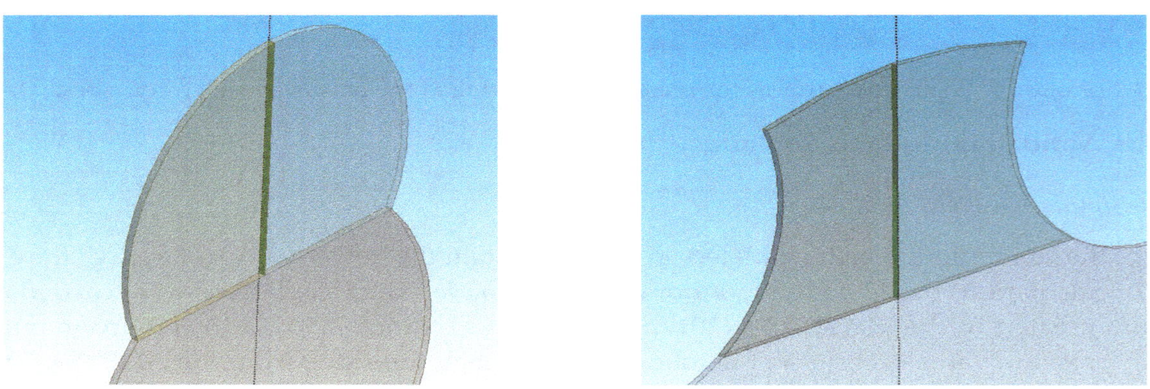

Figure 7. Lobe active surface determination. Active surface marked in green.

After performing the above operations, graphs of the expander power were obtained as a function of time steps (and rotor positions), see Figure 8. On their basis, the average power of the expanders was determined, which was reduced by the mechanical losses specified in Equation (9):

$$\dot{W} = \dot{W}_R \eta_g \eta_{R1} \eta_{R2} \eta_l, \tag{9}$$

where: $\dot{W}$—calculated power of the expander, $\dot{W}_R$—average calculated power on the rotors, η_g—mechanical efficiency of the gearing, η_{R1}—mechanical efficiency of the smaller rotor, and η_{R2}—mechanical efficiency of the larger rotor. Since the simulations were carried out for 1 mm of the expander thickness, the obtained power should be multiplied by the required axial dimension.

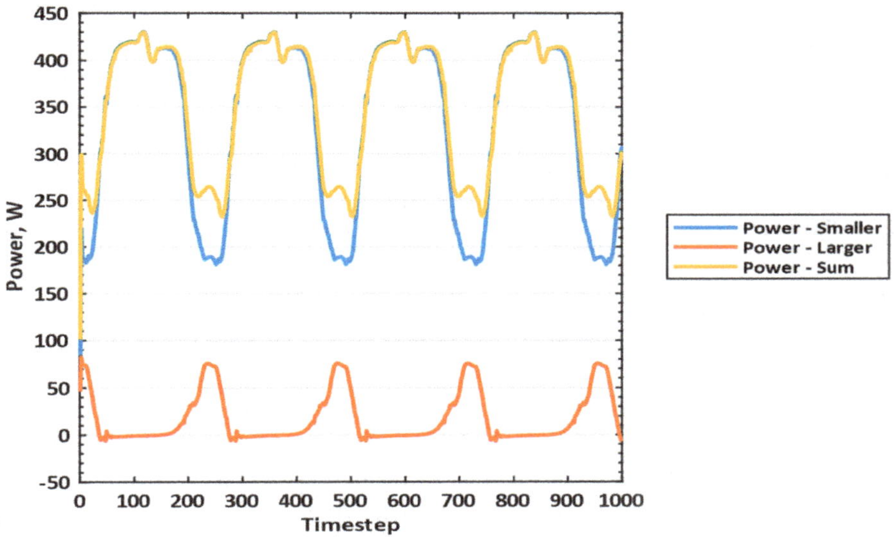

Figure 8. Power as a function of timesteps (rotor position). Simulation No. 2 according to Table 2.

3.2. Results and Validation

The average expander power calculated for geometry No. 1 compared with literature data is presented in the Table 3. The largest error occurred in the third simulation, in which the rotational speed was higher than in other simulations, 1500 and 3750 rpm, respectively, for the larger and smaller rotor. However, it was found that an incomplete expansion may occur at higher rotational speeds, therefore a test simulation was carried out assuming incomplete expansion of the working medium (i.e., up to 1.4 bar), and the results obtained were compared with the literature data. The result of this simulation is also shown in Table 3. Compared with the literature data, the calculation error was surprisingly small (0.18%), which can confirm the incomplete expansion occurring at the higher expander rotation speed.

Table 3. Results comparison with the literature data [21].

Parameter	Unit	No. of Simulation			
		1	2	3	4
Calculated average power of expander	kW	8.55	20.12	12.64	8.46
Expander's power according to literature data	kW	9.53	19.03	11.01	8.29
Error	%	10.28	5.73	14.80	2.05
Calculated power of expander with consideration of incomplete expansion	kW	-	-	10.99	-
Error for incomplete expansion	%	-	-	0.18	-

After analysing the results of the four conduced simulations, a satisfactory accuracy compared with real operational data was obtained. The maximum error was about 10%. The authors of this article concluded that the developed model was ready for use as a tool for conducting tests in terms of the expander geometry and parameters. Then two further simulations were carried out, the first using validated geometry and the second for new geometry No. 2. As the geometrical and operational parameters of the new geometry were imposed and they differ from the four simulations carried out previously, an additional simulation was carried out using geometry No. 1 and the model input parameters were selected as close as possible. The inlet pressure value for both simulations was 6 bar. Required rotational speeds for geometry No. 2 were 1500 and 3000 rpm for the larger and smaller rotor. As the same rotational speeds could not be obtained with geometry No. 1, (which results from the ratio of the rotor pitch diameters) the rotational speeds of 1500 and 3750 rpm were selected. A comparison of the rotary lobe expanders simulation results with two geometries, different rotational speeds, and different number of lobes cannot be unambiguous. Therefore, it was decided to compare the amount of gas consumed to the amount of energy produced. The comparison of calculation results (air consumption and output power) is presented in the Table 4.

Table 4. Results obtained with geometries No. 1 and 2 simulations comparison.

Parameter	Unit	Geometry No. 1	Geometry No. 2
Air consumption	l/s	5.52	0.81
Calculated power	W	19,059	575
Air consumption to output energy ratio	l/J	0.2896	1.4028

The simulation results are presented in the form of velocity vectors in Figures 9 and 10 and pressure projected on streamlines (Figure 11). Following conclusions can be presented after analysing the results. For the second geometry, a rapid decrease in efficiency was noticed. The second geometry was smaller, which with the same width of radial clearance caused a much larger share of gas leaks in the overall consumption. A several-fold decrease in efficiency which can be expressed by the ratio of air consumption to output energy was caused by high leaks. Indeed, the ratio of radial clearance to the width of the inlet channel was also several times lower for the second geometry (Table 1). This was caused by a minimum clearance limitation by the expander's performance which cannot be less than a fixed value of 0.2 mm.

The leaks can be noticed clearly on the velocity vectors (Figure 10) near the clearance between small rotor and expander's chamber as well as on the power trends (Figure 12). It can easily be seen that the power which is generated mainly by the pressure difference acting on the smaller rotor lobes in the case of the first geometry stays at a set high level during a part of rotation, while in the second geometry it dropped quickly after reaching the maximum value.

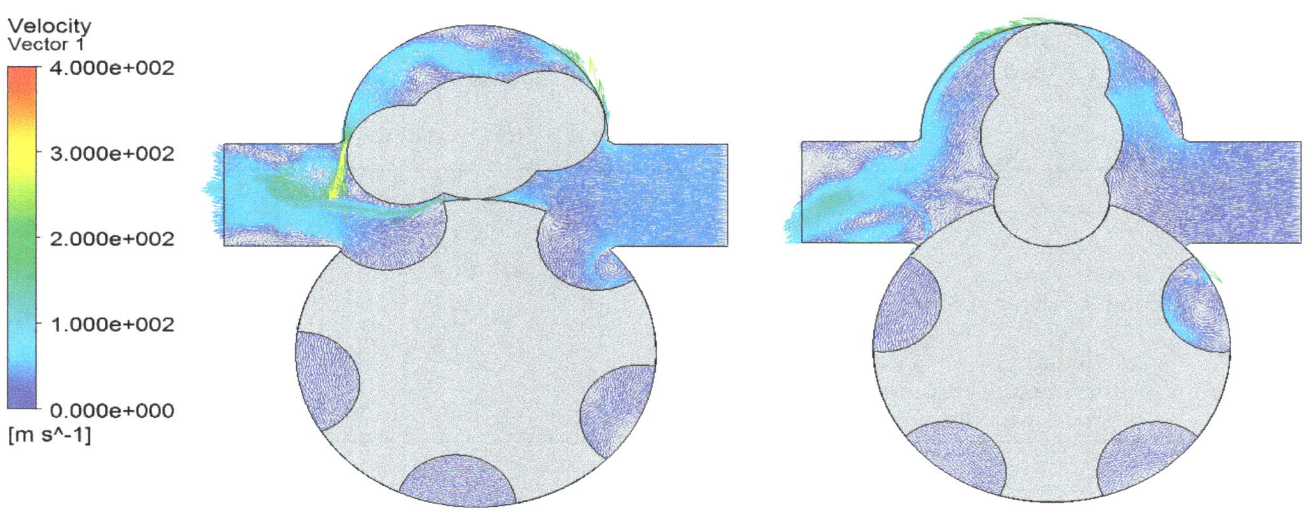

Figure 9. Velocity fields obtained as a result of geometry No. 1 simulation.

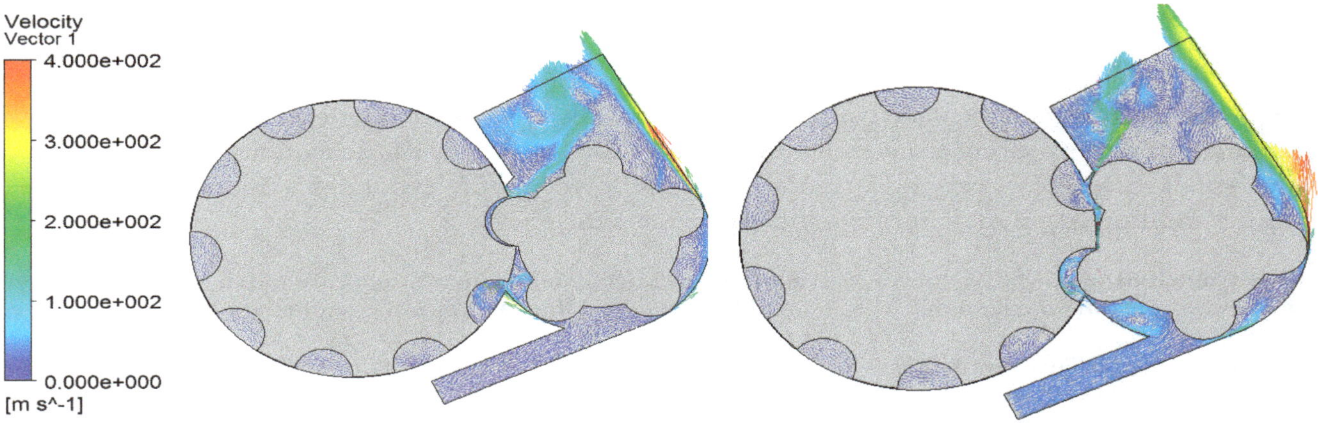

Figure 10. Velocity fields obtained as a result of geometry No. 2 simulation.

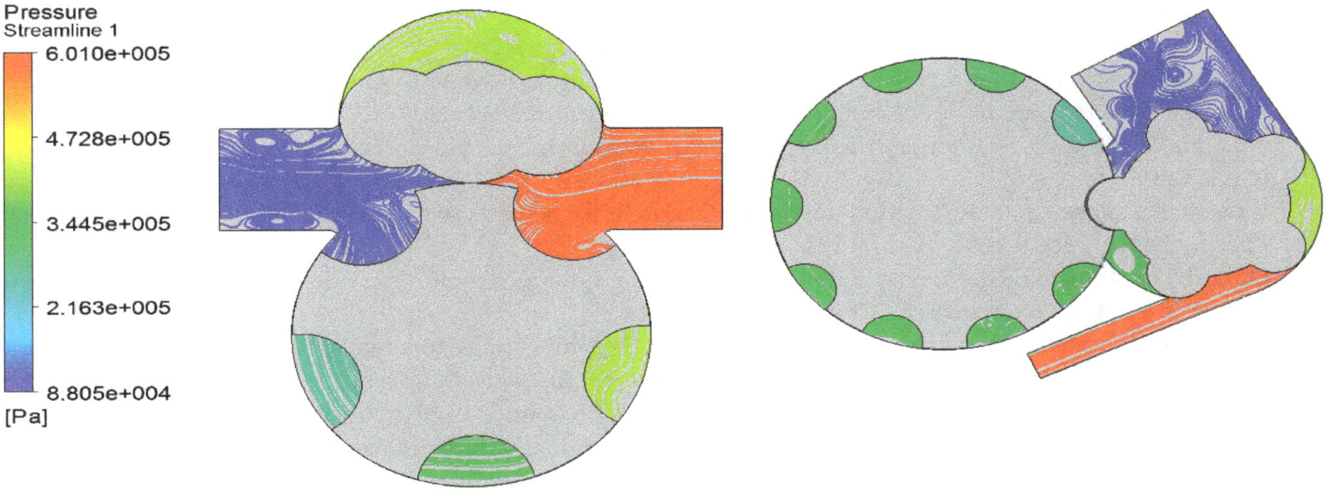

Figure 11. Pressure projected on streamlines.

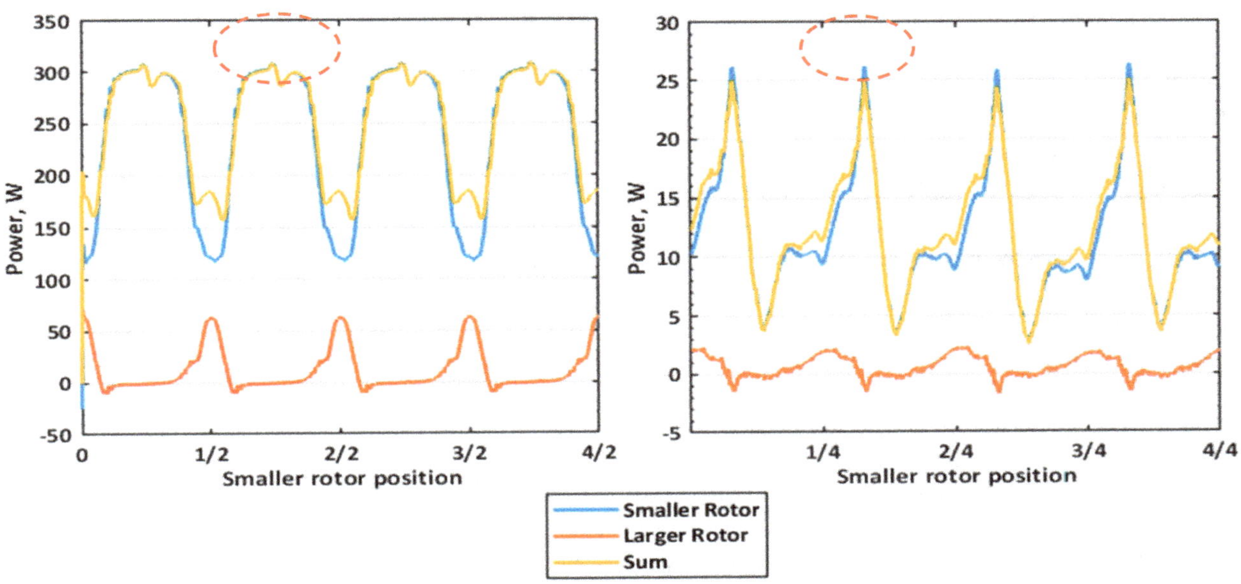

Figure 12. Power as a function of smaller rotor position with a visible leakage impact marked. Geometry No. 1 (left) and geometry No. 2 (right).

4. Conclusions

The conducted research indicates that the construction of the device on a small scale could cause a significant decrease in efficiency, caused by medium leaks in the expander clearances. The increase in the share of losses resulting from leaks undermines the legitimacy of the expander construction on such a small scale. However, the authors claim that lowering the expander's operating pressure or using an additional seal may improve the expanders efficiency.

Author Contributions: M.K.-G. and P.K. conceived and designed the mathematical model; M.K.-G. conceived and designed the numerical model; M.K.-G. and P.K. analysed the data; M.K.-G. wrote the paper.

Acknowledgments: This research is supported by the Programme of Research on the Electricity Sector which is co-financed by the European Union in the framework of the Smart Growth Operational Programme.

References

1. Department for Environment Food & Rural Affairs. *Energy from Waste A Guide to the Debate*; Department for Environment Food & Rural Affairs: London, UK, 2014; Volume 2014.
2. Gumisiriza, R.; Hawumba, J.F.; Okure, M.; Hensel, O. Biomass waste-to-energy valorisation technologies: A review case for banana processing in Uganda. *Biotechnol. Biofuels* **2017**, *10*, 11. [CrossRef] [PubMed]
3. Giacone, E.; Mancò, S. Energy efficiency measurement in industrial processes. *Energy* **2012**, *38*, 331–345. [CrossRef]
4. Worrell, E.; Laitner, J.A.; Ruth, M.; Finman, H. Productivity benefits of industrial energy efficiency measures. *Energy* **2003**, *28*, 1081–1098. [CrossRef]
5. Fiaschi, D.; Manfrida, G.; Maraschiello, F. Thermo-fluid dynamics preliminary design of turbo-expanders for ORC cycles. *Appl. Energy* **2012**, *97*, 601–608. [CrossRef]
6. Weiß, A.P. Volumetric Expander vs. Turbine-Which is the better choice for small ORCC Plants? In Proceedings of the 3rd International Seminar on ORC Power Systems, Brussels, Belgium, 12–14 October 2015.
7. Ghazanfarian, J.; Ghanbari, D. Computational Fluid Dynamics Investigation of Turbulent Flow inside a Rotary Double External Gear Pump. *J. Fluids Eng.* **2014**, *137*, 021101. [CrossRef]
8. Quoilin, S.; Lemort, V.; Lebrun, J. Experimental study and modeling of an Organic Rankine Cycle using scroll expander. *Appl. Energy* **2010**, *87*, 1260–1268. [CrossRef]
9. Qiu, G.; Shao, Y.; Li, J.; Liu, H.; Riffat, S.B. Experimental investigation of a biomass-fired ORC-based micro-CHP for domestic applications. *Fuel* **2012**, *96*, 374–382. [CrossRef]
10. Jradi, M.; Li, J.; Liu, H.; Riffat, S. Micro-scale ORC-based combined heat and power system using a novel scroll expander. *Int. J. Low Carbon Technol.* **2014**, *9*, 91–99. [CrossRef]
11. Kolasiński, P.; Blasiak, P.; Rak, J. Experimental and numerical analyses on the rotary vane expander operating conditions in a micro organic rankine cycle system. *Energies* **2016**, *9*, 606. [CrossRef]
12. He, W.; Wu, Y.; Peng, Y.; Zhang, Y.; Ma, C.; Ma, G. Influence of intake pressure on the performance of single screw expander working with compressed air. *Appl. Therm. Eng.* **2013**, *51*, 662–669. [CrossRef]
13. Twomey, B.; Jacobs, P.A.; Gurgenci, H. Dynamic performance estimation of small-scale solar cogeneration with an organic Rankine cycle using a scroll expander. *Appl. Therm. Eng.* **2013**, *51*, 1307–1316. [CrossRef]
14. Norwood, Z.; Kammen, D.; Dibble, R. Testing of the katrix rotary lobe expander for distributed concentrating solar combined heat and power systems. *Energy Sci. Eng.* **2014**, *2*, 61–76. [CrossRef]
15. Antonelli, M.; Baccioli, A.; Francesconi, M.; Desideri, U.; Martorano, L. Operating maps of a rotary engine used as an expander for micro-generation with various working fluids. *Appl. Energy* **2014**, *113*, 742–750. [CrossRef]
16. Papes, I.; Degroote, J.; Vierendeels, J. New insights in twin screw expander performance for small scale ORC systems from 3D CFD analysis. *Appl. Therm. Eng.* **2015**, *91*, 535–546. [CrossRef]
17. Sadiq, G.A.; Tozer, G.; Al-Dadah, R.; Mahmoud, S. CFD simulations of compressed air two stage rotary Wankel expander–Parametric analysis. *Energy Convers. Manag.* **2017**, *142*, 42–52. [CrossRef]
18. Kurkus-Gruszecka, M.; Krawczyk, P.; Łapka, P. CFD modelling and validation of the rotary lobe compressed air expander. *MATEC Web Conf.* **2018**, *240*, 05015. [CrossRef]
19. GLOBE Benelux GLOBE Benelux. Available online: https://www.globe-benelux.nl (accessed on 18 April 2018).

20. Armak GmbH Armak GmbH-Motoren. Available online: http://www.armakmotor.de (accessed on 18 April 2018).
21. Armak Ltd ARMAK MOTORS. Available online: https://www.armak.co.uk (accessed on 10 June 2019).
22. Vande Voorde, J.; Vierendeels, J.; Dick, E. Flow simulations in rotary volumetric pumps and compressors with the fictitious domain method. *J. Comput. Appl. Math.* **2004**, *168*, 491–499. [CrossRef]
23. Schiffer, J.; Klomberg, S. CFD-Calculation of the Fluid Flow in a Rotary Lobe Pump–Evaluation of a Numerical Model Based on Measurement Results. In Proceedings of the 15th International Conference on Fluid Flow Technologies, Budapest, Hungary, 27 June 2012.
24. Singh, P.J.; Onuschak, A.D. A Comprehensive Computerized Method for Twin Screw Rotor Profile Generation and Analysis. In Proceedings of the International Compressor Engineering Conference, West Lafayette, IN, USA; 1984; pp. 519–527. Available online: https://docs.lib.purdue.edu/cgi/viewcontent.cgi?article=1500&context=icec (accessed on 10 June 2019).
25. Ng, K.C.; Bong, T.Y.; Lim, T.B. A thermodynamic model for the analysis of screw expander performance. *Heat Recover. Syst. CHP* **1990**, *10*, 119–133. [CrossRef]
26. Nilsson, H.R. Rotor Device. U.S. Patent Application No. 2,975,963, 21 March 1961.
27. Kasuya, K.; Fujiwara, M.; Imai, M.; Matsunaga, T.; Takahashi, Y. Male and Female Screw Rotor Assembly with Specific Tooth Flanks. U.S. Patent Application No. 4,401,420, 30 August 1980.
28. Bowman, J.L. Helical Screw Rotor Profiles. U.S. Patent Application No. 4,412,796, 1 November 1981.
29. Lee, H.-T. Screw-rotor Machine with an Ellipse as a Part of its Male Rotor. D.C. Patent Application No. 4,890,992, 2 January 1990.
30. Gulich, J.F. *Centrifugal Pumps*, 3rd ed.; Springer: Berlin, Germany, 2014; Volume 7450.
31. Bianchi, G.; Rane, S.; Kovacevic, A.; Cipollone, R.; Murgia, S.; Contaldi, G. Numerical CFD simulations on a small-scale ORC expander using a customized grid generation methodology. *Energy Procedia* **2017**, *129*, 843–850. [CrossRef]
32. Montenegro, G.; Della Torre, A.; Fiocco, M.; Onorati, A.; Benatzky, C.; Schlager, G. Evaluating the performance of a rotary vane expander for small scale organic rankine cycles using CFD tools. *Energy Procedia* **2014**, *45*, 1136–1145. [CrossRef]
33. Montenegro, G.; Della Torre, A.; Onorati, A.; Broggi, D.; Schlager, G.; Benatzky, C. CFD Simulation of a Sliding Vane Expander Operating Inside a Small Scale ORC for Low Temperature Waste Heat Recovery. *SAE Tech. Paper.* **2014**. [CrossRef]
34. Menter, F.R. Review of the shear-stress transport turbulence model experience from an industrial perspective. *Int. J. Comut. Fluid Dyn.* **2009**, *23*, 305–316. [CrossRef]
35. CFX-Solver, A.N.S.Y.S. ANSYS ANSYS Theory Guide. *Knowl. Creat. Diffus. Util.* **2010**, *15317*, 724–746.
36. ANSYS Inc. *ANSYS CFX*; ANSYS Inc.: Canonsburg, PA, USA, 2017.
37. Rocha, P.A.C.; Rocha, H.H.B.; Carneiro, F.O.M.; Vieira da Silva, M.E.; Bueno, A.V. K-ω SST (shear stress transport) turbulence model calibration: A case study on a small scale horizontal axis wind turbine. *Energy* **2014**, *65*, 412–418. [CrossRef]
38. Zhang, D.; Pan, D.; Shi, W.; Wu, S.; Shao, P. Study on Tip Leakage Vortex in an Axial Flow Pump Based on Modified Shear Stress Transport k-ω Turbulence Model. *Therm. Sci.* **2013**, *17*, 1551–1555. [CrossRef]
39. Shih, T.-H.; Liou, W.W.; Shabbir, A.; Yang, Z.; Zhu, J. A New k-ε Eddy Viscosity Model for High Reynolds Number Turbulent Flows. *Comput. Fluids* **1995**, *24*, 227–238. [CrossRef]

2

Comparative Study on CFD Turbulence Models for the Flow Field in Air Cooled Radiator

Chao Yu *, Xiangyao Xue, Kui Shi, Mingzhen Shao and Yang Liu

Changchun Institute of Optics, Fine Mechanics and Physics, Chinese Academy of Sciences, Changchun 130033, China; my691930402@163.com (X.X.); qw417345721@163.com (K.S.); ou910160194@163.com (M.S.); fg576856647@163.com (Y.L.)

* Correspondence: yuchao@ciomp.ac.cn

Abstract: This paper compares the performances of three Computational Fluid Dynamics (CFD) turbulence models, Reynolds Average Navier-Stokes (RANS), Detached Eddy Simulation (DES), and Large Eddy Simulation (LES), for simulating the flow field of a wheel loader engine compartment. The distributions of pressure fields, velocity fields, and vortex structures in a hybrid-grided engine compartment model are analyzed. The result reveals that the LES and DES can capture the detachment and breakage of the trailing edge more abundantly and meticulously than RANS. Additionally, by comparing the relevant calculation time, the feasibility of the DES model is proved to simulate the three-dimensional unsteady flow of engine compartment efficiently and accurately. This paper aims to provide a guiding idea for simulating the transient flow field in the engine compartment, which could serve as a theoretical basis for optimizing and improving the layout of the components of the engine compartment.

Keywords: computational fluid dynamics; heat transfer; engine compartment; vortex; turbulent model

1. Introduction

The turbulent structure is very important to the flow field distribution of the power compartment of the loader. Its transportation velocity and distribution have important research values. The small-scale turbulent structure is transported at the speed of local flow field, while the large-scale turbulent structure maintains a fixed proportion between the transport speed and the average flow velocity [1].

Bull et al. [2] conducted experiments on subsonic boundary layers, and the results showed that the proportion between the transport velocity and the average flow velocity of the small-scale turbulent structure was 0.53, while that of large-scale turbulent structure was 0.825. Favre et al. [3–5] have extensively studied the spatio-temporal correlation characteristics of boundary layer turbulent fields. The results show that the high and intermediate frequency pulsations contain about 50% turbulent kinetic energy, which is mainly characterized by high viscous dissipation and strong space-time attenuation, while the low frequency pulsations contain another 50% turbulent kinetic energy, of which only 1% is lost due to viscosity. Demetriades [6] conducted an experiment on the compressible supersonic boundary layer, which shows that the spatial correlation between flow direction and radial direction is basically the same, and only after the distance of vortex structure migration exceeds its own scale does the correlation significantly decrease. Ganapathisubramani et al. [7] measured the boundary layer with particle image velocimetry (PIV), and the results showed that the flow direction and the spreading scale in the boundary layer increased with the increase of the distance to the wall, and the change rate was basically linear. In the external flow field, the trend of scale variation is the opposite. The lower speed zone contributes more to the attenuation of flow correlation Ruu. Kline et al. [8] observed that there were a large number of low-velocity strip structures in the boundary layer through the H_2 bubble experiment. Falco et al. [9] studied the quasisturon motion in the turbulent boundary layer based on the smoke spectrum, determined the motion scale, and observed a large number of vortex structures. Moin and Kim [10] carried out a simulation study on turbulent channel flow based on large eddy simulation. Thereafter, numerical studies based on DNS were carried out in large quantities [11–13].

Different from traditional experimental method, numerical technology for the data of the turbulent flow field are measured and stored. The data processing means a great deal of diversity, and it can measure all direct physical quantities, such as pressure, velocity, and deals with predefined indirect quantities. For example, in the previous experiment, single point measurement was carried out based on thermocouple anemometer, so quadrant analysis could only be limited to a few measurement points, and it was concluded that up-jet and down-sweep were upstream and downstream relationships [14]. Numerical studies will not be subject to such restrictions, and not only can we obtain the three-dimensional structure of the quadrant component of specific instantaneous shear product, we can also determine the vortex structure [15–17]. Compared with the experiment, the biggest advantage of numerical study lies in the traceability of specific experimental phenomena, that is, it can trace forward and deduce the quasi-structure and quasi-motion at a certain time point. At the same time, it can also be studied by the means of defining phenomena and using conditional sampling [18]. Calautit J. K. et al. [19] used virtual wind tunnels to replace traditional wind tunnel tests to, respectively, conduct flow field and transmission of the power cabin of passenger cars and loaders. Lu P. et al. [20] adopted the method of one-dimensional and three-dimensional co-simulation to analyze the flow characteristics of the power cabin and the change rules of the internal thermal environment.

However, different turbulence models are used to predict uniformity, which results in different results. Therefore, hybrid meshing inside the engine compartment is used in this research, and a lot of energy and time is spent to improve the grid quality. By comparing it with the experimental results, the prediction performance of different turbulence models on the flow field and temperature field of the engine compartment is evaluated. The temperature field, pressure streamline structure, vorticity, and velocity distribution in the engine compartment are analyzed. Compared with the Reynolds-averaged Navier-Stokes equations (RANS) model, the Large Eddy Simulation (LES) and Detached Eddy Simulation (DES) models can capture transient vorticity characteristics such as the generation, development, and formation of vortices, as well as the trailing edge shedding and fracture. In addition, the DES model has been proven to be a feasible method to accurately and efficiently simulate three-dimensional unsteady flow in complex cabin space.

2. Numerical Simulation

2.1. Turbulence Models

The average time domain of the Navier-Stokes equation to obtain the continuity and momentum equations for incompressible flow is as follows:

$$\begin{aligned} &\frac{\partial \rho}{\partial t} + \frac{\partial(\rho u_i)}{\partial x_i} = 0 \\ &\frac{\partial(\rho u_i)}{\partial t} + \frac{\partial(\rho u_i u_j)}{\partial x_j} = -\frac{\partial p}{\partial x_i} + \frac{\partial}{\partial x_j}(\mu \frac{\partial u_i}{\partial x_j}) + \frac{\partial \tau_{ij}}{\partial x_j} \end{aligned} \tag{1}$$

where ρ denotes the density and u_i denotes velocity. p is the pressure, and μ is the coefficient of viscosity.

Where the Reynolds stress τ_{ij} is:

$$\tau_{ij} = -\rho \overline{u_i u_j}. \tag{2}$$

The most widely used turbulence model is the standard k-ε model. The model is based on the kinetic energy k equation, and then an equation of turbulent dissipation rate ε is introduced. The ε is defined as:

$$k = \frac{1}{2} m v^2, \tag{3}$$

$$\varepsilon = \frac{\mu}{\rho}\overline{\left(\frac{\partial u'_l}{\partial x_k}\right)\left(\frac{\partial u'_l}{\partial x_k}\right)}. \tag{4}$$

The turbulent viscosity μ_t can be expressed as a function of k and ε:

$$\mu_t = \rho C_\mu \frac{k^2}{\varepsilon}. \tag{5}$$

Therefore, the transport equation of the standard k-ε model is [21]:

$$\begin{cases} \frac{\partial(\rho k)}{\partial t} + \frac{\partial(\rho k u_i)}{\partial x_i} = \frac{\partial}{\partial x_j}\left[\left(\mu + \frac{\mu_t}{\sigma_k}\right)\frac{\partial k}{\partial x_j}\right] + G_k + G_b - \rho\varepsilon - Y_M + S_k \\ \frac{\partial(\rho \varepsilon)}{\partial t} + \frac{\partial(\rho k u_i)}{\partial x_i} = \frac{\partial}{\partial x_j}\left[\left(\mu + \frac{\mu_t}{\sigma_\varepsilon}\right)\frac{\partial \varepsilon}{\partial x_j}\right] + C_{1\varepsilon}\frac{\varepsilon}{k}(G_k + C_{3\varepsilon}G_b) - C_{2\varepsilon}\rho\frac{\varepsilon^2}{k} + S_\varepsilon \end{cases} \tag{6}$$

where the kinetic energy caused by the average velocity gradient is:

$$G_k = \mu_t\left(\frac{\partial u_i}{\partial x_j} + \frac{\partial u_j}{\partial x_i}\right)\frac{\partial u_i}{\partial x_j}. \tag{7}$$

The turbulent energy generated by the buoyancy effect G_b is [22]:

$$G_b = \beta g_i \frac{u_t}{\Pr_t}\frac{\partial T}{\partial x_i}. \tag{8}$$

The following values are used for standard k-ε model: $C_{1\varepsilon} = 1.44$, $C_{2\varepsilon} = 1.92$, $C_{3\varepsilon} = 0.09$, $\sigma_k = 1.0$, and $\sigma_\varepsilon = 1.3$.

The basic idea of RANS is to perform time averaging on the N-S equations to transform the unsteady turbulence problem into a steady problem study, at the cost of additional unknown numbers, which are also in the same form as stress, called Reynolds stress. Reynolds stress also needs to be characterized by a model, which is the so-called turbulence model. However, due to the time averaging of the problem, the information contained in the equation itself has been partially lost. It is actually very difficult to give a Reynolds stress model and it is also difficult to apply it to all flows.

LES is a commonly used turbulence simulation method. LES is based on the self-similarity theory and uses a subgrid-scale to simulate small vortices, while the large vortices are based on the geometric calculation. The disadvantage of this model is that it is difficult to calculate the region near the wall. In the LES simulation method, the transport of fluid momentum, mass, energy, and other physical quantities are mainly affected by large-scale vortices. Small-scale vortices with isotropic motion are not affected by geometric and boundary conditions. Therefore, LES uses the N-S equation to calculate the turbulent motion larger than the grid-scale. The LES model obtains a filtered momentum equation by filtering out vortices smaller than the filtered grid in the Fourier equation or spatial domain N-S equation. The filtered N-S equation and the continuity equation are transformed into [22]:

$$\begin{cases} \frac{\partial \rho}{\partial t} + \frac{\partial(\rho \overline{u_i})}{\partial x_i} = 0 \\ \frac{\partial}{\partial t}(\rho\overline{u_i}) + \frac{\partial}{\partial x_j}(\rho\overline{u_i u_j}) = -\frac{\partial \overline{p}}{\partial x_i} + \frac{\partial}{\partial x_j}\left\{\mu\frac{\partial \overline{u_i}}{\partial x_j}\right\} - \frac{\partial \tau_{ij}}{\partial x_j} \end{cases} . \tag{9}$$

The subgrid-scale stress tensor that results from filtering the Navier-Stokes equations consists of three terms:

$$\tau_{ij} = L_{ij} + R_{ij} + C_{ij}, L_{ij} = \overline{\overline{u_i}\overline{u}_j} - \overline{u}_i\overline{u}_j, C_{ij} = \overline{\overline{u_i}u'_j + u'_i\overline{u_j}}, R_{ij} = \overline{u'_i u'_j} \tag{10}$$

where $\overline{u}_i$ denotes the filtered velocity component and $u'_i = u_i - \overline{u_i}$ denotes the SGS component of u_i.L_{ij} is the Leonard term, C_{ij} is the cross term, and R_{ij} is the SGS Reynolds stress.

The sub-grid stress term τ_{ij} is obtained from the vortex viscous model equation:

$$\tau_{ij} - \frac{1}{3}\tau_{kk}\delta_{ij} = -2\mu_t\overline{S_{ij}} \tag{11}$$

where μ_t is the turbulent dynamic viscosity, and $\overline{S_{ij}}$ is the rate-of-strain tensor, which is defined as:

$$\overline{S_{ij}} = \frac{1}{2}\left(\frac{\partial \overline{u_i}}{\partial x_j} + \frac{\partial \overline{u_j}}{\partial x_i}\right). \tag{12}$$

Because the turbulence scale decreases rapidly with the increase of Reynolds number, LES simulation cannot be used in engineering calculation at a high Reynolds number. According to the characteristics of the RANS/LES model, a kind of large-scale separated flow in engineering is reasonably arranged by "model allocation." Since the LES model is not an ideal model for solving near-wall problems, the DES turbulence model was developed. The DES turbulence model combines RANS with LES methods. The basic idea of DES is to use the RANS model within the boundary layer and the LES side in the separation area, which saves computational resources and ensures computational accuracy. The DES model selected SST (Shear Stress Transfer) as the RANS model in this study, and its flow energy diffusion term Y_k is modified to [23]:

$$\begin{cases} Y_k = \rho\beta^* k w F_{DES} \\ F_{DES} = \max\left(\frac{L_t}{C_{DES}\Delta_{\max}}, 1\right) \end{cases} \tag{13}$$

where C_{DES} is a calibration constant in the DES model with a value of 0.61, and $\Delta_{\max}$ is the largest grid gap in the three directions (Δx, Δy, Δz) in the grid.

2.2. Geometric Model Construction and Mesh Generation

In this paper, different computational fluid dynamics (CFD) turbulence models are used to simulate the flow field of a loader's engine compartment. In order to facilitate the calculation, the model is appropriately simplified, and the components such as the bucket, front axle, and wheels that have less influence on the flow field of the engine compartment are omitted. A 3D simulation model of the engine compartment is established to analyze its internal flow field. The simplified geometric model consists of an outer wind tunnel, an inner wind tunnel, and a loader model as shown in Figures 1 and 2.

The GAMBIT software is used to perform hybrid meshing on the radiator group. Due to the irregular structure of the fan rotation domain and engine compartment, the mesh was built by tetrahedron and hexahedral. The grid of the air intake, inlet fan, and air duct area of the engine compartment are refinement. In order to reduce the total mesh size of the model, a two-layer wind tunnel nested structure is used. The size-Function and Hex-core functions are used to reduce the total mesh number in the outer and inner wind tunnels. The grid distribution in the engine compartment is dense. The outer wind tunnel and the inner wind tunnel mesh are sparse, and the virtual wind tunnel model mesh is shown in Figure 3. The minimum grid size of the simulation model is 1 mm. The total number of grids in the entire calculation model is 800,000.

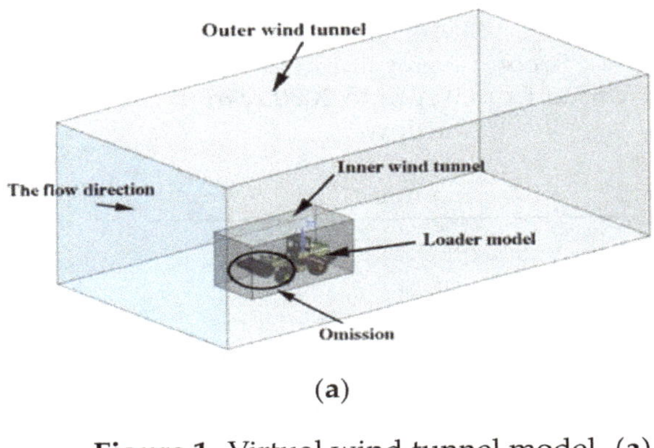

(a)

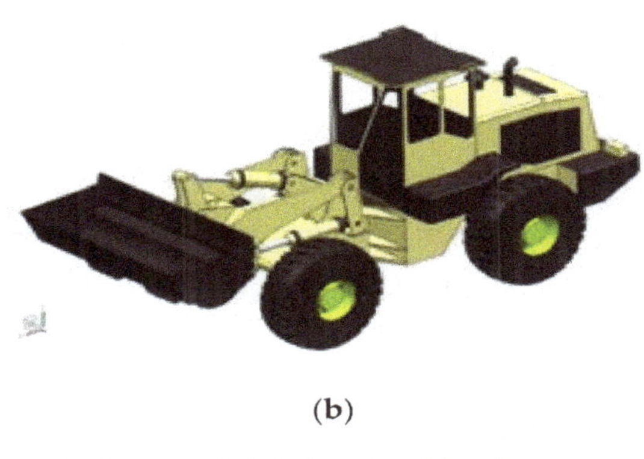

(b)

Figure 1. Virtual wind tunnel model. (**a**) The whole model; (**b**) the wheel loader.

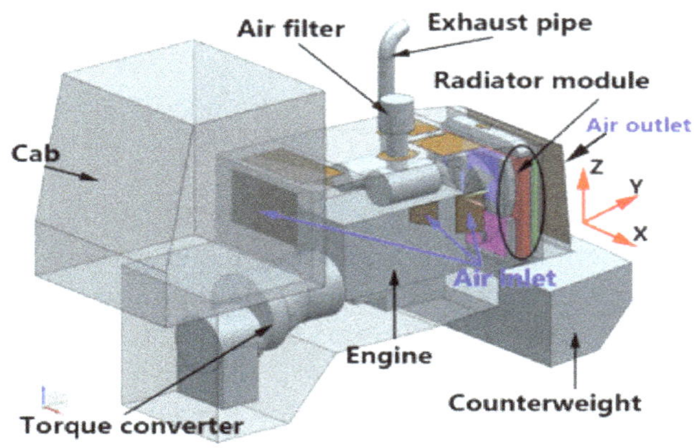

Figure 2. Simplified engine compartment.

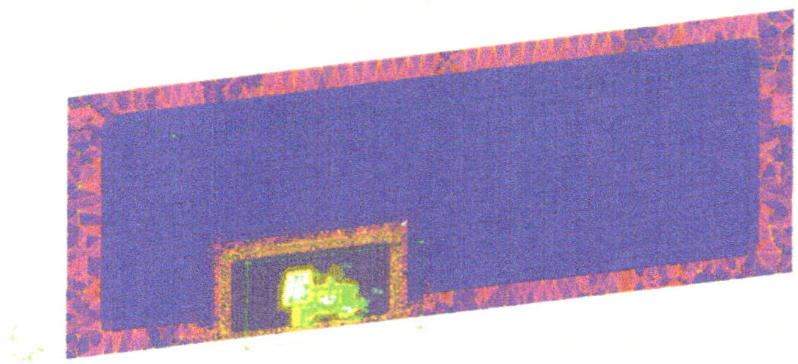

Figure 3. Engine compartment virtual wind tunnel model meshing (x = 0 section).

2.3. Boundary Condition Setting

The incompressible N-S equation is solved by ANSYS Fluent software. During the calculation, the fan rotates along the mesh section at a certain time step (0.01 s). When the change of mass flow rate in two consecutive circulatory iterations and all normalized residuals in mass and momentum conservation equations were less than 10^{-4} for the differences, the solution was assumed to be converged [24–27]. The fan speed is the same as the experimental value. The mesh interface is set between the various modules of the engine compartment. The boundary conditions and settings of the CFD solver are shown in Table 1. The radiators in the engine compartment are simulated by a porous medium model [28].

Table 1. CFD (computational fluid dynamics) model description.

Analysis Type	Transient State
Turbulence Model	LES, DES, and Standard $k-\varepsilon$ model
Pressure-Velocity Coupling	SIMPLEC
Transient Formulation	Second-order implicit
Fan Status	Varied from 0 rpm to 2000 rpm
Vent Status	Outflow
Environment Temperature	25 °C
Inlet velocity	1.5 m/s

3. Results and Analysis

3.1. Experiment Analysis

The credibility of the computational model is verified by experiments. There are 9 monitoring points set at the exit of the engine compartment. When the fan speed reaches 1000 rpm, they collect the wind speed value. The monitoring point position is shown in Figure 4. Figure 5 shows the comparison of the three simulation results at 9 monitoring points with the experimental data collected at 1000 rpm. Compared with the experimental results, the error of the calculation results of each simulation model is less than 15%, and the simulation model can solve the actual fluid flow of the loader's engine compartment relatively correctly.

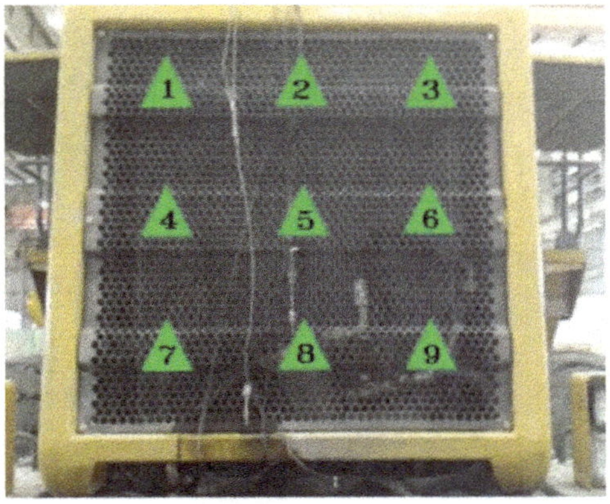

Figure 4. Flow rate collection points of experiment.

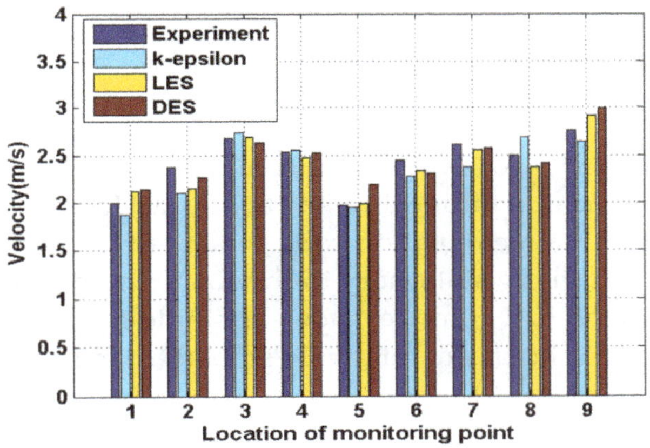

Figure 5. Simulation calculation and experimental data comparison (at 2000 rpm).

3.2. Pressure Distribution

The arrangement of various heat dissipating components in the loader plays an important role in the pressure distribution of the engine compartment. Figure 6 shows the evolution of pressure in the engine compartment with fan speed. As the fan speed increases, the velocity of the fluid in the cabin gradually increases. A low-pressure region appears at the fan shaft center and a high-pressure region appears at the outer edge of the fan. At low speeds, the surface unevenness of the radiator is consistent with the dynamic pressure of the fan outlet. Due to the large space in the upper part of the engine, the fluid inflow is large at high speeds. Under the commonality of the viscous force and pressure gradient, the radiator gradually approaches the equilibrium pressure.

There is a slight difference in the pressure gradient between the RANS, LES, and DES models. Compared with the other two simulation models, the LES model has a significantly low-pressure region at the fan shaft center when fan speeds are high. The pressure trends described in the three turbulence models are the same, indicating that all simulations can effectively capture the mainstream structure.

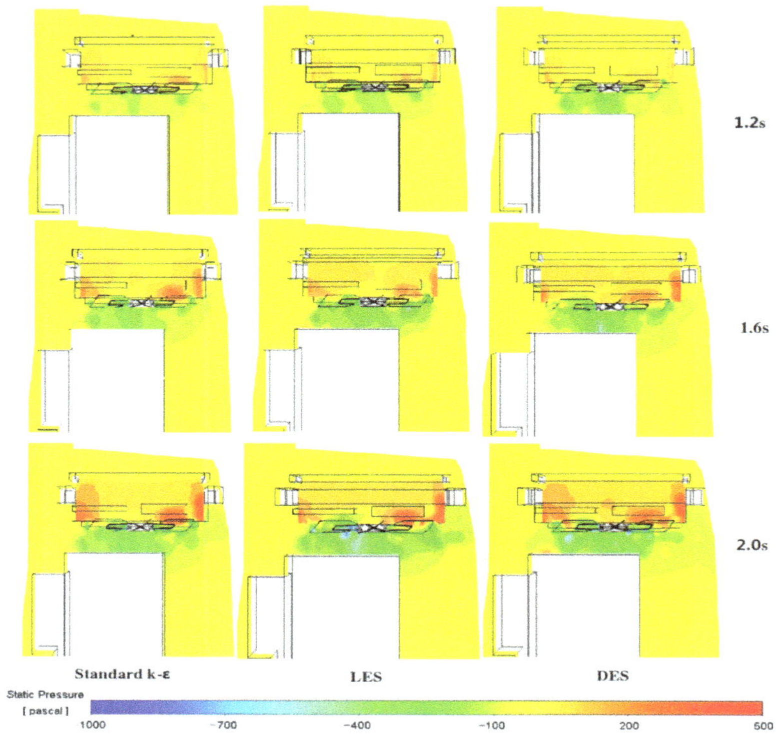

Figure 6. 2D contours of pressure magnitude at YZ plane.

3.3. Velocity Distribution

The evolution of the fluid flow field in the engine compartment is shown in Figure 7. All simulation models are good at predicting the magnitude and level of flow. However, after a closer look at the YZ section, it can be seen that there is a recirculation zone generated by the intake air in the LES and DES models, while it remains substantially stationary in the RANS model. Figure 8 shows the flow field distribution of the XY section. The RANS model shows multiple small turbulences between the engine and fan, while the LES and DES models are relatively smooth in this area.

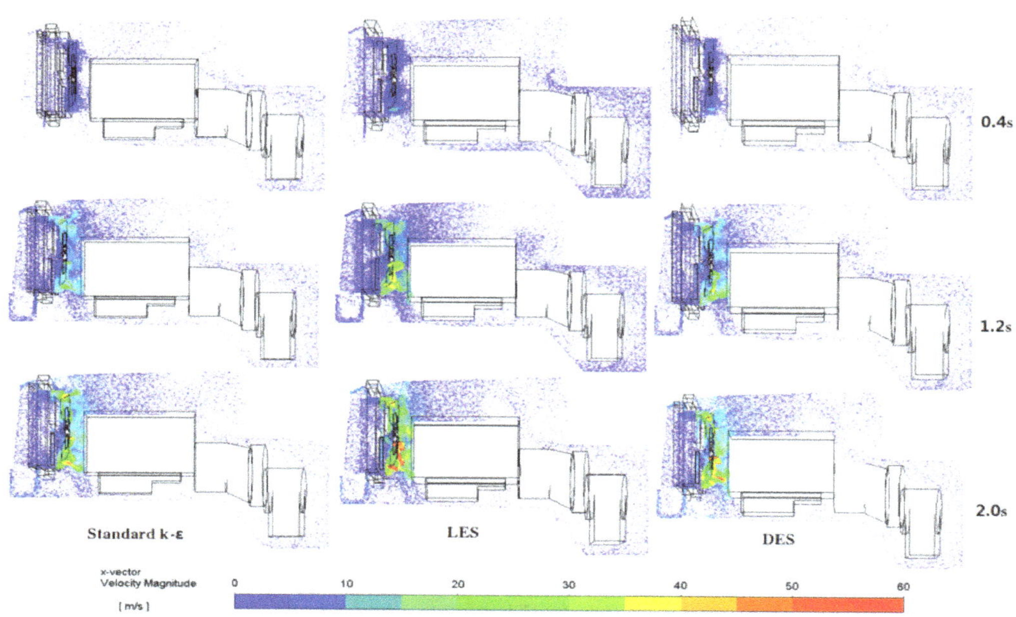

Figure 7. 2D contours of velocity magnitude at YZ plane.

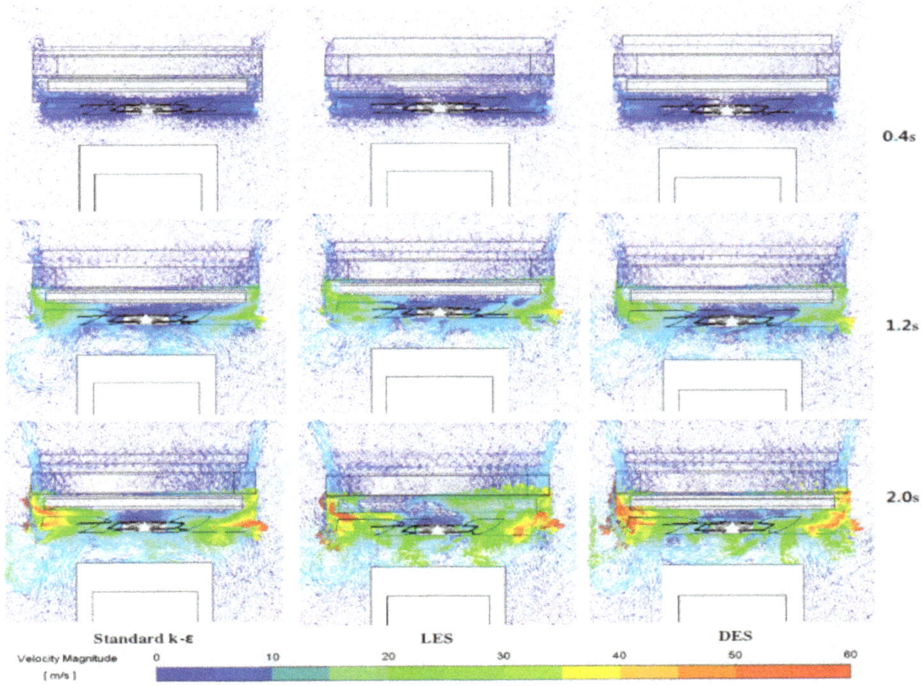

Figure 8. 2D contours of velocity magnitude at XY plane.

According to Figures 7 and 8, as the rotational speed increases, the high-speed zone on the outer edge of the fan matches the pressure distribution described above. A high flow rate region occurs at the boundary of the radiator, and the seal between the radiators of each layer allows the cold air to pass sufficiently. There is a retardation zone between adjacent radiators, and the efficient radiator has a relatively small pressure drop retardation zone. In the simulation of the engine compartment, it can be seen that the cold air mainly flows from above and below the power compartment.

3.4. *Vortex Structure*

The vortex was shown by a special isosurface (called vortex core). Based on the Q criterion, the Q criterion index is used to describe it. The recipe is defined as follows [29,30]:

$$Q = \frac{1}{2}(\Omega^2 + S^2) \tag{14}$$

where Ω and S represent the vorticity tensor and strain rate tensor, respectively.

Figure 9 is a vortex core diagram of the rotating region. As we can see in the standard transient structure for the Q standard, the vortex structure among RANS, LES, and DES models are completely different. LES and DES can reflect more detailed vortex cores.

It can be seen from the figure that the vortex core of the fan is mainly composed of two parts; the vortex core is generated from the blade root vortex and tip vortex formed by the leading edge of the blade. The vortex sheets remain attached on the pressure side while separated at the tip. A high-velocity vortex core appears on the windward surface of the fan. As the rotational speed increases, the fan blade surface gradually forms a vortex. Due to the faster transition between DES and LES, the vortex is also present at the root of the blade, which is gradually developed more abundantly and orderly.

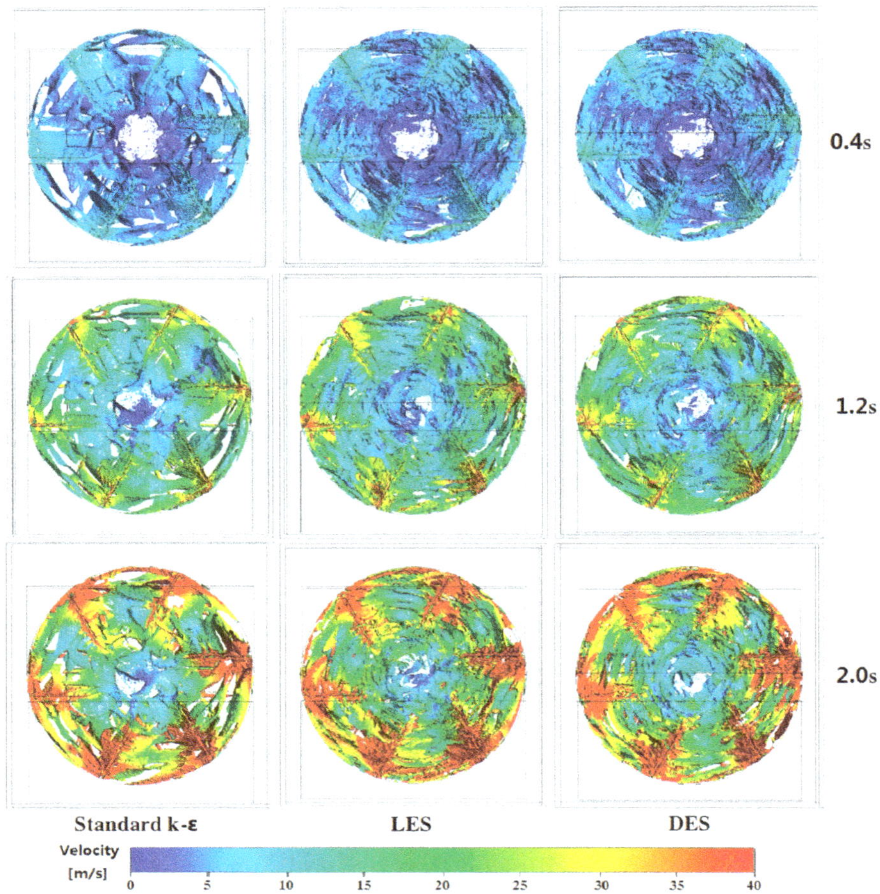

Figure 9. Vortex structures of the fan.

3.5. Prediction Performance of the Three Turbulent Models

As shown in Figure 10, the simulation data are in good agreement with the experimental data, and the wind speed at each monitoring point increases gradually. A low-speed zone occurs at the detection point 5, which is consistent with the unevenness of the fan flow rate. The prediction speed of the Standard k-ε model and LES model is relatively larger than the DES model. However, the DES model is more accurate than the Standard k-ε model and LES model, whose absolute error is under 20%.

The calculation time of the three simulation models is shown in Table 2. The RANS simulation is about 2.8 times faster than the LES simulation. In industrial design, the RANS model is used for CFD simulation. Although the calculation cost of the RANS model is low, its accuracy is lower than that of the LES model. The LES model has high precision, but the calculation time is long and it is difficult to apply to actual engineering. Besides, the DES model, which combines the advantages of LES and RANS models, is an ideal choice for CFD simulation with short computation time and high accuracy.

Table 2. Comparison of computation time with different models (3.4 GHz, 2 CPU).

Model	CPU Time (h)
RANS	12.3
DES	22.1
LES	34.4

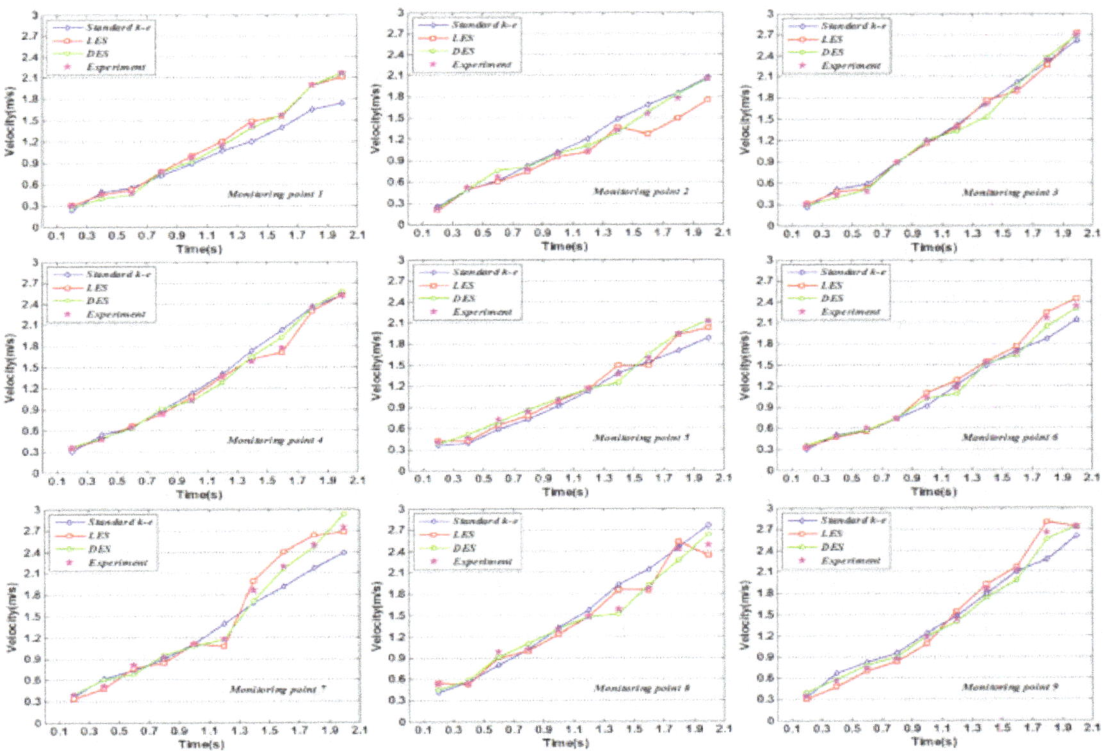

Figure 10. Prediction performance of three models.

4. Conclusions

In this paper, the hybrid grid model is used to simulate the unsteady flow of the engine power compartment. The ability of the RANS, DES, and LES models to predict the turbulence distribution is compared and analyzed. The engine compartment propulsion simulation using RANS, DES, and LES is proposed. Scholars can use the RANS model to estimate the flow velocity and structure reasonably and accurately. However, LES or DES models should be used when it is necessary to obtain more accurate turbulence distributions with complex geometric structures. Under normal working conditions, the natural wind energy in the environment improves the heat dissipation performance of the engine compartment of the loader. In this study, considering the influence of natural wind on heat dissipation can improve the accuracy of performance prediction. The DES model combines the advantages of the RANS model and LES model, which can accurately simulate a complex eddy current field. At the same time, the LES model can simulate large-scale eddy currents in the boundary layer. The DES model has proven to be an accurate and efficient model for simulating three-dimensional unsteady turbulent flow in complex channels. Meanwhile, theoretical research on this aspect will be the focus of CFD simulation in the future.

Author Contributions: Conceptualization, C.Y.; methodology, C.Y.; software, X.X.; validation, K.S.; formal analysis, K.S.; investigation, X.X.; resources, M.S.; data curation, M.S.; writing—original draft preparation, C.Y.; writing—review and editing, C.Y.; visualization, Y.L.; supervision, X.X.; project administration, X.X.; funding acquisition, X.X. All authors have read and agreed to the published version of the manuscript.

Nomenclature

k	Turbulent kinetic energy [m^2/s^2]	ε	Turbulent dissipation rate [m^2/s^3]
N	Rotating speed [rpm]	ω	Relative velocity [m/s]

u_i	Time-averaged velocity [m/s]	ν	Turbulent viscosity [m^2/s]
X_i	Coordinate [m]	ρ	density [kg/m^3]
t	Time [s]	μ	Circumferential velocity [m/s]
P	Fluid pressure [Pa]	μ_t	Turbulence viscosity coefficient
		τ_{ij}	Reynolds stress

References

1. Del Álamo, J.C.; Jiménez, J. Estimation of turbulent convection velocities and corrections to Taylor's approximation. *J. Fluid Mech.* **2009**, *640*, 5–26.
2. Bull, M. Wall-pressure fluctuations associated with subsonic turbulent boundary layer flow. *J. Fluid Mech.* **1967**, *28*, 719–754.
3. Favre, A. Review on space-time correlations in turbulent fluids. *J. Appl. Mech.* **1965**, *32*, 241–257.
4. Favre, A.; Gaviglio, J.; Dumas, R. Further space-time correlations of velocity in a turbulent boundary layer. *J. Fluid Mech.* **1958**, *3*, 344–356.
5. Favre, A.; Gaviglio, J.; Dumas, R. Structure of Velocity Space-Time Correlations in a Boundary Layer. *Phys. Fluids* **1967**, *10*, S138–S145.
6. Demetriades, A. Turbulence correlations in a compressible wake. *J. Fluid Mech.* **1976**, *74*, 251–267.
7. Ganapathisubramani, B.; Hutchins, N.; Hambleton, W.T.; Longmire, E.K.; Marusic, I. Investigation of large-scale coherence in a turbulent boundary layer using two-point correlations. *J. Fluid Mech.* **2005**, *524*, 57–80.
8. Kline, S.J.; Reynolds, W.C.; Schraub, F.A.; Runstadler, P.W. The structure of turbulent boundary layers. *J. Fluid Mech.* **1967**, *30*, 741–773.
9. Falco, R.E. Coherent motions in the outer region of turbulent boundary layers. *Phys. Fluids* **1977**, *20*, S124.
10. Moin, P.; Kim, J. Numerical investigation of turbulent channel flow. *J. Fluid Mech.* **1982**, *118*, 341–377.
11. Kim, J.; Moin, P.; Moser, R. Turbulence statistics in fully developed channel flow at low Reynolds number. *J. Fluid Mech.* **1987**, *177*, 133–166. [CrossRef]
12. Moser, R.D.; Kim, J.; Mansour, N.N. Direct numerical simulation of turbulent channel flow up to $Re_\tau = 590$. *Phys. Fluids* **1999**, *11*, 943–945. [CrossRef]
13. Abe, H.; Kawamura, H.; Matsuo, Y. Direct numerical simulation of a fully developed turbulent channel flow with respect to the Reynolds number dependence. *J. Fluids Eng. Trans. ASME* **2001**, *123*, 382–393. [CrossRef]
14. Wallace, J.M.; Eckelmann, H.; Brodkey, R.S. The wall region in turbulent shear flow. *J. Fluid Mech.* **1972**, *54*, 39–48. [CrossRef]
15. Lozano-Durán, A.; Flores, O.; Jiménez, J. The three-dimensional structure of momentum transfer in turbulent channels. *J. Fluid Mech* **2012**, *694*, 100–130. [CrossRef]
16. Robinson, S.K. Coherent motions in the turbulent boundary layer. *Annu. Rev. Fluid Mech.* **1991**, *23*, 601–639. [CrossRef]
17. Adrian, R.J. Hairpin vortex organization in wall turbulence. *Phys. Fluids* **2007**, *19*, 041301. [CrossRef]
18. Jiménez, J. Coherent structures in wall-bounded turbulence. *J. Fluid Mech.* **2018**, *842*, 1. [CrossRef]
19. Calautit, J.K.; Chaudhry, H.N.; Hughes, B.R.; Sim, L.F. A validated design methodology for a closed-loop subsonic wind tunnel. *J. Wind Eng. Ind. Aerodyn.* **2014**, *125*, 180–194. [CrossRef]
20. Lu, P.; Gao, Q.; Wang, Y. The simulation methods based on 1D/3D collaborative computing for the vehicle integrated thermal management. *Appl. Therm. Eng.* **2016**, *104*, 42–53. [CrossRef]
21. Galvan, S.; Reggio, M.; Guibault, F. Assessment study of k-ε turbulence models and near-wall modeling for steady state swirling flow analysis in draft tube using fluent. *Eng. Appl. Comput. Fluid Mech.* **2011**, *5*, 459–478. [CrossRef]
22. Yang, X.; Gupta, S.; Kuo, T.W.; Gopalakrishnan, V. RANS and large Eddy simulation of internal combustion engine flows-a comparative study. *J. Eng. Gas Turbines Power* **2014**, *136*, 051507. [CrossRef]
23. Xu, C.; Chen, L.; Lu, X. Large-eddy and detached-eddy simulations of the separated flow around a circu-lar cylinder. *J. Hydrodyn.* **2007**, *19*, 559–563. [CrossRef]
24. Wood, H.G.; Yan, Q.; Wei, W. Parametric analysis and optimization of inlet deflection angle in Torque converters. *J. Fluid Eng.* **2015**, *137*, 031101.
25. Chen, X.; Liu, Y. *Finite Element Modeling and Simulation with ANSYS Workbench*; CRC Press: Boca Raton, FL, USA, 2014.

26. Yu, C.; Qin, S.; Liu, Y.; Chai, B. Heat exchange performance optimization of a wheel loader cooling system based on computational fluid dynamic simulation. *Adv. Mech. Eng.* **2018**, *10*, 1687814018803984. [CrossRef]
27. Yu, C.; Qin, S.; Chai, B.; Huang, S.; Liu, Y. The Effect of Compressible Flow on Heat Transfer Performance of Heat Exchanger by Computational Fluid Dynamics (CFD) Simulation. *Entropy* **2019**, *21*, 829. [CrossRef]
28. Yang, L.; Sicheng, Q. The optimization design of Off-Highway machinery radiator based on genetic algorithm and ε-NTU. *Acta Tech. CSAV* **2017**, *62*, 465–476.
29. Ji, B.; Wang, J.; Luo, X.; Miyagawa, K.; Xiao, L.Z.; Long, X.; Tsujimoto, Y. Numerical simulation of cavitation surge and vortical flows in a diffuser with swirling flow. *J. Mech. Sci. Technol.* **2016**, *30*, 2507–2514. [CrossRef]
30. Hunt, J.C.R.; Wray, A.A.; Moin, P. Eddies, streams, and convergence zones in turbulent flows. *Stud. Turbul. Using Numer. Simul. Databases* **1988**, *2*, 193–208.

3

Numerical Study on the Influence of Mass and Stiffness Ratios on the Vortex Induced Motion of an Elastically Mounted Cylinder for Harnessing Power

Vidya Chandran [1], Sekar M. [2], Sheeja Janardhanan [3,*] and Varun Menon [4]

1 Department of Mechanical Engineering, Karunya Institute of Technology and Sciences, Coimbatore, Tamil Nadu 600018, India; vidya.rudn@gmail.com

2 Department of Mechanical Engineering, AAA College of Engineering and Technology, Sivakasi, Tamil Nadu 600018, India; mailtosekar@gmail.com

3 Department of Mechanical Engineering, SCMS School of Engineering and Technology, Ernakulam, Kerala 673307, India

4 Department of Computer Science and Engineering, SCMS School of Engineering and Technology, Ernakulam, Kerala 673307, India; varungmenon46@gmail.com

* Correspondence: sheejajanardhanan@scmsgroup.org.

Abstract: Harnessing the power of vortices shed in the wake of bluff bodies is indeed a boon to society in the face of fuel crisis. This fact serves as an impetus to develop a device called a hydro vortex power generator (HVPG), comprised of an elastically mounted cylinder that is free to oscillate in the cross-flow (CF) direction even in a low velocity flow field. The oscillatory motions in turn can be converted to useful power. This paper addresses the influence of system characteristics viz. stiffness ratio (k^*) and mass ratio (m^*) on the maximum response amplitude of the elastically mounted cylinder. Computational fluid dynamics (CFD) simulations have been used here to solve a two way fluid–structure interaction (FSI) problem for predicting the trend of variation of the non-dimensional amplitude Y/D with reduced velocity U_r through a series of simulations. Maximum amplitude motions have been attributed to the lowest value of m* with $U_r = 8$. However, the maximum lift forces correspond to $U_r = 4$, providing strong design inputs as well as indicating the best operating conditions. The numerical results have been compared with those of field tests in an irrigation canal and have shown reasonable agreement.

Keywords: computational fluid dynamics (CFD); flow around cylinder; fluid structure interaction (FSI); hydrodynamic response; numerical methods; simulation and modeling; vortex induced vibration (VIV) ratio

---

1. Introduction

As the sources of fossil fuels are depleting at a faster pace, energy scientists all over the world are keen on the search for new technologies that can provide renewable and clean energy. Hydroelectric power generation is of course a clean source of energy, but considering the capital investment and the effects of dams on natural ecosystems, the need for a much cleaner energy source becomes more important. This paper discusses the design and manufacture of the hydro vortex power generator (HVPG) model, which when scaled up can be viewed as one such cleaner source of electricity. Also, the paper discusses a numerical method to optimize the design parameters of HVPG model. The principle behind the working of HVPG is vortex shedding in the wake of bluff bodies in fluid flow. The phenomenon of vortex shedding behind bluff bodies has been an extensively researched topic [1,2]. The presence of such vortex shedding has been considered undesirable and researchers had been in search of methods to suppress vortex shedding [3]. Vortex power was proved useful to mankind by the researchers at Michigan University, who first converted vortex power into electricity [4,5]. HVPG works on the principle of vortex induced vibration of bluff bodies subjected to fluid current. The power of vortices shed in the wake of these bluff bodies are converted into vibration energy and then into electricity. HVPG can be made useful as a single standing power unit that can provide electricity to remote locations; and also as a multiunit module which can supply power to the grid [6].The paper discusses the design optimization of a single, power-generating, scaled-down module, harnessing power from vortices. In this paper, an attempt has been made to optimize the design based on the major influencing parameters, oscillating mass ratio (m^*) and stiffness ratio (k^*). Griffin consolidated experimental results and plotted them to show the dependence of maximum amplitude on system characteristics [7]. However, the drawback of Griffin's plot was its considerable scatter which could be attributed to the inclusion of mode shapes also as an influencing parameter. Later researchers could successfully reduce the scatter in Griffin's plot and establish a simple relationship between maximum amplitude and mass damping parameter by eliminating mode shapes from the list of variables [7,8]. The simplified mass damping parameter as in [8] was applicable for cylinders of high and low mass ratios equally. Many researchers have formulated empirical formulae to express the parametric relationship and experimentally verified the correlations [9,10]. Later in the experiments, however, it has been observed that with variation in the damping ratio, U_r and the Strouhal number (St) varies [11]. Also, these experiments revealed relatively larger response amplitude compared to other studies. Recently researchers have also succeeded in theoretically proving the effectiveness of harnessing VIV energy for powering underwater mooring platforms [12].

The present work studies the influence of k^* and m^* on maximum response amplitude of an elastically mounted cylinder with a single DOF. It also provides a detailed insight into the vortex shedding pattern at various U_r. The simulations are carried out at Re of the order 10^4 which corresponds to the realistic flow regime encountered by power generating vortices [13]. The numerical results have been verified using field tests conducted at Palissery irrigation canal (Palissery irrigational canal is one of the irrigation projects by Government of Kerala located in Thrissur district, Kerala, India).

Previous studies in the same domain had only considered the effect of complex and coupled parameters such as the Skop Griffin parameter [14]. The novelty in the present work is the effort made to represent the influence of tangible parameters viz the stiffness coefficient and mass of the cylinder. These parameters are easily controllable from a designer's point of view. A non-dimensional approach has been used here to generalize the results for the design of power harnessing devices of any scale.

2. The Concept of HVPG

HVPG works on the principle of vortex induced vibration (VIV). If a bluff body is not completely secured with at least one degree of freedom motion, and the frequency of vortex shedding matches the natural frequency of the structure, the structure begins to resonate, vibrating with harmonic oscillations of large amplitude. This phenomenon is known as 'lock-in'. During lock-in, vortex shedding frequency shifts to the natural frequency of the structure, leading to large amplitude vibrations. Vortex shedding in the wake of a cylinder is shown in Figure 1.

The vortex shedding occurs at a discrete frequency and is a function of the Reynolds number (Re), defined by Equation (1)

$$Re = \frac{\rho V D}{\mu} \tag{1}$$

The dimensionless frequency of the vortex shedding, $St = f_v\ D/V$, is approximately equal to 0.2 when the Reynolds number is greater than 1000 [15]. When vortices are shed from the cylinder, uneven pressure distribution develops around the upper and lower surfaces of the cylinder, generating an oscillating hydrodynamic lift force on the cylinder. This unsteady force given by Equation (2) can induce significant cross flow vibrations on a structure, especially if the resonance condition is met.

$$F_L = C_L \frac{1}{2} \rho A V^2. \tag{2}$$

where F_L is the lift force and C_L is the coefficient of lift. ρ is the density of water, A the projected area in the direction of flow, and V is the velocity of flowing water. The cylinder also experiences a net force along the flow direction and is called the drag force and is given by the Equation (3).

$$F_D = C_D \frac{1}{2} \rho A V^2 \tag{3}$$

where F_D is the drag force and C_D is the drag coefficient.

The oscillating lift force acting on the cylinder makes the cylinder oscillate in the cross flow (CF) direction at the frequency of vortex shedding. For the making of HVPG, the cylinder has been mounted elastically, which enables the entire module to be considered as a spring-mass system with the cylinder considered as the mass and the elastic supports as springs. When the natural frequency of spring-mass system matches the vortex shedding frequency, the cylinder oscillates with large amplitudes. The linear motion of the mass can then be converted to rotary motions through a slider-crank mechanism and the crank rotations can be used to drive a generator unit.

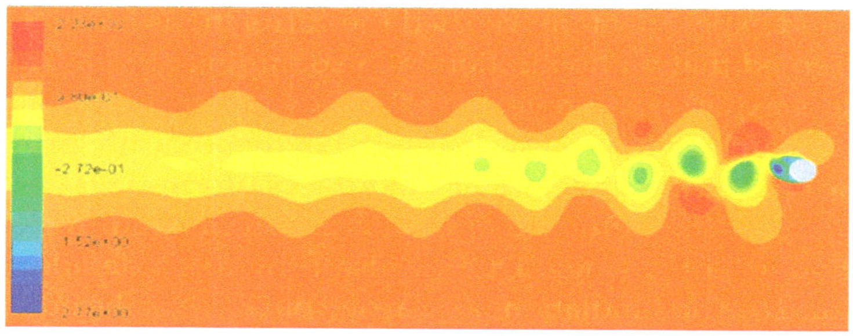

Figure 1. Shedding of alternate vortices behind a cylinder represented as pressure (N/m^2) contours. (Von Kármán Vortex Street).

3. Mathematical Model

A single power module of HVPG has been modeled as a spring mass system undergoing instability induced vibration. The instability is caused by the shedding of vortices in the wake of the cylinder when the flow encounters a bluff body. Alternate vortex shedding causes oscillatory forces that induce structural vibrations, where the rigid cylinder is now similar to a spring-mass system with a harmonic forcing term. This phenomenon is referred to as heave motion [14]. Equation of motion [16] for this system can be written as

$$m\ddot{Y} + c\dot{Y} + kY = F(t) \tag{4}$$

where Y is the displacement of the cylinder under VIV in the cross-flow direction; m is the sum of mass of the oscillating system or body mass, m_b and added mass of the system, m_a. m_a is defined as $m_a = C_A m_b$, where C_A is the added mass coefficient. c is the damping coefficient and k is the coefficient of stiffness of the spring mass system. *F(t)* is the time varying force acting on the cylinder due the flow instability. Relatively small oscillation amplitudes are approximated by

$$F(t) = F_L sin(\omega_v \, t + \varphi) \tag{5}$$

where ω_v is the circular frequency of vortex shedding and φ the phase difference between the force and cylinder displacement. F_L is the maximum value of oscillating hydrodynamic lift force acting on the cylinder and is given by Equation (2).

The amplitude of oscillation of the system depends on the mass (m^*) ratio of the oscillating cylinder given by

$$m^* = \frac{m}{m_{fd}} \tag{6}$$

where m_{fd} is the mass of fluid displaced by the oscillating mass. The maximum possible response amplitude at any Re, Y_{max} can be calculated from the empirical relation between non-dimensional amplitude ($A_y = \frac{Y_{max}}{D}$) and Re [10] as given by Equation (7).

$$A_Y = -0.4435\left[log\frac{\alpha}{Re}\right] - 1.5 \tag{7}$$

where α is defined as

$$\alpha = (m^* + C_A)\xi \tag{8}$$

where C_A is the added mass coefficient and ξ is the damping ratio.

Maximum amplitude of oscillation occurs when shedding frequency locks on to the natural frequency of the oscillating system (f_n). This condition is known as lock-in. Amplitude of oscillation of a spring mass system can also be obtained from Equation (9)

$$Y = \frac{F_L}{k}\left[\frac{1}{\sqrt{(1-\eta^2)^2 + (2\xi\eta)^2}}\right] \tag{9}$$

where η is the frequency ratio represented by Equation (10)

$$\eta = \frac{f_n}{f_v} \tag{10}$$

where f_v is the vortex shedding frequency. During lock-in $\eta = 1$ and Equation (9) simplifies as shown by Equation (11).

$$Y = \frac{F_L}{2k\xi} \tag{11}$$

The paper discusses effect of mass and stiffness ratios of the oscillating cylinder on the maximum response amplitude. The structural damping variations are not considered as it is observed to be less significant compared to its inertia and elastic counterparts. Also, mass and stiffness of the system are more tangible parameters from design point of view compared to damping. Moreover, the study focuses on non-dimensionalizing the influencing parameters so that results hold applicable for prototypes and models equally. Many researchers have considered the combined effect of mass and damping through mass damping parameter $m^*\xi$. In such analysis also ξ is kept constant and m^* is varied independently [17].

4. Numerical Determination of Hydrodynamic Lift Forces and Motions

Vortex induced vibration, a two-way fluid structure interaction phenomenon, is complex in nature due to the fact that the cylinder displacement is capable of changing the vortex shedding pattern behind it leading to a variation in the hydrodynamic load acting on the cylinder. Parameters k^*, m^*, and c have significant influence on the oscillation amplitude. A significant amount of research has been carried out to bring clarity on the influence of these parameterson the maximum cylinder response amplitude. Experiments conducted by [18] could capture lock-in phenomenon for a cylinder with a single degree of freedom. It was observed that cylinder oscillation frequency matches with vortex shedding frequency for cylinders having low mass ratios [19]. For mass ratios below critical mass ratio, m^*_{cr}, the range of U_r over which resonance occurs tend to extend towards infinity. The following section of the paper is an effort to understand the effect of mass and stiffness coefficient on the response of cylinder numerically in a simpler and economic way, and the results of the numerical study is verified using a field test.

4.1. Modeling the Flow

Numerically this problem has been treated as a case of two-way fluid structure interactions (two-way FSI). Modeling and meshing has been performed in ANSYS ICEM CFD (version 12) [20] and solving using ANSYS FLUENT (version 12) [21]. Flow around the cylinder is modeled using the transient, incompressible Navier–Stokes equation based RANS solver with k–ω SST as the turbulence model. The RANS solver does the virtual averaging of velocities over an interval of time and hence for a specific interval the velocity vector appears to be constant in a RANS solver. In the present work, an optimized fine grid is used to compensate for this drawback of the solver enabling it to capture the physics of von Kármán street eddies. RANS solver for transient two-dimensional analysis can be explained as follows [21]

$$\frac{\partial \rho}{\partial t} + \frac{\partial}{\partial x_i}(\rho u_i) = 0 \tag{12}$$

$$\frac{\partial}{\partial t}(\rho u_i) + \frac{\partial}{\partial x_i}(\rho u_i u_j) = -\frac{\partial p}{\partial x_i} + \frac{\partial}{\partial x_j}\left[\mu\left(\frac{\partial u_i}{\partial x_j} + \frac{\partial u_j}{\partial x_i} - \frac{2}{3}\delta_{ij}\frac{\partial u_l}{\partial x_l}\right)\right] + \frac{\partial}{\partial x_j}\left(-\rho\overline{u_i' u_j'}\right) \tag{13}$$

where u_i and u_i' are mean and fluctuating velocity components for $i = 1, 2$, and 3.

The velocities and other solution variables in the above equation represent the time averaged values. Equation (12) is solved by modeling Reynolds stresses $\rho\overline{u_i' u_j'}$, effectively using k–ω SST as turbulence model [22]. k–ω SST is capable of accurately predicting the commencement and the intensity of flow separation at fixed boundaries while the standard k–ε model has proved its efficacy in predicting the wake characteristics accurately. This fact has been established after extensive studies conducted by the authors. k–ωSST turbulence model demands a very high near wall grid resolution and hence the maximum element size is fixed to be less than 1×10^{-4}, which satisfies CFL criterion and near wall $y+$ values.

The above governing equations are discretized using finite difference method. Non iterative time advancement (NITA) scheme with fractional time stepping method (FSM) has been chosen for pressure-velocity coupling of the grid. A Least Squares Cell Based scheme is used for gradient in spatial discretization and a second order upwind scheme as convective scheme.

NITA

For capturing the physics of the flow with accuracy in the boundary as well as in the wake of the cylinder, the computational grid needs to be extremely fine. Solving a dynamic mesh case with such extreme grid fineness using iterative time advancement scheme demands a considerable number of iterations to be performed using a very small time step size to satisfy the dynamic mesh criteria. This in turn leads to huge computational cost and effort. As an alternative and computationally economic method, in the present work, NITA with FSM has been implemented. NITA scheme assures the same time accuracy by reducing the splitting error, which occurs while solving the discretized Navier–Stokes equation to the same order as the truncation error. Splitting error need not be reduced to zero in NITA scheme, saving a lot of computational effort.

4.2. Structural Modeling

An elastically mounted cylinder can be mathematically represented by Equation (4). This equation of motion is solved using six degrees of freedom solver (6DOF), an integral part of the main solver by defining the cylinder as an object with one degree of freedom (1DOF) in transverse direction. A user defined function (UDF) compiled in C programming language has been hooked to the cylinder dynamic boundary conditions. The governing equation for the motion of the center of gravity of the cylinder in the transverse direction is solved in the inertial coordinate system. Velocity in the transverse direction is obtained by performing integration on Equation (14).

$$\ddot{Y} = \frac{1}{m}\sum F \tag{14}$$

where $\ddot{Y}$, is the translational acceleration in the transverse direction, m is the mass of the cylinder and F, resultant fluid force acting on the cylinder. Position of the center of gravity of the cylinder (CG) is updated after solving the equation of motion of a spring mass system.

$$m\ddot{Y} + c\dot{Y} + kY = F(t) \tag{15}$$

The inertial force term on the left-hand side of Equation (15) is computed by the 6DOF solver for each time step from Equation (14) and the UDF hooked to the moving cylinder inputs the restoring force term as 6DOF load acting on the cylinder. Mass of the cylinder is given in the UDF as

$$m = m_b + m_a \tag{16}$$

$$m_a = (1 + C_A)m_b \tag{17}$$

Added mass coefficient C_A for the aspect ratio of the present model is found to be equal to 0.7 [13].

4.3. Mesh Deformation

Mesh motion to adapt to the movement of the cylinder is achieved by using displacement based smoothing algorithm. The governing equation for mesh motion is represented by Equation (18).

$$\nabla\cdot(\gamma\nabla\vec{u}) = 0 \tag{18}$$

where $\vec{u}$ is the velocity of mesh displacement. The boundary conditions for Equation (18) are computed by the 6DOF solver and the boundary mesh motion diffuses into the interior of the deforming mesh according to the Laplace equation, Equation (18). Diffusion coefficient, γ is calculated using boundary distance formulation given by Equation (19).

$$\gamma = \frac{1}{d^{\tau}} \tag{19}$$

where τ is the diffusion parameter and d is the normalized boundary distance. Diffusion parameter is set as unity to avoid excessive deformation of the near cylinder elements.

4.4. Fluid Structure Interaction

In this paper, a two-way implicit approach is used to study the effect of m^* and k^* on the response of cylinder under VIV. Flow equations and structural equations are solved simultaneously in iterations with a time step. A flow chart for the solution procedure is shown in Figure 2.

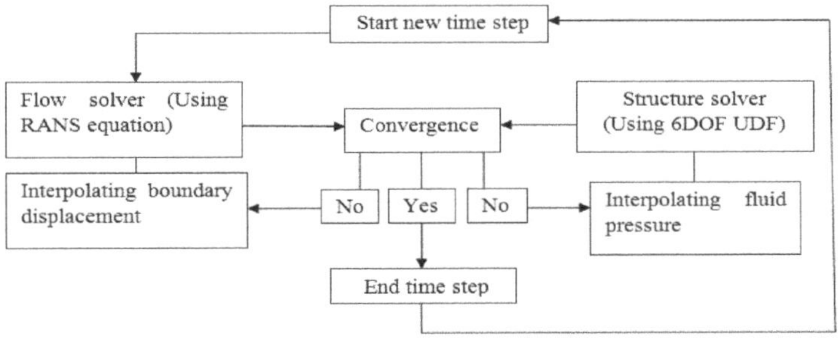

Figure 2. Flow chart of two-way implicit FSI solution procedure.

5. Problem Description

5.1. Simulation Parameters

In the present study, geometrically identical cylinders of different masses are considered for VIV of an elastically mounted horizontal cylinder. Based on the present study, a basic model of HVPG as in [5] was fabricated and a field study was carried out on it for verifying the numerical results. The main parameters of the model are summarized in Table 1. The model has an aspect ratio of 13.12 and an outer diameter of 0.0762 m. The structural damping is considered to be zero in the present study. The computational domain for the study is shown in Figure 3a. The representation of the present problem as a spring-mass system is depicted in Figure 3b. The mesh fineness in indicated in Figure 3c and the grid used for the present computation is represented by Figure 3d.

Table 1. Properties of the elastically mounted cylinder model.

Properties	Values	Units
Diameter of the cylinder (D)	0.0762	m
Aspect ratio of the cylinder (L/D)	13.12	-
Flow velocity (V)	0.5	m/s
Reynolds Number of flow (Re)	3.8 × 104	-
Mass ratio (m^*)	0.66	-

A simple representation of the two-dimensional computational domain for the cylinder is displayed in Figure 3a. The flow direction is parallel to the global x-axis and the flow velocity is set to be 0.5 m/s. Domain size is fixed based on previously published analysis and allowance is given to accommodate the vertical motion of cylinder boundary [23]. Simulations are performed for three different values of mass ratio, $m^* = 0.66$, 1.32, and 1.98 which correspond to moderate mass ratios above m^*_{cr}, for which maximum response in confined to the range of $U_r = 4$–12. For fixed mass, influence of stiffness ratio k^* is studied by varying the reduced velocity value which is defined as

$$U_r = \frac{V}{f_n D} \tag{20}$$

$$f_n = \frac{1}{2\pi}\sqrt{\frac{k}{m}} \tag{21}$$

Each case has been analyzed over a range of reduced velocity, $U_r = 4$–12, over which the cylinder is predicted to have maximum amplitude of oscillation [24]. $U_r = 5$ has also been analyzed since the case corresponds to $\eta = 1$ where one might expect resonance. The incoming flow velocity is fixed as 0.5 m/s to maintain the flow regime uniform at $Re = 3.8 \times 10^4$. The mass ratio, stiffness ratio, and other parameters for each case are summarized in Table 2. Stiffness coefficient of the cylinder has been non-dimensionalized to generalize the applicability of the analysis. Stiffness ratio is defined as

$$k^* = \frac{k}{mg/L} \tag{22}$$

Table 2. Reduced velocity, stiffness ratio, and frequency ratio at $Re = 3.8 \times 10^4$, for $m^* = 0.66$, 1.32, and 1.98.

U_r	k^*	η
4	11.17	1.3
5	6.9	1.0
6	4.81	0.84
8	2.7	0.63
10	1.73	0.51
12	1.21	0.42

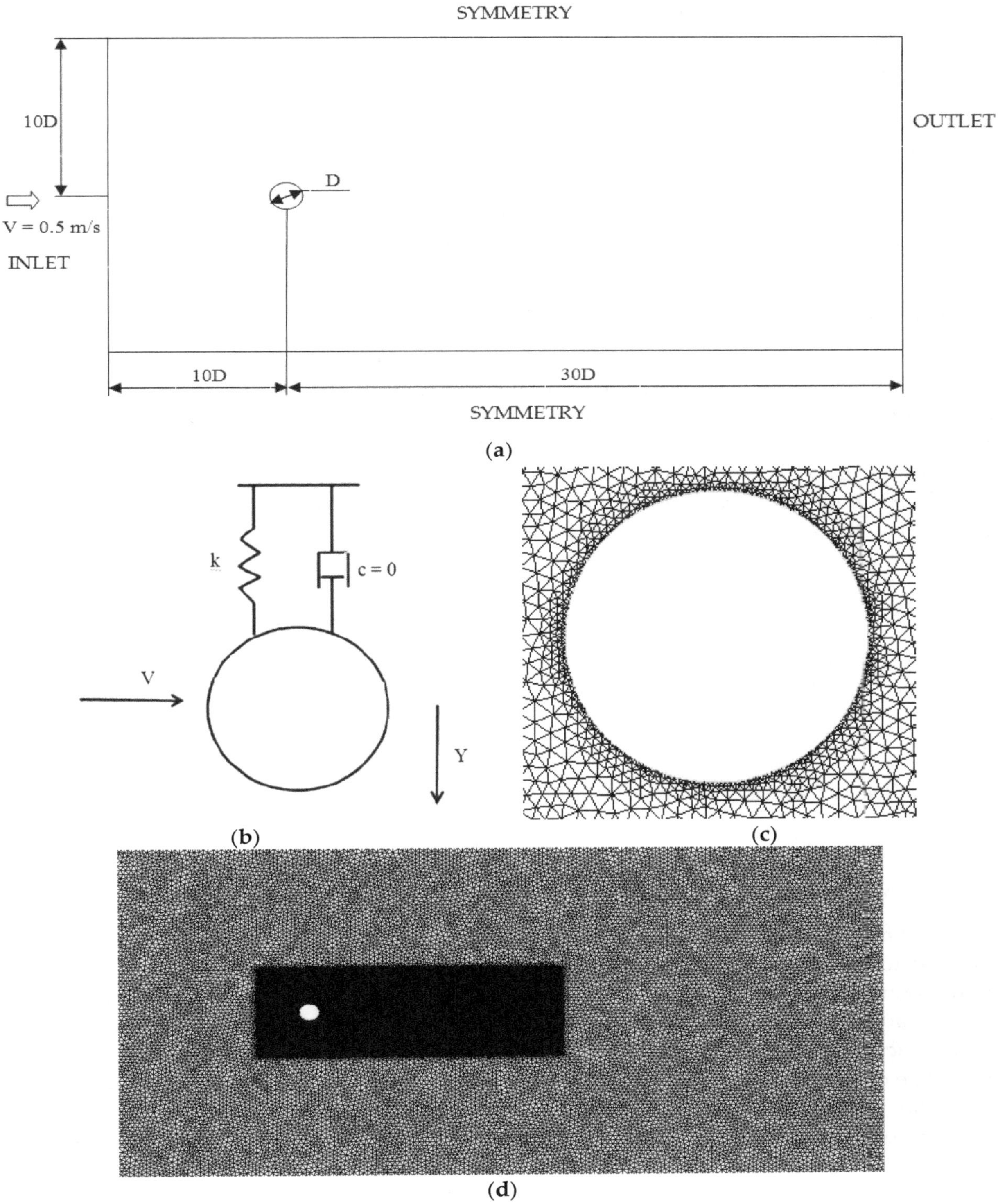

Figure 3. (**a**) Computational domain; (**b**) representation of elastically mounted cylinder model; (**c**) mesh around the cylinder; (**d**) computational mesh.

5.2. *Fluid Domain and Boundary Conditions*

Figure 3a shows the computational domain for the CFD simulation of VIV of an elastically mounted horizontal cylinder. The origin of the Cartesian coordinate system is located at the center of the cylinder. The length of the domain is 40D with the cylinder located at 10D away from the inlet boundary. The cross-flow width of the domain in 20D with the center of the cylinder at the middle. Detailed views of the mesh around the cylinder along with the computational domain after meshing have been shown in Figure 3c,d respectively. There are 307 nodes around the circumference of the cylinder and the minimum element size near the rigid wall boundary has been computed from boundary layer theory to be $0.0001D$. The non-dimensional element size represented as $y+$, next to the cylinder surface is found to be less than unity. For the cylinder wall, a no slip boundary condition has been applied assuming the surface to be smooth. Inlet boundary has been treated has velocity-inlet with inflow velocity, $V = 0.5$ m/s. Outlet boundary has been treated as pressure outlet, the gradients of fluid velocity are set to zero and the pressure with zero reference pressure. On the two transverse boundaries a symmetry boundary condition has been applied.

5.3. Mesh Independence Study

An unstructured 2D mesh has been used in the present CFD simulation to facilitate computationally economic platform for dynamic mesh simulation. The three dimensionality of the wake reduces as a result of the motion of the cylinder [16]. Hence it is possible to get a reasonably accurate result from a 2D analysis saving much computational cost and effort. The meshing strategy is that finer mesh is used in the vicinity of the moving cylinder with extra fine meshing in the boundary layer. Boundary layer thickness and the near wall element size have been calculated from boundary layer theory. The thickness of laminar sub-layer is obtained from Equation (23) [25]

$$\delta' = \frac{11.6\vartheta}{V^*} \tag{23}$$

where V^* is the frictional velocity given by

$$V^* = \sqrt{\frac{\tau_0}{\rho}} \tag{24}$$

and τ_0, the wall shear stress is obtained as

$$\tau_0 = \frac{0.664}{\sqrt{Re_D}}.\frac{\rho V^2}{2} \tag{25}$$

Increased mesh density has been adopted in the near cylinder and its wake in order to capture the physics of vortex shedding accurately. For ensuring that the results are independent of the grid size, a mesh independence study has been carried out. Three different grids have been used to simulate a specific case $m^* = 2.45$ and $U_r = 8$ and has been verified using results of experiments conducted at a towing tank facility in the Department of Ocean Engineering, Indian Institute of Technology Madras, India [10]. The details of the mesh independency study are given in Table 3. The last three grids give almost similar results and experiments, and [10] shows a 4.9% deviation from the present results. This can be assumed to be due to not accounting for damping in the present study. It can be concluded that the variation in the numerical results given by Meshes II, III, and IV are in the acceptable range and considering the computational economy Grid II with 49,995 nodes has been chosen for further analysis.

Table 3. Mesh independency study results

	Re	m^*	U_r	Nodes	Y_{max}/D
Grid I	3.8 × 104	2.45	8	35,487	1.241
Grid II				49,995	1.220
Grid III				70,857	1.219
Grid IV				98,475	1.219
Narendran et al. (2015)	0.3–2.4 × 105	2.45	8	-	1.160

While due steps like mesh independence studies have been carried out in this investigation, we acknowledge that the use of eddy-viscosity based turbulence models introduces a small amount of discrepancy in the final results. The errors of the k–ωSST model in flow separation and wakes has been documented comprehensively in prior studies [26,27].

6. Results and Discussion

Numerical simulations have been carried out for three different mass ratios over reduced velocities ranging from 4 to 12. The cylinder is modeled to be having only single degree of freedom (SDOF) in the transverse or CF direction. Influence of m^* and k^* on both hydrodynamic force coefficient in the CF direction and the response of the cylinder have been studied in detail. The CF response results can be verified with experimental [17,28]. Time history of coefficient of lift, C_L and non-dimensional CF response Y/D over the range of U_r for $m^* = 0.66$, 1.32, and 1.98 are displayed in Figures 4–6 respectively.

6.1. Case I

Under Case I, the response of cylinder is studied for $m^* = 0.66$. The mass of the cylinder is taken as 3 kg and added mass coefficient $C_A = 0.7$. U_r is varied in the analysis by varying the coefficient of stiffness k and in turn the natural frequency f_n of the oscillating system. Details of the simulation parameters have been given in Table 2. For a stationary cylinder, it was observed from numerous experimental and numerical works [29–31] that C_L oscillates in a symmetrical fashion about zero due to vortex shedding [13]. In the present study for lower values of U_r, C_L is not oscillating about zero symmetrically. However, the response of the cylinder is observed to be symmetrical for all values of U_r at $m^* = 0.66$. At $U_r = 8$, C_L becomes almost symmetrical with an effective lift coefficient 0.18. Beat phenomenon is captured in the response of the cylinder at $U_r = 8$. Results for different cases are presented in Table 4. Maximum lift force is observed at $U_r = 4$ with $C_L = 1.12$ and maximum response at $U_r = 8$ with $Y/D = 1.26$. With increase in U_r response amplitude of the cylinder decreases beyond $U_r = 8$. Time histories of C_L and Y/D for $m^* = 0.66$ is presented in Figure 4.

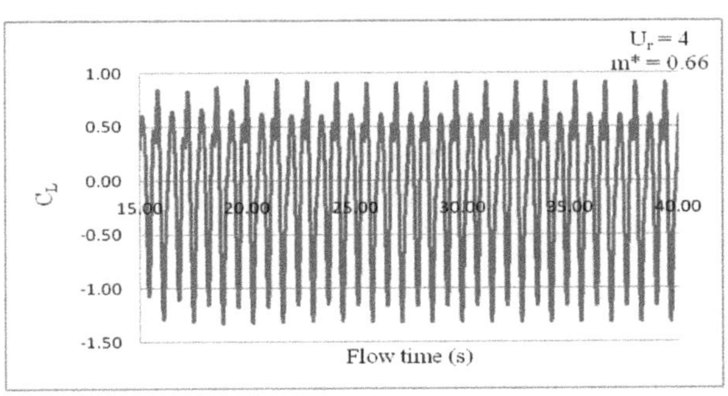

(a)

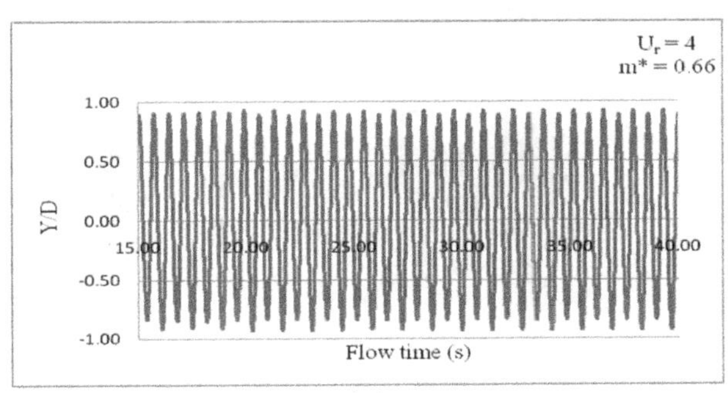

(g)

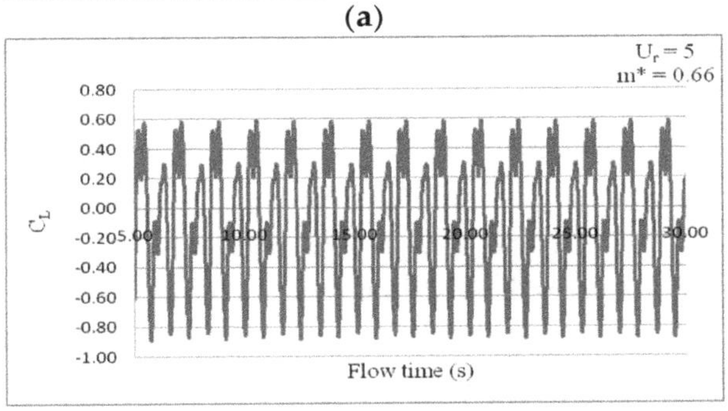

(b)

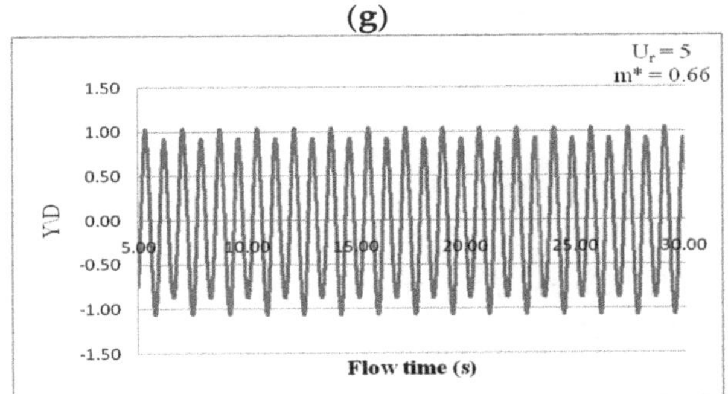

(h)

Figure 4. *Cont.*

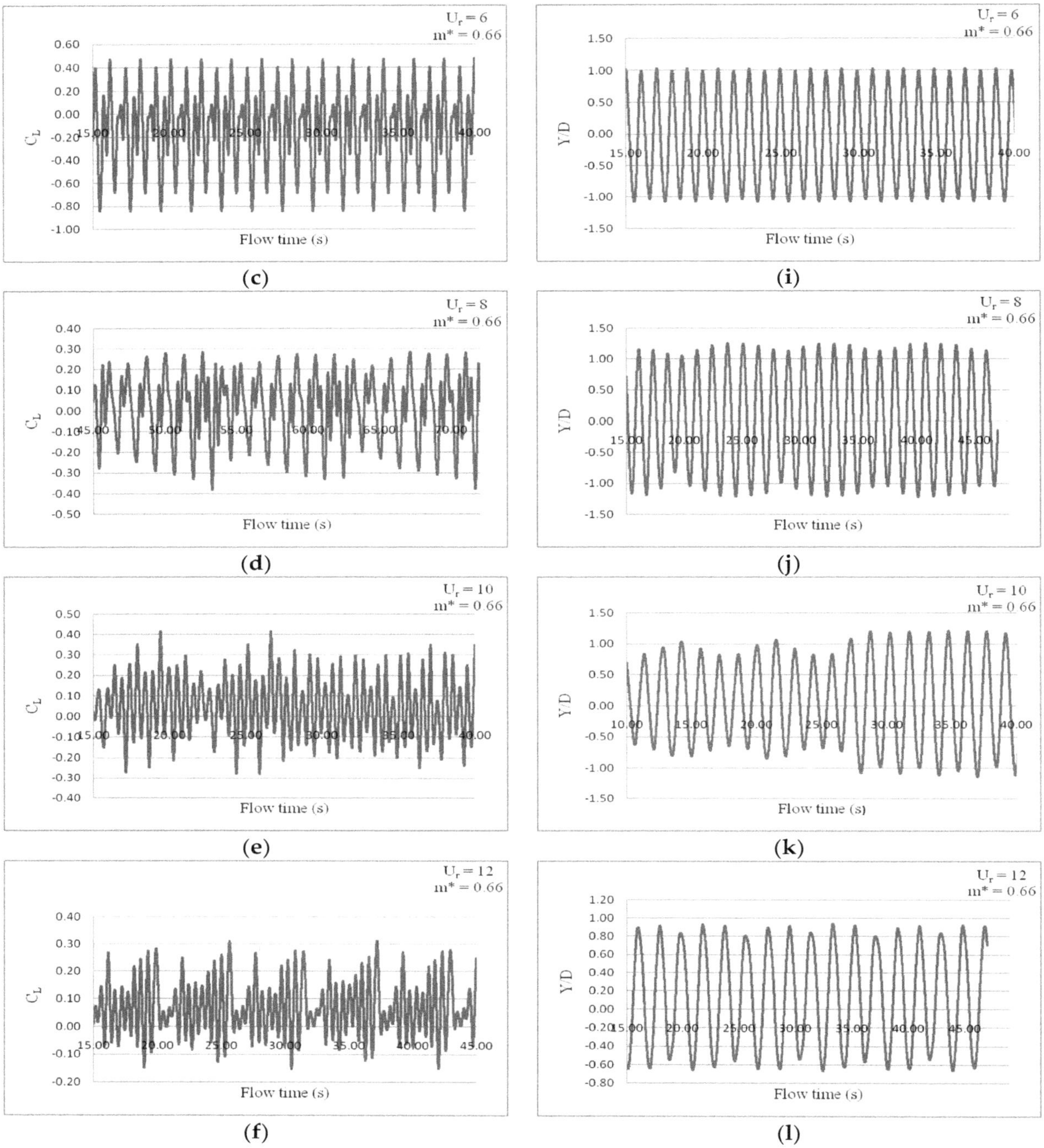

Figure 4. Time histories of hydrodynamic and structural response parameters at $m^* = 0.66$: (**a**) C_L for $U_r = 4$; (**b**) C_L for $U_r = 5$; (**c**) C_L for $U_r = 6$; (**d**) C_L for $U_r = 8$; (**e**) C_L for $U_r = 10$; (**f**) C_L for $U_r = 12$; (**g**) Y/D for $U_r = 4$; (**h**) Y/D for $U_r = 5$; (**i**) Y/D for $U_r = 6$; (**j**) Y/D for $U_r = 8$; (**k**) Y/D for $U_r = 10$; (**l**) Y/D for $U_r = 12$.

6.2. Case II

Under Case II, the response of cylinder is studied for $m^* = 1.32$. Mass of the cylinder is taken as 6 kg and added mass coefficient $C_A = 0.7$. Time histories of C_L and Y/D for $m^* = 1.32$ is presented in Figure 5. At $m^* = 1.32$, the cylinder exhibits similar response characteristics as in the previous case with maximum response at $U_r = 8$ with $Y/D = 1.17$. As the mass ratio increases, a slight decrease in the cross flow response amplitude is observed. Unlike the previous case, beat phenomenon is observed at $U_r = 10$.

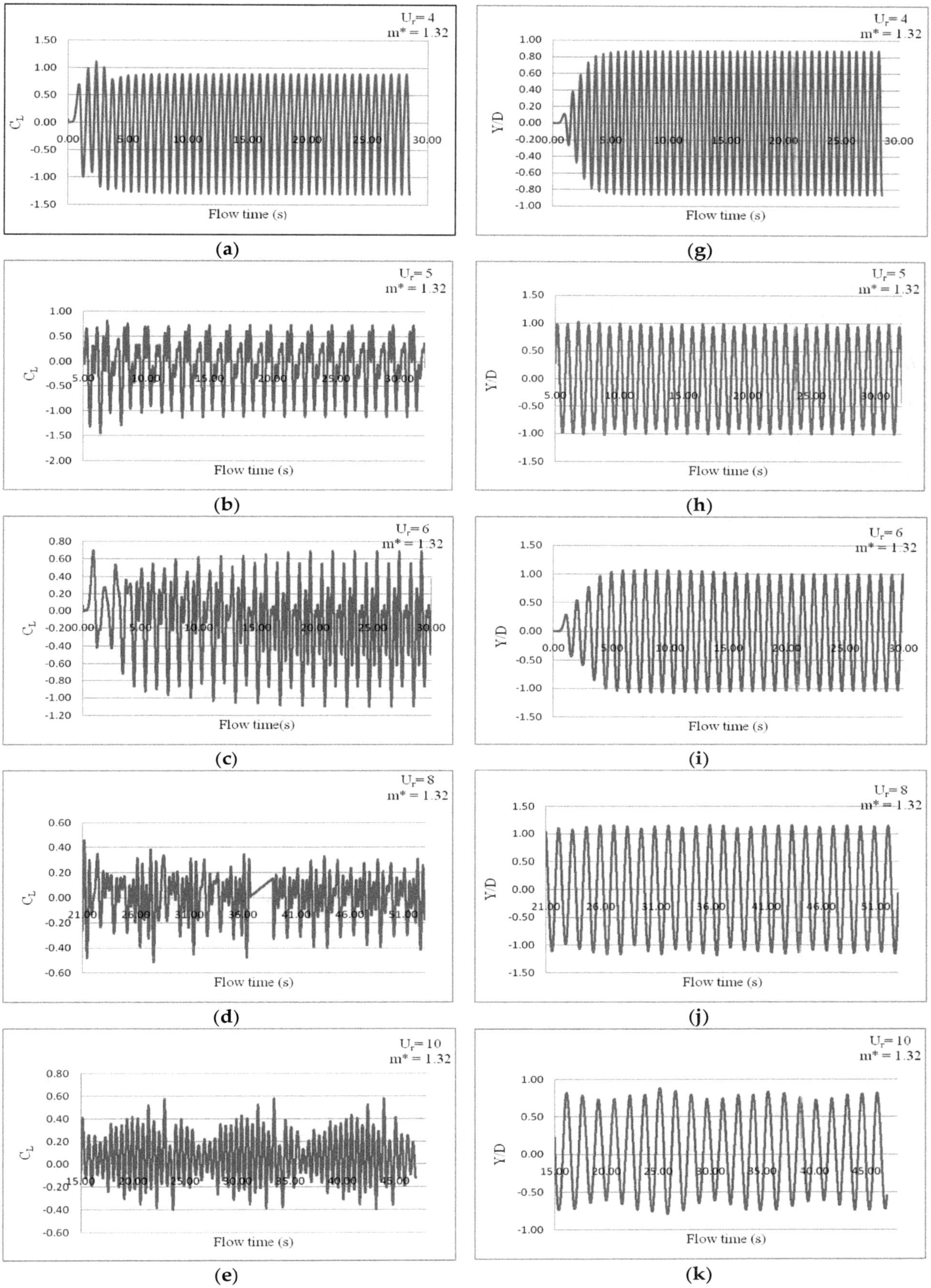

Figure 5. *Cont.*

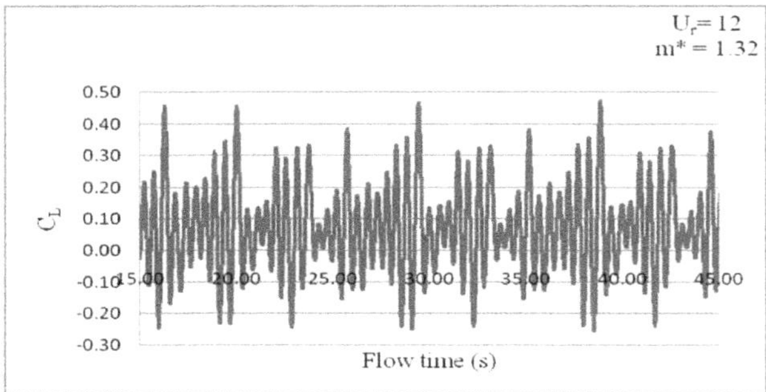

(f)

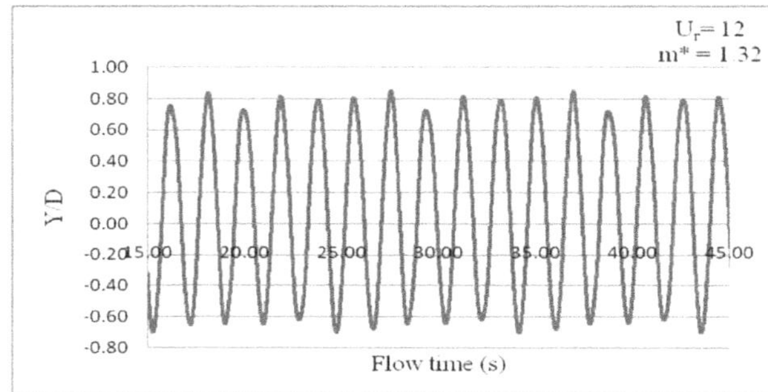

(l)

Figure 5. Time histories of hydrodynamic and structural response parameters at $m^* = 1.32$: (**a**) C_L for $U_r = 4$; (**b**) C_L for $U_r = 5$; (**c**) C_L for $U_r = 6$; (**d**) C_L for $U_r = 8$; (**e**) C_L for $U_r = 10$; (**f**) C_L for $U_r = 12$; (**g**) Y/D for $U_r = 4$; (**h**) Y/D for $U_r = 5$; (**i**) Y/D for $U_r = 6$; (**j**) Y/D for $U_r = 8$; (**k**) Y/D for $U_r = 10$; (**l**) Y/D for $U_r = 12$.

6.3. Case III

Under Case III, the response of cylinder is studied for $m^* = 1.98$. The mass of the cylinder is taken as 9 kg and added mass coefficient $C_A = 0.7$. Time histories of C_L and Y/D for $m^* = 1.98$ is presented in Figure 6. Maximum amplitude is observed at $U_r = 8$ with $Y/D = 1.13$ which is slightly less than the previous two cases. Here also prominent beat is observed at $U_r = 10$.

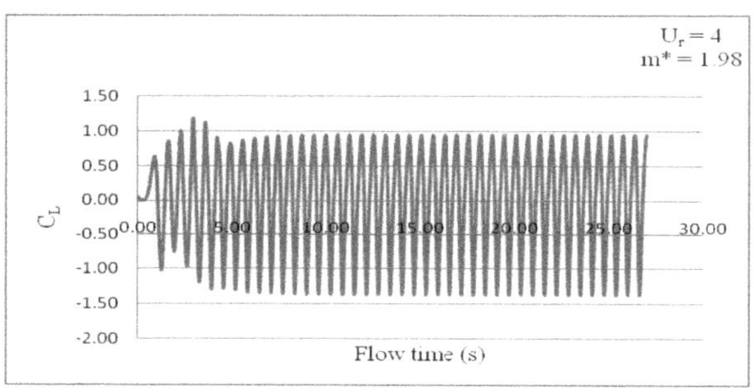

(a)

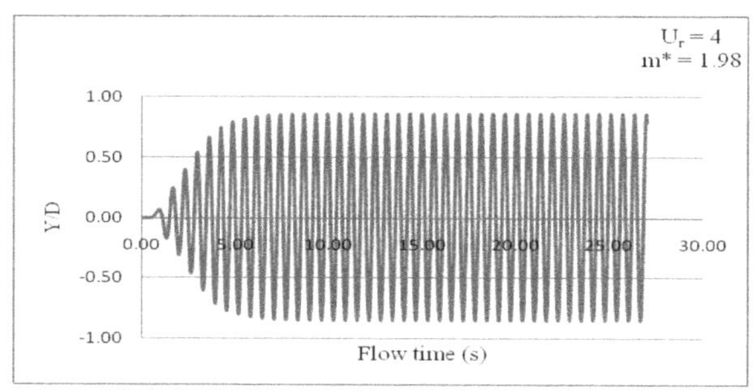

(g)

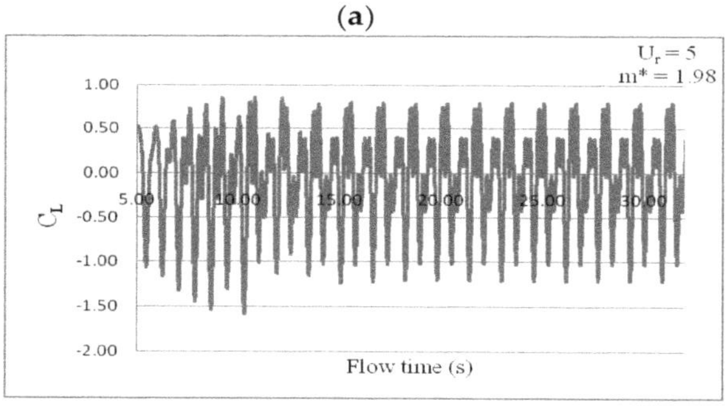

(b)

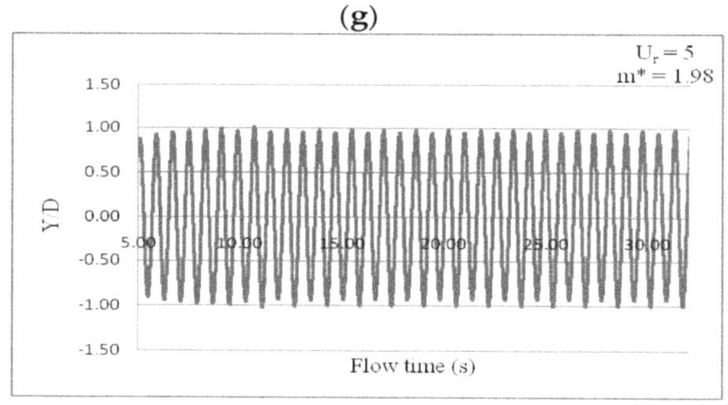

(h)

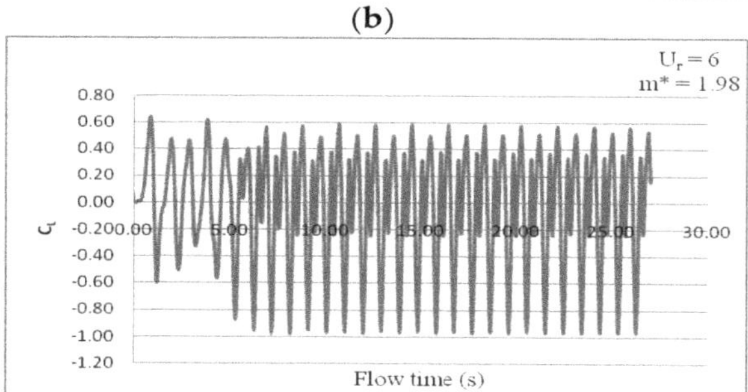

(c)

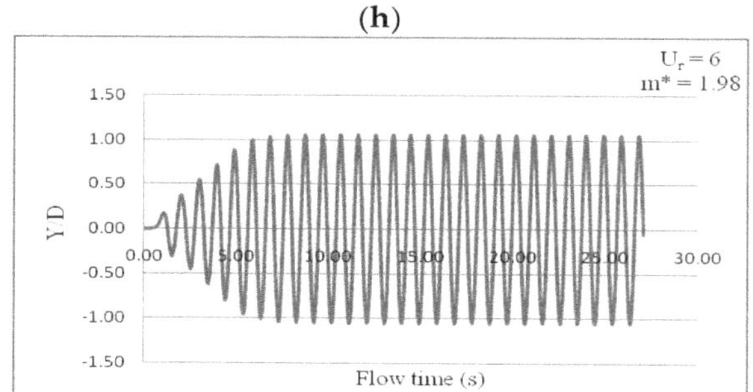

(i)

Figure 6. *Cont.*

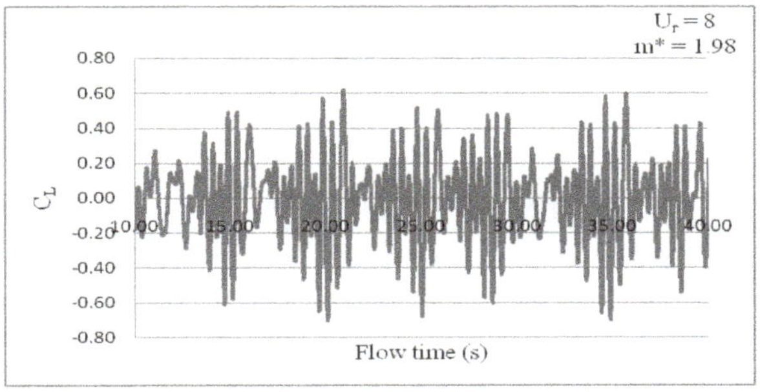

(d)

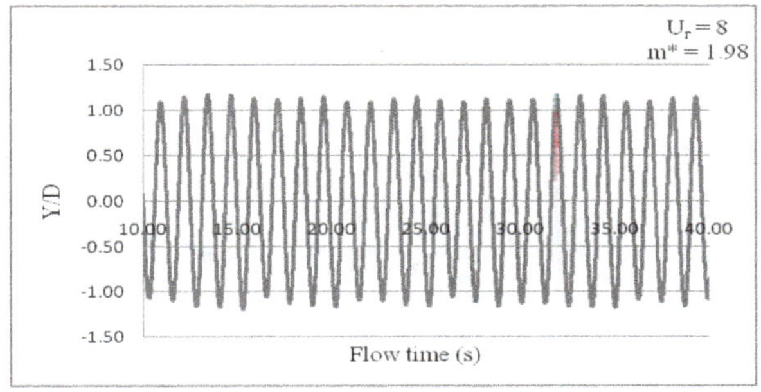

(j)

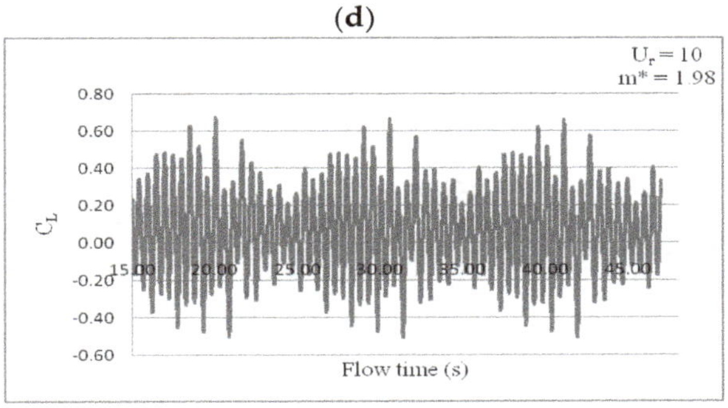

(e)

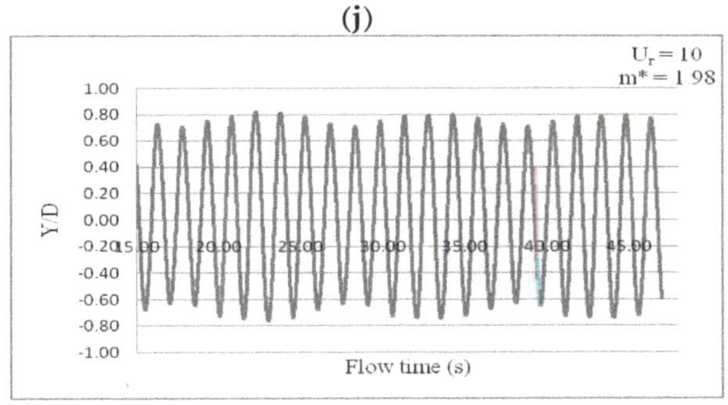

(k)

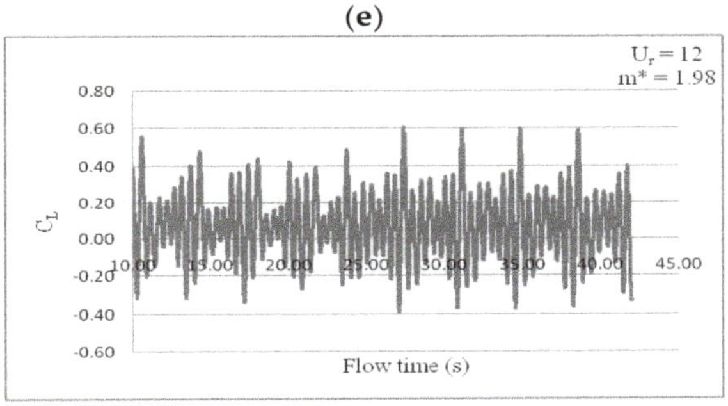

(f)

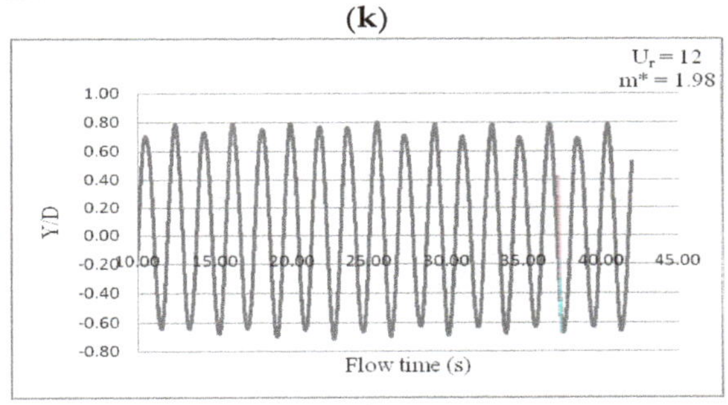

(l)

Figure 6. Time histories of hydrodynamic and structural response parameters at $m^* = 1.98$: (**a**) C_L for $U_r = 4$; (**b**) C_L for $U_r = 5$; (**c**) C_L for $U_r = 6$; (**d**) C_L for $U_r = 8$; (**e**) C_L for $U_r = 10$; (**f**) C_L for $U_r = 12$; (**g**) Y/D for $U_r = 4$; (**h**) Y/D for $U_r = 5$; (**i**) Y/D for $U_r = 6$; (**j**) Y/D for $U_r = 8$; (**k**) Y/D for $U_r = 10$; (**l**) Y/D for $U_r = 12$.

6.4. Shedding Characteristics

Numerical simulations for each case show that the shedding pattern and the characteristic variation of C_L strongly depend on the natural frequency of the oscillating mass. A more detailed history of C_L and Y/D for $m^* = 0.66$, 1.32, and 1.98 are presented in Figures 7–9 respectively.

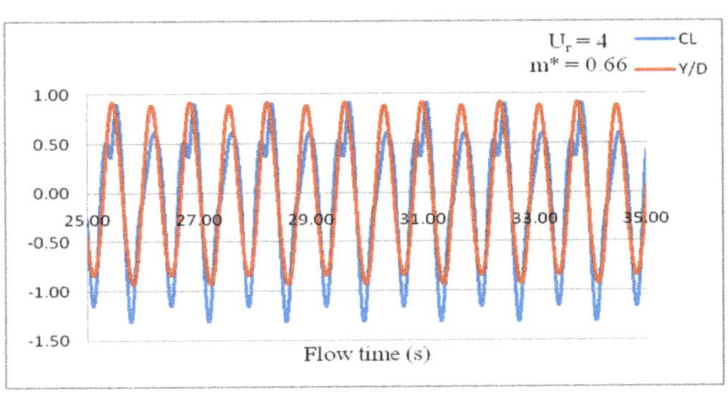

(a)

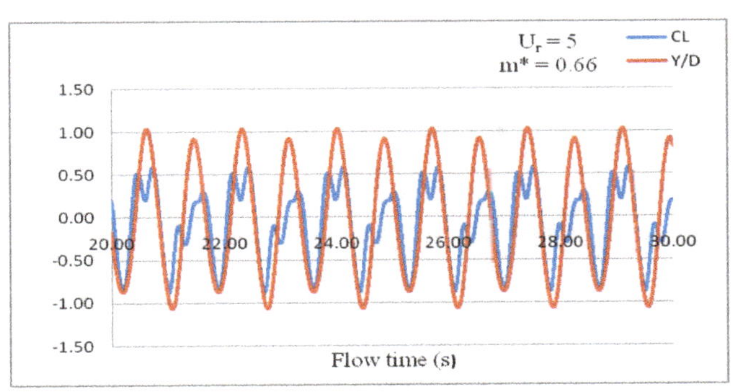

(b)

Figure 7. *Cont.*

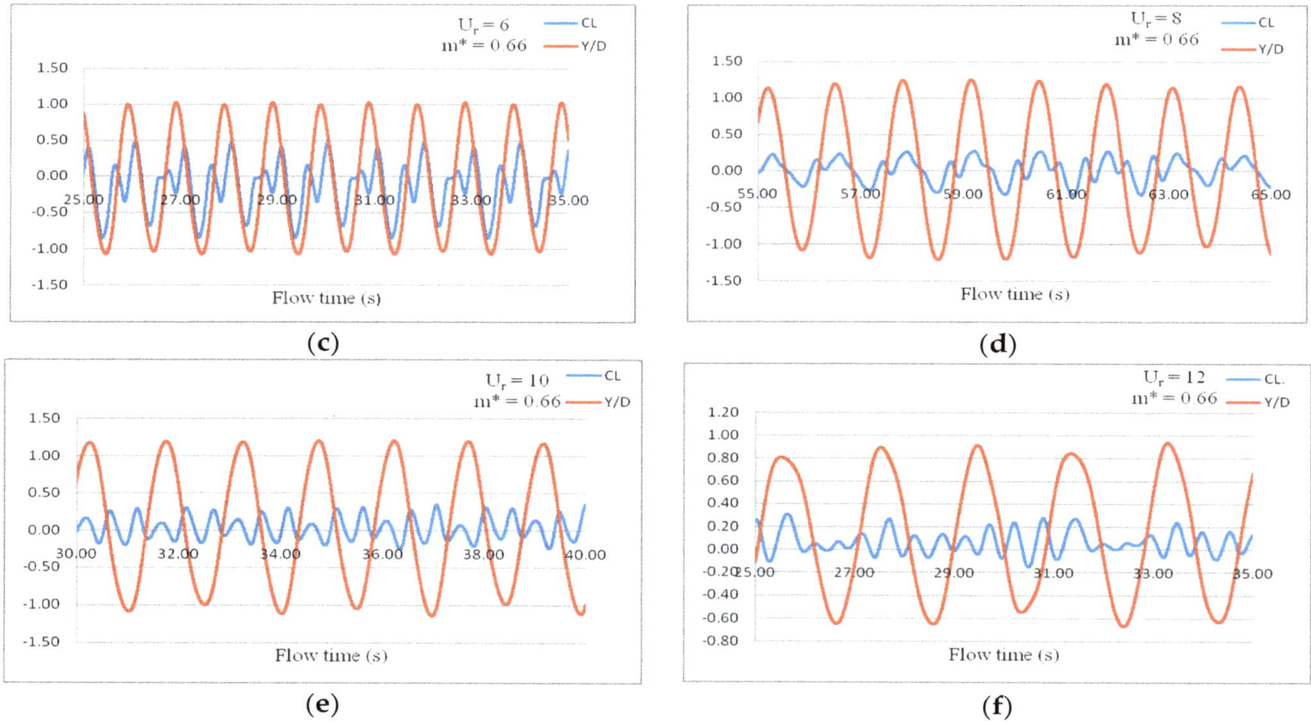

Figure 7. Variation of C_L with Y/D for m^* = 0.66 (**a**) U_r = 4; (**b**) U_r = 5; (**c**) U_r = 6; (**d**) U_r = 8; (**e**) U_r = 10; (**f**) U_r = 12.

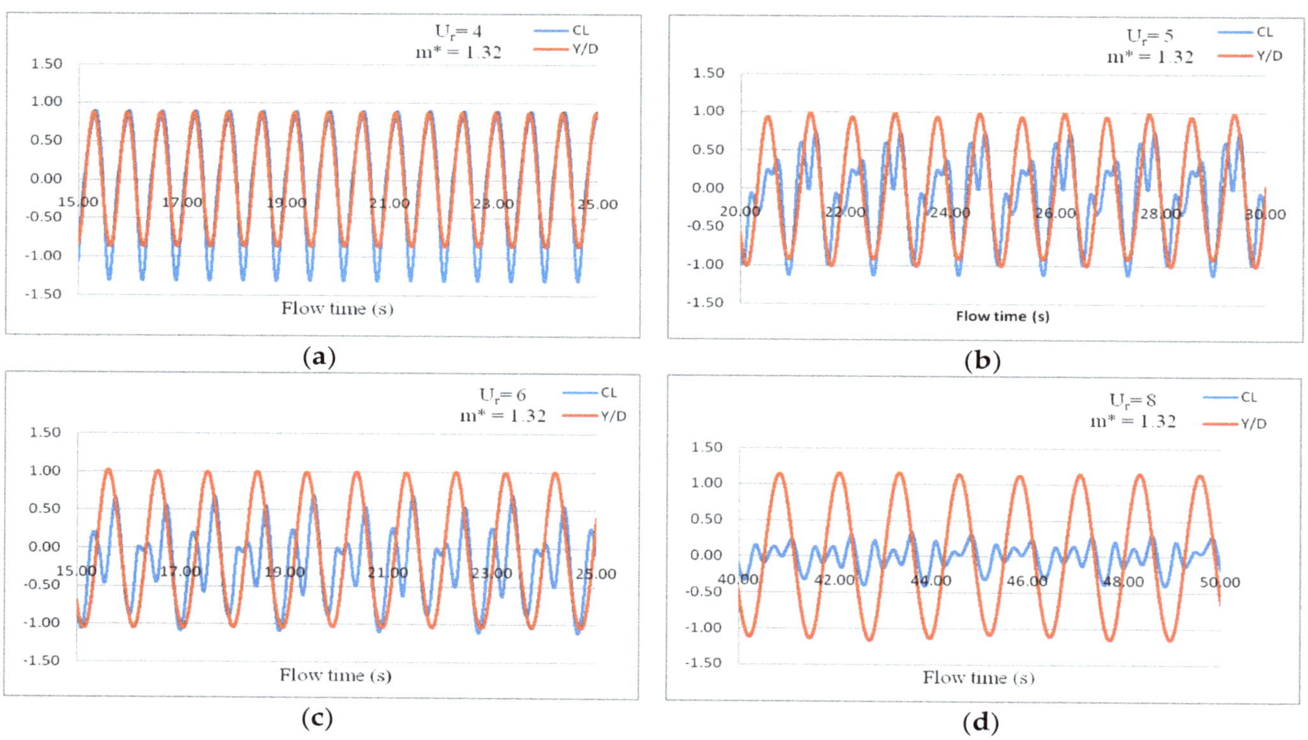

Figure 8. *Cont.*

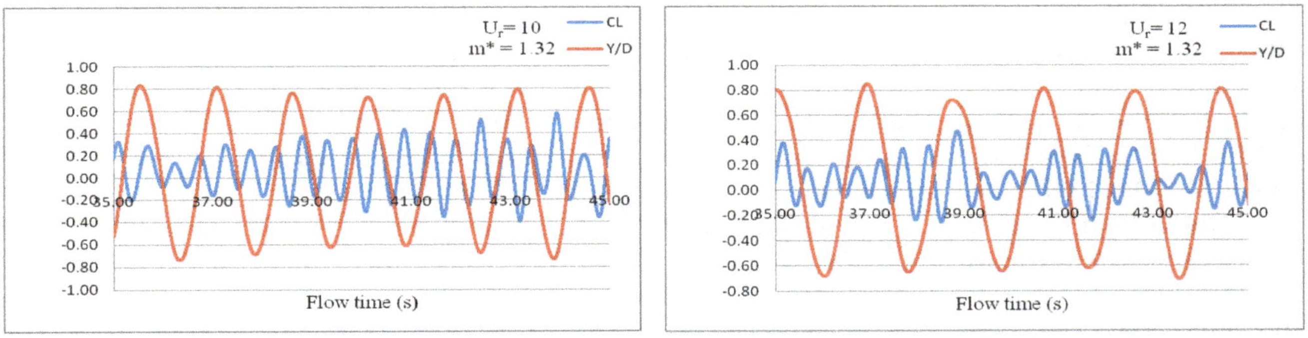

(e) (f)

Figure 8. Variation of C_L with Y/D for $m^* = 1.32$: (**a**) $U_r = 4$; (**b**) $U_r = 5$; (**c**) $U_r = 6$; (**d**) $U_r = 8$; (**e**) $U_r = 10$; (**f**) $U_r = 12$.

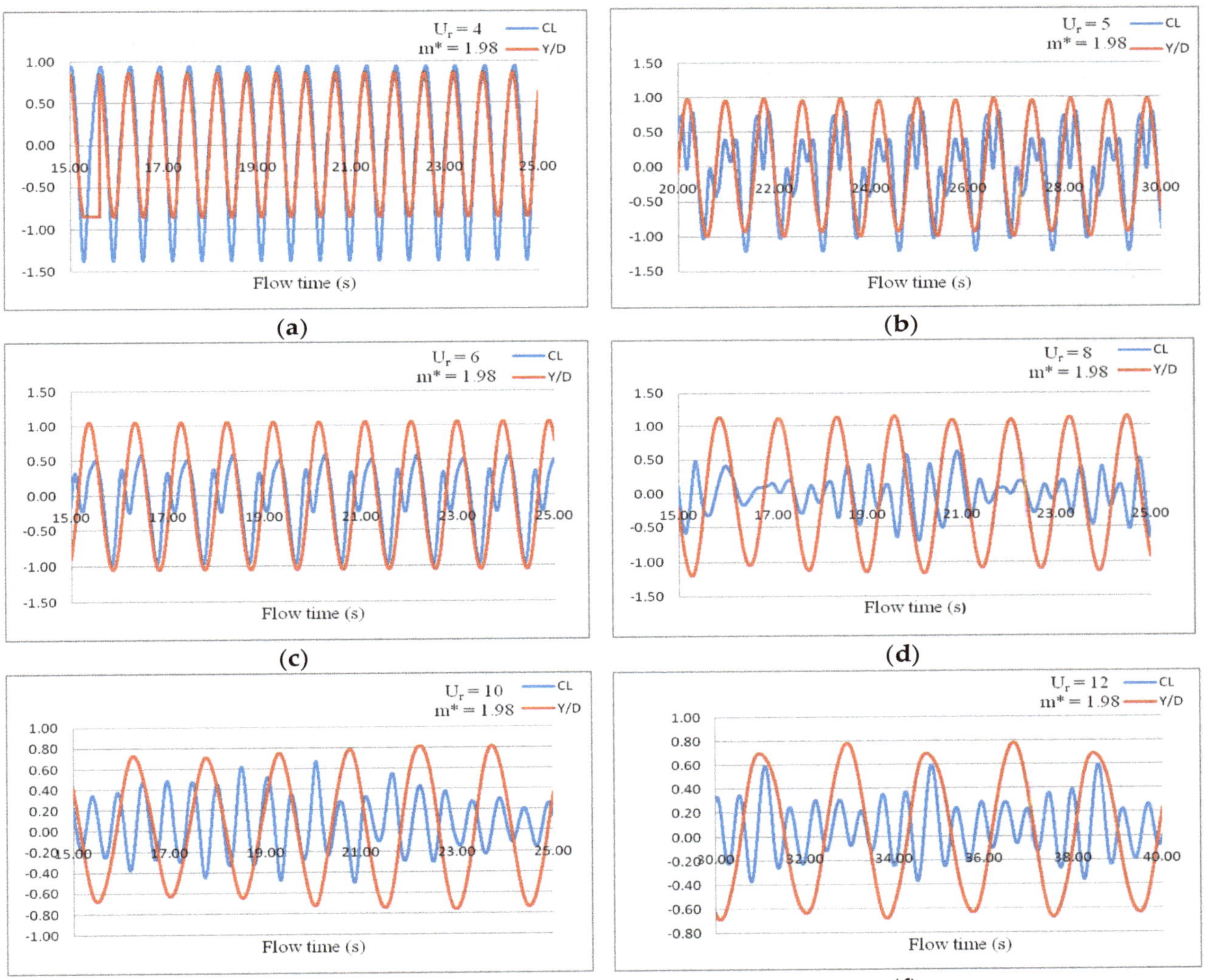

(e) (f)

Figure 9. Variation of C_L with Y/D for $m^* = 1.98$: (**a**) $U_r = 4$; (**b**) $U_r = 5$; (**c**) $U_r = 6$; (**d**) $U_r = 8$; (**e**) $U_r = 10$; (**f**) $U_r = 12$.

From a detailed analysis of amplitude of variation of lift coefficient, it is observed that the characteristics of hydrodynamic load causing the cylinder to oscillate varies with U_r and in turn with frequency ratio, η. For all the three cases, a similarity can be observed in the shedding characteristics which are reflected in the pattern of lift coefficient variation. In each case, a single oscillation of the cylinder is actuated by a single oscillation of lift force when $U_r = 4$. For $U_r = 6$, double oscillation of lift coefficient is observed for each oscillation of Y/D. The first oscillation is of relatively low amplitude and the second one of larger amplitude. $U_r = 8$ and 10 which is more prone to beating phenomenon displays three separate C_L oscillations for each oscillation of the cylinder. $U_r = 12$ is observed to have an additional half oscillation. For all mass ratios, a 2P mode of vortex shedding is observed in the range of $U_r = 4$–12 which corresponds to synchronization.

A single beat sustains for a time period of 10.53 and 10.42 seconds for $m^* = 1.98$ and 1.32, respectively. Differently for $m^* = 0.66$, single beat sustains at $U_r = 8$ for 7.84 s which is significantly lesser than the other two cases, showing the highest value of cross flow response amplitude. Also, at the lowest mass ratio, $m^* = 0.66$, it is observed from the response amplitude history that the cylinder initially shows a tendency to beat at $U_r = 10$ with a significantly larger response amplitude with $Y/D = 1.08$ compared to 0.82 and 0.80 for $m^* = 1.32$ and 1.98 respectively at same U_r. Further study on the influencing parameters led to a trend showing a relationship between m^* and U_r, through which they have proved that the range of reduced velocity over which cylinder under VIV shows maximum cross flow response widens as m^* decreases [10]. The relationship put forward by [14] is shown in Figure 11. Results of the present study also show that Y/D increases with decreasing mass ratio and the range of U_r over which the maximum amplitude of response to be expected widens.

The effect of U_r on various simulations parameters is illustrated in Figure 10. A comparison of various hydrodynamic and structural parameters at different k* values for each case is given in Figure 11a,b. Present simulation results matches well with the values represented in Figure 11b. In all cases maximum response occurs at $U_r = 8$. Even though the response amplitude is high in the range $U_r = 4$–12, as it deviates away from $U_r = 8$, amplitude decreases. A similar result was reported by [19] is represented in the modified Griffin's plot.

The numerical simulation could successfully reproduce the response behavior depicted by the lower branch of response. Figure 11b shows the response of a cylinder of mass ratio $m^* = 2.4$ which is comparable with the present study.

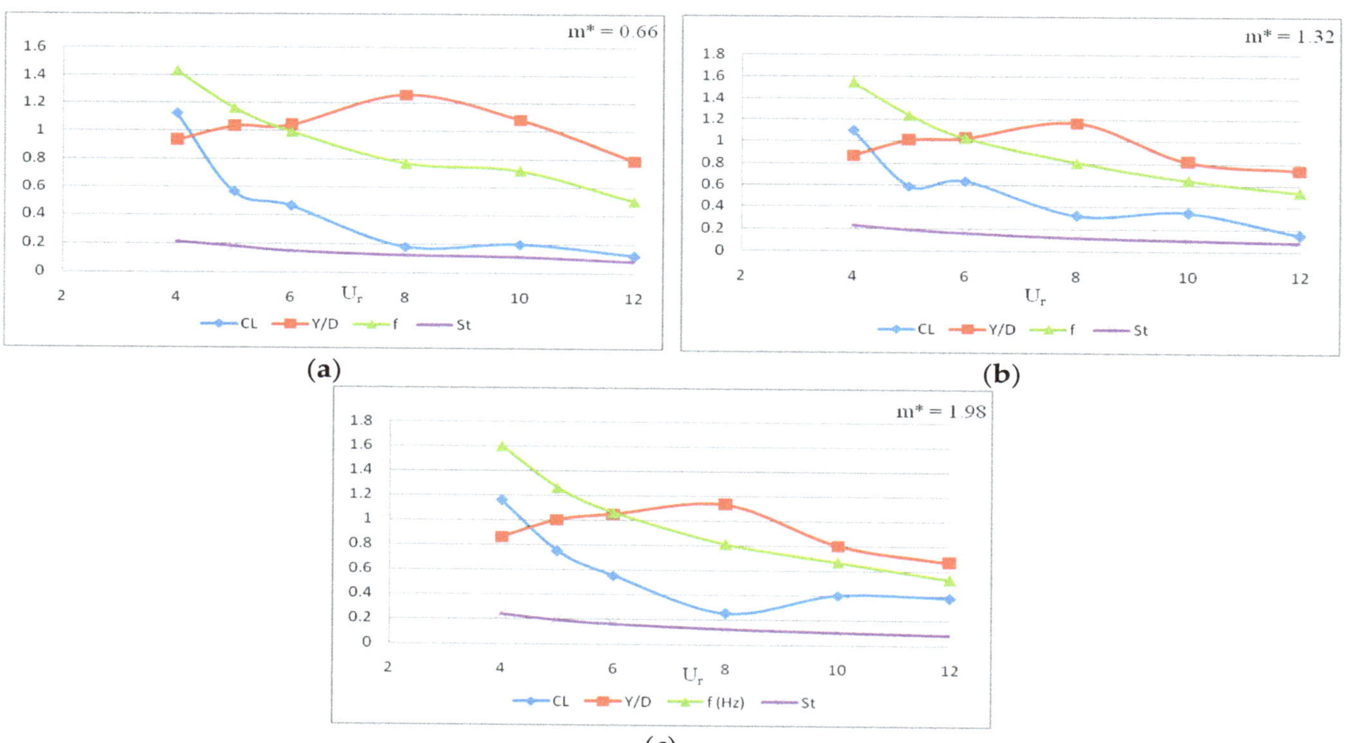

Figure 10. Effect of U_r on various hydrodynamic and structural parameters for VIV of a cylinder (**a**) $m^* = 0.66$; (**b**) $m^* = 1.32$; (**c**) $m^* = 1.98$.

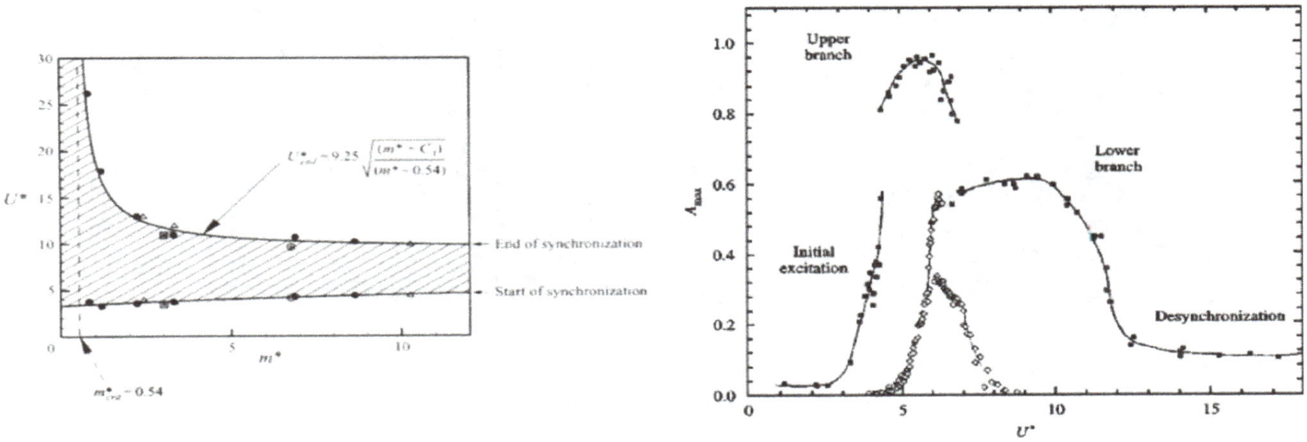

(a) (b)

Figure 11. (**a**) Relationship between U_r and m^* presented by Williamson and Govardhan [23]; (**b**) maximum response amplitudes A_{max} as functions of the reduced velocity U_r for $m^* = 2.4$ and $m^* = 248$ [19].

It was previously observed that 2P mode shedding is the reason for synchronized response [20]. Since detailed analysis of physics of the flow is not in the scope of this paper, shedding pattern is not further discussed. A representation of 2P mode of vortex shedding in the synchronization range of U_r is shown in Figure 12. A detailed study to analyze this phenomenon is pending. Shedding characteristics at different m^* and k are listed in Table 4.

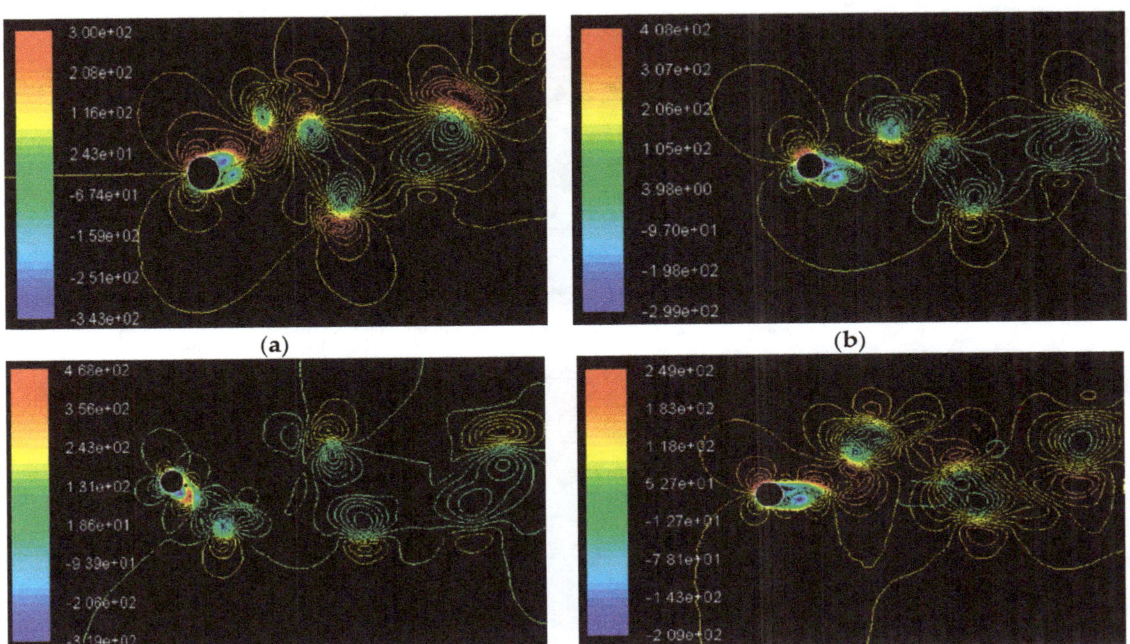

(c) (d)

Figure 12. *Cont.*

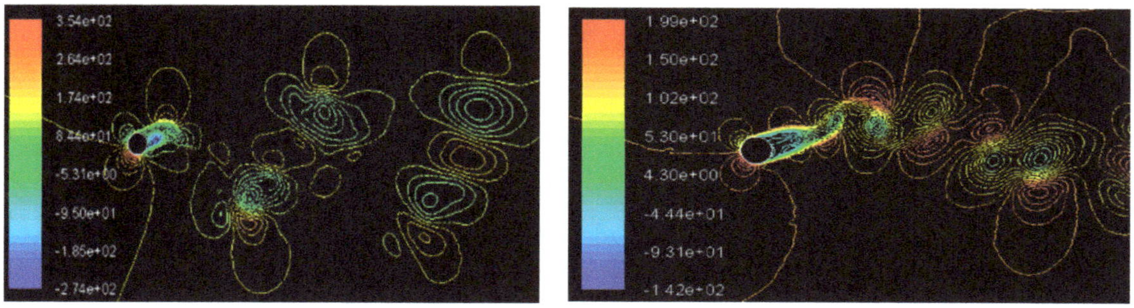

(e) (f)

Figure 12. Representation of pressure (N/m^2) contours in 2P mode of vortex shedding for $m^* = 0.66$ (**a**) $U_r = 4$; (**b**) $U_r = 5$; (**c**) $U_r = 6$; (**d**) $U_r = 8$; (**e**) $U_r = 10$; (**f**) $U_r = 12$.

Table 4. Hydrodynamic and structural response characteristics for $m^* = 0.66$, 1.32, and 1.98.

Mass Ratio, m^*	U_r	C_L	Y/D	f_v (Hz)	η	St	Shedding Characteristics
$m^* = 1.98$	4	1.16	0.86	1.6	1.04	0.24	2P Lift force oscillates about zero value once during one time period of oscillation of the cylinder.
	5	0.75	1.0	1.26	1.04	0.19	2P Lift force oscillates twice during one time period of oscillation of the cylinder.
	6	0.55	1.05	1.06	1.03	0.16	2P Lift force oscillates twice during one time period of oscillation of the cylinder.
	8	0.25	1.13	0.81	1.01	0.12	2P Lift force oscillates thrice during one time period of oscillation of the cylinder.
	10	0.4	0.8	0.67	0.98	0.1	2P Lift force oscillates thrice during one time period of oscillation of the cylinder. Beat phenomenon is observed with time period 10.53 s.
	12	0.38	0.67	0.53	1.04	0.08	2P Lift force oscillates 3.5 times during one time period of oscillation of the cylinder.
$m^* = 1.32$	4	1.1	0.87	1.55	1.07	0.23	2P Lift force oscillates about zero value once during one time period of oscillation of the cylinder.
	5	0.59	1.01	1.24	1.06	0.19	2P Lift force oscillates twice during one time period of oscillation of the cylinder.
	6	0.64	1.03	1.03	1.06	0.16	2P Lift force oscillates twice during one time period of oscillation of the cylinder.
	8	0.32	1.17	0.81	1.01	0.12	2P Lift force oscillates thrice during one time period of oscillation of the cylinder.
	10	0.35	0.82	0.65	1.02	0.1	2P Lift force oscillates thrice during one time period of oscillation of the cylinder. Beat phenomenon is observed with time period 10.42 s.
	12	0.15	0.74	0.54	1.01	0.082	2P Lift force oscillates 3.5 times during one time period of oscillation of the cylinder.
$m^* = 0.66$	4	1.12	0.93	1.43	1.14	0.21	2P Lift force oscillates about zero value once during one time period of oscillation of the cylinder.
	5	0.56	1.03	1.16	1.13	0.18	2P Lift force oscillates twice during one time period of oscillation of the cylinder.
	6	0.46	1.04	0.99	1.10	0.15	2P Lift force oscillates twice during one time period of oscillation of the cylinder.
	8	0.18	1.26	0.77	1.06	0.12	2P Lift force oscillates thrice during one time period of oscillation of the cylinder. Beat phenomenon is observed with time period 7.84 s.
	10	0.2	1.08	0.72	1.1	0.11	2P Lift force oscillates thrice during one time period of oscillation of the cylinder. No beat is observed.
	12	0.12	0.79	0.5	1.1	0.08	2P Lift force oscillates 3.5 times during one time period of oscillation of the cylinder.

Energy possessed by a spring mass system undergoing oscillation can be represented as the sum of its kinetic and potential energies as given by Equation (26).

$$E = \frac{1}{2}m\dot{Y}^2 + \frac{1}{2}kY^2 \tag{26}$$

When the position of the mass corresponds to maximum amplitude, the entire kinetic energy of the system will be converted into potential energy and Equation (26) reduces to

$$E = \frac{1}{2}kY^2 \tag{27}$$

At zero amplitude position of the mass, the entire potential energy is converted into kinetic energy. Since the total energy of the system is conserved energy balance can be written as Equation (28).

$$E = \frac{1}{2}kY^2 = \frac{1}{2}m\dot{Y}^2 \tag{28}$$

Hence the maximum possible velocity with which the system oscillates can be expressed as

$$\dot{Y} = \sqrt{\frac{k}{m}}Y \tag{29}$$

Which may be also represented in non-dimensional form represented by Equation (30)

$$\dot{Y} = \sqrt{\frac{Dk^*g}{L/D}\frac{Y}{D}} \tag{30}$$

For the optimum condition proven numerically, V_{max} is obtained as 0.5 m/s. Power associated with the oscillatory motion can be expressed by Equation (31)

Maximum velocity has been calculated using the above expression and a comparison of average power output estimated is represented in Table 5. Even though the maximum amplitude of oscillation is obtained for each m^* at $U_r = 8$, power output is maximum at $U_r = 4$, suggesting the best operating conditions.

$$P_{avg} = F_L\dot{Y} = C_L\frac{1}{2}\rho AV^2\dot{Y} \tag{31}$$

Table 5. Cylinder velocity and calculated average power for different configurations.

Re	m^*	U_r	k^*	C_L	Y/D	$\dot{Y}$	P_{avg} (W)
3.8×10^4	0.66	4	11.17	1.12	0.93	0.74	7.90
		5	6.9	0.56	1.03	0.65	3.44
		6	4.81	0.46	1.04	0.54	2.38
		8	2.7	0.18	1.26	0.49	0.85
		10	1.73	0.2	1.08	0.34	0.64
		12	1.21	0.12	0.79	0.21	0.24
	1.32	4	11.17	1.1	0.87	0.69	7.26
		5	6.9	0.59	1.01	0.63	3.55
		6	4.81	0.64	1.03	0.54	3.28
		8	2.7	0.32	1.17	0.46	1.40
		10	1.73	0.35	0.82	0.26	0.86
		12	1.21	0.15	0.74	0.19	0.28
	1.98	4	11.17	1.16	0.86	0.69	7.57
		5	6.9	0.75	1	0.63	4.47
		6	4.81	0.55	1.05	0.55	2.87
		8	2.7	0.25	1.13	0.44	1.05
		10	1.73	0.4	0.8	0.25	0.95
		12	1.21	0.38	0.67	0.18	0.64

7. Field Test Validation of Numerical Results

The HVPG module with specifications as given in Table 1 has been used for a field test in the Palissery irrigation canal. The flow velocity in the canal was measured to be 0.5 m/s. The model has been tested for different k^* values. Cylinder displacement was measured by attaching a pantograph to the side vanes. As the cylinder oscillates, the marking pencil attached to the spring side of the module marks its impression on the paper attached on the vane side. The results are tabulated in Table 6. $U_r = 4$, 8, and 12, but the effect of k^* is observed to follow the same trend as predicted numerically. An average deviation of 30% is observed between numerical method and field tests. Over prediction of the response amplitude by the numerical method may be attributed to not accounting for structural damping. Also, friction in the sliding parts of the guide vane contributes to lowering the response. Components of the HVPG and field test set up are shown in Figure 13.

Table 6. Results of field test conducted at Palissery irrigation canal.

Re	m^*	U_r	k^*	Y_{field} (cm)	Y/D_{fileld}	$Y/D_{numerical}$
		4	11.17	5.5	0.72	0.93
3.8×10^4	0.66	8	2.7	7.5	0.98	1.26
		12	1.21	4.5	0.59	0.79

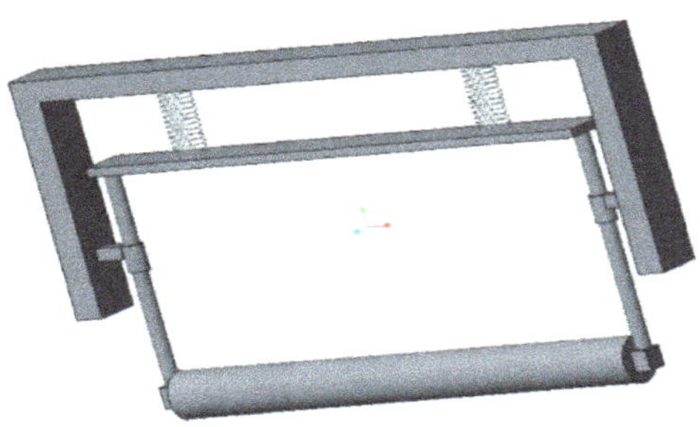

(a)

(b)

Figure 13. (**a**) Model of HVPG; (**b**) field test at Palissery irrigation canal for $m^* = 0.66$ and $U_r = 4$, 8, and 12.

8. Conclusions

Extensive research carried out to understand and interpret the intrinsic vortex shedding phenomenon has brought out several correlations to estimate response amplitude Y/D [32], but most of these expressions define the non-dimensional oscillation amplitude Y/D as a function of complex parameters like the Skop–Griffin parameter [14]. The present study is an effort to understand the hydrodynamic response of the cylinder from a designers' perspective by considering the effect of tangible system parameters only. Optimum response is obtained at $m^* = 0.66$ and $U_r = 8$, but optimum estimated power output for the same mass ratio is obtained at $U_r = 4$. It is observed that maximum power output can be derived from an HVPG operating at low mass ratio and in the lowest regime of reduced velocity in the synchronization range irrespective of amplitude of cylinder response. The reduced power output at $U_r = 8$ is due to the lower value of C_L which in turn is due to several oscillations of lift force within one time period of oscillation of the cylinder. The true values of frequency ratio, η obtained from the simulations are indicative of synchronized response for the range of U_r considered. Hence it is observed that the developed numerical method could successfully simulate the flow around an oscillating cylinder. The numerical method developed is capable of predicting the trend of variation of Y/D which is verified using the results of field test. At $m^* = 0.66$ where maximum response is observed, η significantly exceeds over unity compared to the higher mass ratio cases. It can be concluded that maximum amplitude of response is observed at mass ratios corresponding to η values greater than unity rather than at unity η values. Even though hydrodynamic lift force acting on the cylinder is proportional to the incoming flow velocity, C_L strongly depends on the natural frequency of the oscillating system and the shedding pattern. Response amplitude of the cylinder also depends on stiffness ratio k^* and in turn on natural frequency of the oscillating system f_n. Response amplitude increases with decreasing mass ratio and the range of U_r over which the response is high (resonance) widens. Occurrence of beat phenomenon also depends on m^* and k^* and the relationship is more pronounced at lower values of m^*. The mode of vortex shedding depends only on U_r at a constant flow velocity and is independent of m^*. On the whole, the work provides strong design inputs to the construction of the envisaged HVPG model.

9. Scope for Future Research

The discussions and conclusions from Sections 7 and 8 respectively indicate an immense scope for future research. Authors are in the process of improving the present design by incorporating structural damping in the studies. Inclusion of a greater number of rollers in the guide vanes is also being aimed at for eliminating the friction between sliding parts. The influence of these additional parameters in the equation of motion can reduce the errors in the subsequent numerical prediction, leading the design parameters closer to those of practical values. Improvised designs are also being devised for minimizing the transmission losses during power generation.

Author Contributions: Conceptualization, V.C., S.M., and S.J.; Data curation, V.C. and S.J.; Formal analysis, V.C. and S.J.; Funding acquisition, S.J. and V.M.; Investigation, V.C.; Methodology, S.M. and S.J.; Project administration, S.M. and S.J.; Resources, S.J. and V.M.; Software, V.M.; Supervision, S.M and S.J.; Validation, V.C. and S.J.; Visualization, V.C. and S.J.; Writing—original draft, V.C. and S.J.; Writing—review & editing, V.C., S.J., and V.M.

Acknowledgments: The authors would like to thank the management and principal of SCMS School of Engineering and Technology for all the support offered for the research and development. The authors would also like to thank Energy Management Center, Government of Kerala for all the support.

References

1. Gerrard, J.H. The mechanics of the formation region of vortices behind bluff bodies. *J. Fluid Mech.* **1966**, *25*, 401–413. [CrossRef]
2. Gao, Y.; Fu, S.; Xiong, Y.; Zhao, Y.; Liu, L. Experimental study on response performance of vortex-induced vibration on a flexible cylinder. *Ships Offshore Struct.* **2016**, *12*, 116–134. [CrossRef]
3. Bimbato, A.M.; Pereira, L.A.; Hirata, M.H. Suppression of vortex shedding on a bluff body. *J. Wind Eng. Ind. Aerodyn.* **2013**, *122*, 16–18. [CrossRef]
4. Bernitsas, M.; Raghavan, K.; Ben-Simon, Y.; Garcia, E. VIVACE (Vortex Induced Vibration Aquatic Clean Energy): A New Concept in Generation of Clean and Renewable Energy from Fluid Flow. *J. Offshore Mech. Arct. Eng.* **2006**, *130*, 041101. [CrossRef]
5. An, X.; Song, B.; Tian, W.; Ma, C. Design and CFD Simulations of a Vortex-Induced Piezoelectric Energy Conve rter (VIPEC) for Underwater Environment. *Energies* **2018**, *11*, 330. [CrossRef]
6. Janardhanan, S.; Chandran, V.; Varghese, C.; Achuth, D.; Devassy, D.; Mathews, D.C. Hydro vortex power generator design and construction. In Proceedings of the Kerala Technological CONGRESS, KETCON 2018—Human Computer Interface, Thrissur, India, 24 February 2018.
7. Griffin, O.M. Vortex-Excited Cross-Flow Vibrations of a Single Cylindrical Tube. *ASME J. Press. Vessel Technol.* **1980**, *102*, 158–166. [CrossRef]
8. Khalak, A.; Williamson, C.H.K. Dynamics of a hydroelastic cylinder with very low mass and damping. *J. Fluids Struct.* **1996**, *10*, 455–472. [CrossRef]
9. Narendran, K.; Murali, K.; Sundar, V. Vortex-induced vibrations of elastically mounted circular cylinder at Re of the O(105). *J. Fluids Struct.* **2015**, *54*, 503–521. [CrossRef]
10. Govardhan, R.; Williamson, C.H.K. Defining the 'modified Griffin plot' in vortex-induced vibration: Revealing the effect of Reynolds number using controlled damping. *J. Fluid Mech.* **2006**, *561*, 147–180. [CrossRef]

11. Bernitsas, M. Out of the Vortex. *Mech. Eng.* **2010**, *132*, 22–27. [CrossRef]
12. Tian, W.; Mao, Z.; Zhao, F. Design and Numerical Simulations of a Flow Induced Vibration Energy Converter for Underwater Mooring Platforms. *Energies* **2017**, *10*, 1427. [CrossRef]
13. Khan, N.B.; Ibrahim, Z.; Tuan, L.; Javed, M.F.; Jameel, M. Numerical investigation of the vortex-induced vibration of an elastically mounted circular cylinder at high Reynolds number (Re = 104) and low mass ratio using the RANS code. *PLoS ONE* **2017**, *12*, e0185832. [CrossRef] [PubMed]
14. Williamson, C.H.K.; Govardhan, R. Vortex induced vibrations. *Annu. Rev. Fluid Mech.* **2004**, *36*, 413–455. [CrossRef]
15. Achenbach, E.; Heinecke, E. On Vortex Shedding from Smooth and Rough Cylinders in the Range of Reynolds Numbers 6 × 103 to 5 × 106. *J. Fluid Mech.* **1981**, *109*, 239–251. [CrossRef]
16. Blevins, R.D. *Flow-Induced Vibration*, 2nd ed.; Van Nostrand Reinhold: New York, NY, USA, 1990; pp. 163–164, ISBN 1-57524-183-8.
17. Gabbai, R.D.; Benaroya, H. An overview of modeling and experiments of vortex-induced vibration of circular cylinders. *J. Sound Vib.* **2005**, *282*, 575–616. [CrossRef]
18. Anagnostopoulos, P.W.; Bearman, P.W. Response characteristics of a vortex exited cylinder at low Reynolds number. *J. Fluids Struct.* **1992**, *6*, 39–50. [CrossRef]
19. Khalak, A.; Williamson, C.H.K. Investigation of the relative effects of mass and damping in vortex induced vibration of a circular cylinder. *J. Wind Eng. Ind. Aerodyn.* **1997**, *69–71*, 341–350. [CrossRef]
20. IcemCfd, A. *12.0 User's Ma*; Ansys Inc.: Canonsburg, PA, USA, 2009; Volume 5.
21. Fluent, A. *12.0 Theory Guide*; Ansys Inc.: Canonsburg, PA, USA, 2009; Volume 5.
22. Menter, F.R. Two-equation eddy-viscosity turbulence models for engineering applications. *AIAA J.* **1994**, *32*, 1598–1605. [CrossRef]
23. Anton, G. Analysis of Vortex-Induced Vibration of Risers. Master's Thesis, Applied Mechanics, Chalmers University of Technology, Gothenburg, Sweden, 2012.
24. Vandiver, J.K. Damping parameters for flow-induced vibration. *J. Fluids Struct.* **2012**, *35*, 105–119. [CrossRef]
25. Schlichting, H. *Boundary Layer Theory*, 8th ed.; McGraw-Hill Book Company: New York, NY, USA, 1979; ISBN 13 978-3540662709.
26. Iaccarino, G.; Mishra, A.A.; Ghili, S. Eigenspace perturbations for uncertainty estimation of single-point turbulence closures. *Phys. Rev. Fluids* **2017**, *2*, 024605. [CrossRef]
27. Mishra, A.A.; Gianluca, I. Uncertainty Estimation for Reynolds-Averaged Navier Stokes Predictions of High-Speed Aircraft Nozzle Jets. *AIAA J.* **2017**, *55*, 3999–4004. [CrossRef]
28. Govardhan, R.; Williamson, C.H.K. Modes of vortex formation and frequency response for a freely-vibrating cylinder. *J. Fluid Mech.* **2000**, *420*, 85–130. [CrossRef]
29. Chen, W.; Zhang, Q.; Li, H.; Hu, H. An experimental investigation on vortex induced vibration of a flexible inclined cable under a shear flow. *J. Fluids Struct.* **2015**, *54*, 297–311. [CrossRef]
30. Feng, C.C. The Measurements of Vortex Induced Effects in Flow Past a Stationary and Oscillating Circular and D-Section Cylinders. Master's Thesis, The University of British Columbia, Vancouver, BC, Canada, 1968.
31. Naudascher, E.; Rockwell, D. *Flow Induced Vibration—An Engineering Guide*; Dover Publications Inc.: Mineola, NY, USA, 2005; pp. 37–38, ISBN 13 978-0-486-44282-2.
32. Domal, V.; Sharma, R. An experimental study on vortex-induced vibration response of marine riser with and without semi-submersible. *J. Eng. Marit. Environ.* **2017**, *232*, 176–198. [CrossRef]

4

Possibilities and Limitations of CFD Simulation for Flashing Flow Scenarios in Nuclear Applications

Yixiang Liao [1,2,*] and Dirk Lucas [2]

1 Gesellschaft für Anlagen-und Reaktorsicherheit, Boltzmannstraße 14, 85748 Garching bei München, Germany
2 Helmholtz-Zentrum Dresden-Rossendorf, Institute of Fluid Dynamics, Bautzner Landstraße 400, 01328 Dresden, Germany; D.Lucas@hzdr.de
* Correspondence: Yixiang.Liao@grs.de or Y.Liao@hzdr.de.

Academic Editor: Bjørn H. Hjertager

Abstract: The flashing phenomenon is relevant to nuclear safety analysis, for example by a loss of coolant accident and safety release scenarios. It has been studied intensively by means of experiments and simulations with system codes, but computational fluid dynamics (CFD) simulation is still at the embryonic stage. Rapid increasing computer speed makes it possible to apply the CFD technology in such complex flow situations. Nevertheless, a thorough evaluation on the limitations and restrictions is still missing, which is however indispensable for reliable application, as well as further development. In the present work, the commonly-used two-fluid model with different mono-disperse assumptions is used to simulate various flashing scenarios. With the help of available experimental data, the results are evaluated, and the limitations are discussed. A poly-disperse method is found necessary for a reliable prediction of mean bubble size and phase distribution. The first attempts to trace the evolution of the bubble size distribution by means of poly-disperse simulations are made.

Keywords: flashing; computational fluid dynamics (CFD) simulation; two-fluid-model; mono-disperse; poly-disperse

---

1. Introduction

Flash boiling (flashing) is a process of phase change from liquid to vapor. It distinguishes itself from traditional boiling by the way the liquid gets superheated. It is caused by depressurization or pressure drop instead of heating. For this reason, flash boiling is sometimes also called "adiabatic boiling", namely without external heat sources. Another more familiar phenomenon of the same type is "cavitation". Under the term of cavitation, we often understand an isothermal process, where the growth or collapse of pre-existing vapor nuclei is an inertia-controlled process driven by the pressure difference across the gas-liquid interface. The phase change is controlled by pressure drop, and its recovery while liquid temperature remains constant. In this case, the vapor generation rate can be approximated by using the Rayleigh-Plesset equation. On the other hand, the phase change process in flash boiling is similar to that in traditional boiling with external heat sources. It is characterized by nucleation and the inter-phase heat transfer rate, namely by thermal non-equilibrium. The superheated liquid is cooled down by giving up surplus energy to vapor generation. The effect of pressure non-equilibrium between the inside and outside of a vapor bubble is often neglected. Nevertheless, an actual phase change process taking place in superheated liquid due to depressurization is controlled by both thermal (temperature difference) and mechanical (i.e., pressure difference) effects. The terms of "cavitation" and "flashing" above represent only simplification assumptions, i.e., neglecting thermal or mechanical effects, in theoretical treatments. The ratio of thermal and mechanical non-equilibrium contribution to bubble growth and phase change is dependent on the temperature (or pressure) level at which the process is taking place. At a high pressure level, especially in low depressurization rate situations, the phase change process is dominated by the inter-phase thermal heat transfer and is often treated as a flash boiling process.

The flash boiling phenomenon has a fundamental and decisive presence in many industrial and technical applications, where significant pressure drop is present. In nuclear engineering, as an example, it can take place in the following scenarios:

1. Large break loss-of-coolant accidents (LOCA) of pressurized water nuclear reactors;
2. Pressure release through blow-off valves at the pressurizer or steam generator;
3. Two-phase critical flow problem through nozzles;
4. Flashing-induced instability in natural circulation systems.

Under these circumstances, the properties of the flashing flow, such as the discharge flow rate, vapor generation rate, void hold-up, as well as two-phase morphology, are of key safety and economic importance. All of these quantities are influenced by the degree of non-equilibrium substantially.

Since the middle of the last century there have arisen many theoretical and experimental investigations on two-phase flashing flow owing to great concern in nuclear safety. With respect to the theoretical study, simplifying and empirical assumptions usually have to be adopted due to the complexity of its nature. The degree of non-equilibrium is accounted for partially or neglected fully. Therefore, the general validity of available methods is largely limited. So far, one-dimensional approaches or system codes are routinely applied to deal with such kinds of issues. However, the flashing two-phase mixture has a strongly three-dimensional nature, which is accompanied with a large heterogeneous gaseous structure and high gas volume fraction. All of these features along with the micro-scale bubble dynamic processes require a more sophisticated prediction tool with high time and space resolution, such as the CFD (computational fluid dynamics) technology. Furthermore, the drawback of increased computational effort in CFD simulations is being offset by the rapidly increasing speed and decreasing hardware costs of parallel computers. Therefore, there exists a need to assess the restrictions and update the closures for flashing situations.

2. State of the Art of CFD Simulation of Flashing Flow

Recently, promising CFD research on flashing flows has been published, such as in [1–9]. All of these simulations are based on the framework of a simplified two-fluid-model for bubbly flow, where the interfacial area density for inter-phase exchange is obtained by the particle model. Furthermore, they all assume a mono-disperse interface morphology. That means that the bubble size in each computational cell has a single value instead of a spectrum at any given time.

Laurien and his co-workers studied water evaporation and re-condensation phenomena caused by steady-state pressure variation inside a three-dimensional complex pipeline [1–4]. Frank [5] simulated the well-known Edwards pipe blow-down test [10] using a one-dimensional simplified mesh in CFX (a commercial CFD code of the company ANSYS). Both of them employed a five-equation model including two continuity equations, two momentum equations for liquid and vapor, respectively, and one energy equation for liquid. The vapor was assumed to remain always at the saturation condition corresponding to local pressure, which is uniform inside and outside the bubble. The assumption is reasonable in the case of a small depressurization rate. For the computation of interfacial area density, a constant bubble diameter, e.g., $d_g = 1$ mm, is prescribed in the whole domain. In addition, the momentum interaction between the gas and liquid phases is modelled only as a drag force while the effect of non-drag forces is ignored. However, Liao [11] showed that non-drag forces have a significant effect on the spatial distribution of phases.

Later on, Laurien [4] suggested that a model presuming bubble number density instead of bubble size, which allows bubble size to grow, is more close to the physical picture of boiling flow. This assumption is acceptable when the nucleation zone is sufficiently narrow and bubble dynamics, such as coalescence and break-up, is negligible. Otherwise, an additional transport equation for the bubble number density with appropriate source terms or even a poly-disperse method is necessary.

Maksic and Mewes [6] simulated flashing flows in pipes and nozzles by using a four-equation model, where a common velocity field is assumed for both phases. The inter-phase heat transfer is assumed to be dominated by conduction. However, it has been shown that in most flashing expansion cases, the convective contribution due to the relative motion of bubbles dominates the heat transfer [12]. Neglecting of inter-phase velocity slip obviously under-predicts the vapor generation rate [11]. Wall nucleation was considered as a unique source of bubble number density in the additional transport equation. The Jones model [13–16] was used to determine the nucleation rate. Inter-phase mass, momentum and energy transfer due to nucleation were ignored.

Marsh and O'Mahony [7] simulated the nozzle flashing flow using a six-equation model in FLUENT (a commercial CFD code of the company ANSYS), with separate mass, momentum and enthalpy balance equations for liquid and vapor. Inter-phase mass and momentum, as well as energy transfer resulting from both nucleation and phase change were accounted for. However, the effects of non-drag forces on momentum exchange and the heat transfer between vapor and vapor-liquid interface were neglected. A modified version of the Blander and Katz nucleation model [17] was employed to compute the source of bubble number density. The original model was found to create large numerical instability, which is based on the classical homogeneous nucleation theory.

Mimouni et al. [8] simulated the cavitating flow using a six-equation model in NEPTUNE_CFD. The vapor temperature was ensured to be very close to the saturation temperature by using a special heat transfer coefficient. Besides the drag, added mass and lift force were included in the inter-phase momentum transfer. The contribution of nucleation to the vapor generation rate, as well as momentum and energy transfer were considered by the slightly modified Jones model [13–16]. The original model was shown to be insufficiently general by the authors. Nevertheless, the effect of nucleation and vaporization on the mean bubble size was ignored, namely, a constant bubble size was assumed.

Janet [9] studied the performance of various nucleation models in a flashing nozzle flow by using the five-equation model in CFX Version 14.5. It was found that predictions obtained with the Jones model are more reliable than the RPI (Rensselaer Polytechnic Institute) [18] and Riznic models [19].

Besides closure models for heat transfer coefficient and nucleation rate, basic differences among the above model setups can be summarized as follows.

Although in [7,8], the energy equation of the steam phase was solved, special treatments on heat the transfer coefficient were needed to maintain numerical stability. As a consequence, steam was either near saturated or no heat transfer to the interface. Based on these considerations, the five-equation model is chosen for the present work, which is proven to be sufficient. Various mono-disperse particle models are applied to flashing scenarios relevant to nuclear safety analysis. The limitations of each approach are discussed, and subsequently, a poly-disperse model free of open parameters is established for flow conditions with a broad spectrum of bubble size.

The rest of this paper is structured as follows. Section 3 begins with a brief mathematical description of applied transport equations, as well as the main closure relations. Results and discussions of the mono-simulations are found in Section 4. Section 5 presents a poly-disperse method and the simulation results. Suggestions for future work in Section 6 conclude the paper.

3. Physical Setup

3.1. Fundamental Transport Equations

The ensemble-averaged mass, momentum and energy conservation equations for individual phases are given as follows. Steam is assumed to stay always at the saturated state corresponding to local absolute pressure, and a common pressure field p is assumed for both phases. Pressure non-equilibrium at the steam-water interface is neglected.

Water:

$$\frac{\partial}{\partial t}(\alpha_l \rho_l) + \nabla \cdot \left(\alpha_l \rho_l \vec{U}_l\right) = \Gamma_{lg}, \tag{1}$$

$$\frac{\partial}{\partial t}\left(\alpha_l \rho_l \vec{U}_l\right) + \nabla \cdot \left(\alpha_l \rho_l \vec{U}_l \vec{U}_l\right) = -\alpha_l \nabla p + \nabla \cdot \left(\alpha_l \vec{\tau}_l\right) + \alpha_l \rho_l \vec{g} + \vec{F}_{lg} + \Gamma_{lg} \vec{U}_i, \tag{2}$$

$$\frac{\partial}{\partial t}\left(\alpha_l \rho_l \left(H_{tot,l} - \frac{p}{\rho_l}\right)\right) + \nabla \cdot \left(\alpha_l \rho_l \vec{U}_l H_{tot,l}\right) = \nabla \cdot (\alpha_l \lambda_l \nabla T_l) + \Gamma_{lg} H_{l,i} + E_l. \tag{3}$$

Steam:

$$\frac{\partial}{\partial t}(\alpha_g \rho_g) + \nabla \cdot \left(\alpha_g \rho_g \vec{U}_g\right) = -\Gamma_{lg}, \tag{4}$$

$$\frac{\partial}{\partial t}\left(\alpha_g \rho_g \vec{U}_g\right) + \nabla \cdot \left(\alpha_g \rho_g \vec{U}_g \vec{U}_g\right) = -\alpha_g \nabla p + \nabla \cdot \left(\alpha_g \vec{\tau}_g\right) + \alpha_g \rho_g \vec{g} - \vec{F}_{lg} - \Gamma_{lg} \vec{U}_i, \tag{5}$$

The unknown terms Γ_{lg}, F_{lg} and E_l describe the mass, momentum and heat flux to the liquid phase from the interface, which have to be modelled by constitutive relations or closure models. U_i and $H_{l,i}$ represent interfacial values of velocity and enthalpy carried into or out of the phase due to phase change. They are determined according to an upwind formulation:

$$\vec{U}_i = \begin{cases} \vec{U}_g, & \text{if } \Gamma_{lg} > 0 \text{ (condensation)} \\ \vec{U}_l, & \text{if } \Gamma_{lg} < 0 \text{ (evaporation)} \end{cases}, \tag{6}$$

$$H_{l,i} = \begin{cases} H_{l,sat}, & \text{if } \Gamma_{lg} > 0 \text{ (condensation)} \\ H_{l,tot}, & \text{if } \Gamma_{lg} < 0 \text{ (evaporation)} \end{cases}, \tag{7}$$

where $H_{l,sat}$ is liquid saturation enthalpy. The saturation parameters are interpolated from the published IAPWS-IF97 steam-water property tables corresponding to local pressure.

3.2. Main Closure Models

3.2.1. Inter-Phase Mass Transfer

As discussed in the Introduction, phase change in a flashing process is assumed to be induced by inter-phase heat transfer, which is called "thermal phase change model" in CFX. The volumetric mass transfer rate is related to heat flux density as follows:

$$\Gamma_{lg} = \frac{E_l \cdot A_i}{H_{g,sat} - H_{l,i}}, \tag{8}$$

where A_i is the interfacial area density and $H_{g,sat}$ is the vapor saturation temperature.

3.2.2. Interfacial Area Density

The particle model is adopted for interfacial transfer between two phases, which assumes that one of the phases (here water) is continuous and the other (steam) is dispersed. The surface area per unit volume is then calculated by assuming that steam is present as spherical particles of mean diameter d_g. Using this model, the interfacial area density is:

$$A_i = 6\frac{\alpha_g}{d_g} \text{ or } (6\alpha_g)^{2/3}(\pi n_g)^{1/3}. \tag{9}$$

Either mean diameter d_g or the number density n_g has to be known. They are prescribed or obtained by solving additional transport equations; see Table 1. Depending on the way of calculating d_g or n_g, mono-disperse and poly-disperse methods are derived. In the latter case, d_g represents the Sauter mean diameter of a size spectrum.

Table 1. Model setup available for CFD simulation of flashing flow.

Papers	Number of Conservation Equations	CFD Software	Particle Model for Interfacial Area Density with
[1–3]	Five	CFX 4.2	prescribed size
[4]	Five	CFX 4.2	prescribed number density
[6]	Four	CFX 4.2	additional transport equation for number density
[7]	Six	FLUENT 6.2.16	additional transport equation for number density
[8]	Six	NEPTUNE_CFD	prescribed size
[9]	Five	CFX 14.5	additional transport equation for number density

3.2.3. Inter-Phase Heat Transfer

The sensible heat flux transferring to water from the steam-water interface, E_l, is given by:

$$E_l = h_l(T_{sat} - T_l) \tag{10}$$

where h_l is the overall heat transfer coefficient. It is approximated by the Ranz–Marshall correlation [20,21], which is applicable for spherical particles. It has been widely used such as in [1–5,8,9].

3.2.4. Nucleation Model

The effect of nucleation is investigated by considering two kinds of nucleation mechanisms, namely wall nucleation and bulk nucleation. The wall nucleation rate is computed according to the Jones model [16].

$$J_{het,1,W} = 0.25 \times 10^{-7} \frac{R_d^2}{R_{cr}^4} \cdot C_{dp}(T_l - T_{sat})^3, \tag{11}$$

where R_d and R_{cr} are bubble departure radius and critical radius, respectively. The constant $C_{dp} = 10^4$ ($K^{-3} \cdot s^{-1}$).

Heterogeneous nucleation due to impurities in the bulk flow is accounted for with the model given by Rohatgi and Reshotko [22].

$$J_{nuc,1,B} = N_{im,B} \cdot \sqrt{\frac{2\sigma}{\pi m_W}} \cdot \exp\left(-\frac{16\pi\sigma^3\varphi}{3k_B T_l}\left(\frac{T_{sat}}{T_l - T_{sat}} \frac{{\rho_g}^{-1} - {\rho_l}^{-1}}{\Delta H_L}\right)^2\right), \tag{12}$$

where m_w is the mass of a single liquid molecule, k_B the Boltzmann constant and ΔH_L latent heat. $N_{im,B}$ is the number density of impurities (nucleation sites) in the bulk, and φ is the heterogeneous factor. Both $N_{im,B}$ and φ are treated as adjustable constants; $N_{im,B} = 5 \times 10^3$ (m^{-3}) and $\varphi = 10^{-6}$ are adopted in this work.

3.2.5. Turbulence Model

Turbulence in the liquid phase is described by a standard shear stress transport (SST) model augmented by the addition of more source terms. These source terms describe the effect of bubbles, i.e., the bubble-induced turbulence (BIT).

Concerning the source for the k-equation, there is a general agreement that the bubbles' contribution to turbulence generation comes from the work done by interphase drag force, i.e.,:

$$\varphi_k = 0.75\rho_l \frac{C_D}{d_g} \alpha_g \left|\vec{U}_l - \vec{U}_g\right|^3, \tag{13}$$

Similar to single-phase dissipation, the ε-equation source is obtained by scaling φ_k with a time scale:

$$\varphi_\varepsilon = C_{\varepsilon B} \frac{\varphi_k}{\tau}, \tag{14}$$

where $C_{\varepsilon B}$ is an adjustable constant and set to one. The time scale τ is approximated with the ratio between bubble size and turbulence intensity [23]:

$$\tau = \frac{d_g}{\sqrt{k_l}}. \tag{15}$$

3.2.6. Inter-Phase Momentum Transfer

Inter-phase momentum transfer occurs due to interfacial forces acting on each phase by interaction with the other phase. In the present work, both drag force and non-drag forces like, lift force, wall lubrication force, virtual mass force and turbulent dispersion force, are considered, i.e.,:

$$\vec{F}_{lg} = \vec{F}_{lg,D} + \vec{F}_{lg,L} + \vec{F}_{lg,W} + \vec{F}_{lg,VM} + \vec{F}_{lg,TD}. \tag{16}$$

These forces have to be computed according to appropriate models. So far, identical relations are commonly used for air-water and steam-water dispersed flows. The baseline-closure models defined in the previous work [24] are adopted.

In the following text, the above transport equations and closure models will be used in the CFD simulation of various flashing scenarios. Results obtained with different assumptions for interfacial area density are presented below.

3.3. Numerical Schemes and Convergence Criteria

In the simulation the coupled volume fraction algorithm was used, which allows the implicit coupling of the velocity, pressure and volume fraction equations. The high resolution scheme was selected to calculate the advection terms in the discrete finite volume equations. The discretization algorithm for the transient term was the second order backward Euler. The upwind advection and the first order backward Euler transient scheme were used in the turbulence numerics. The convergence of the simulation was monitored by using the criterion of the root mean square residual throughout the domain of less than 10^{-4}.

4. Mono-Disperse Simulation Results

4.1. Edwards and O'Brien Blowdown Test

4.1.1. Description of the Test

In nuclear safety analysis, the Edwards and O'Brien test [10] represented a standard problem for the verification and assessment of computational programs [25,26]. It consisted of fluid depressurization studies in a horizontal straight pipe 4.096 m long with an inner diameter of 0.073 m; see Figure 1. One end of the pipe is a fixed wall, while at the other end, a glass disc was mounted, which was designed to burst to initiate the depressurization transient.

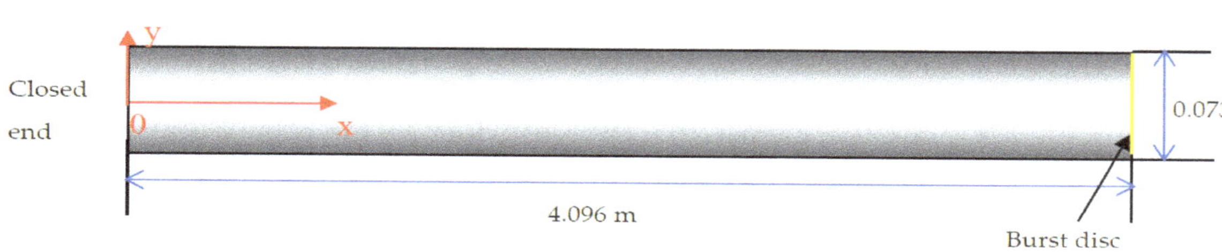

Figure 1. Edwards and O'Brien blowdown test.

The pipe was filled with sub-cooled water, whose initial conditions ranged from 3.55 MPa to 17.34 MPa and from 514.8 K to 616.5 K. At the burst of the glass disc, the depressurization wave propagates through the pipe, and water becomes superheated and vaporizes suddenly. Measurements were performed at several horizontal positions.

4.1.2. Simulation Results

The problem investigated in the present work has initial conditions of 7.00 MPa and 513.7 K. A pressure boundary was assumed for the outlet, and the background pressure was 1 atm. The simulation is done on a 3D cylinder grid, and furthermore, the five-equation model mentioned above with the prescribed bubble diameter (d_g = 1 mm) as in [3] is applied. The results of absolute pressure and void fraction at the position x = 1.469 m are presented in Figure 2a,b and compared with the experimental data.

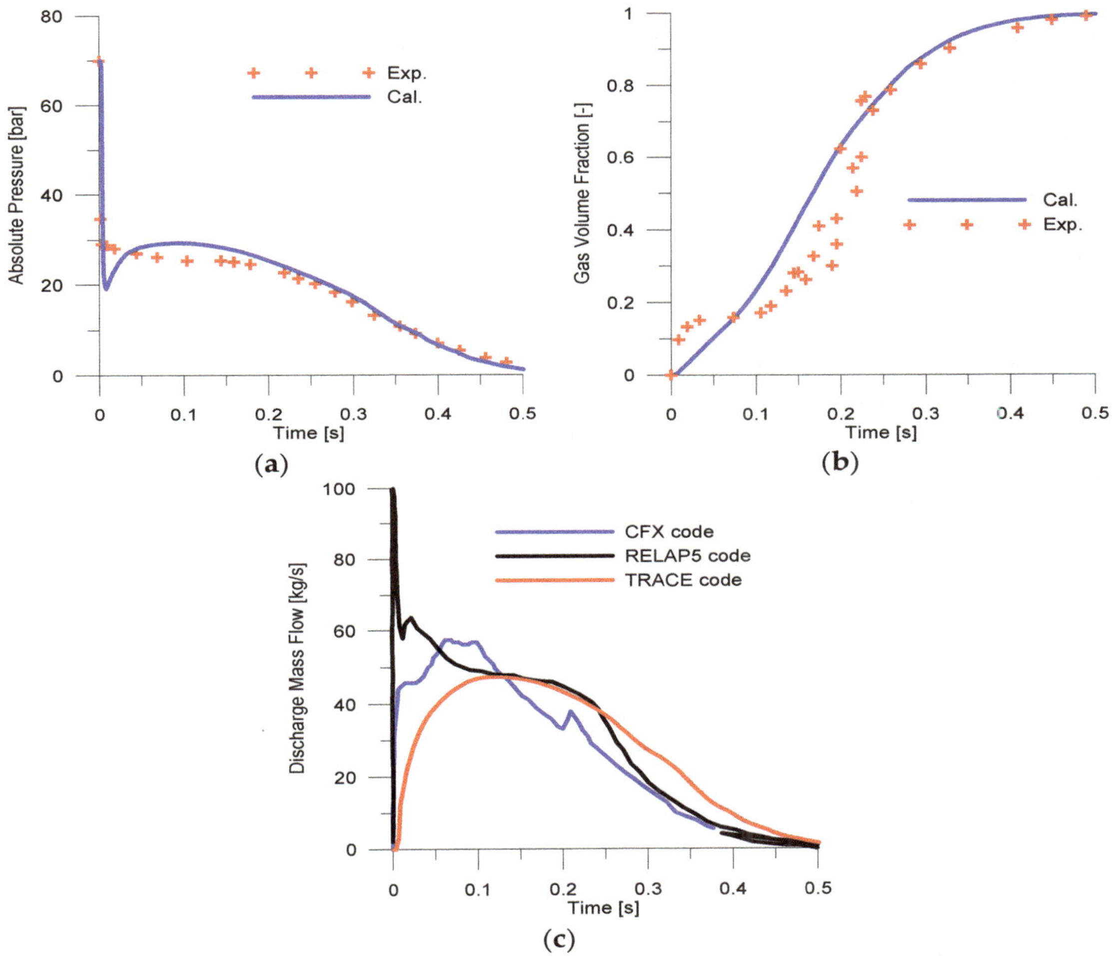

Figure 2. Evolution of flow parameters at x = 1.469 m after the disc burst: (**a**) absolute pressure; (**b**) void fraction; (**c**) discharged mass flow rate change with the time.

One can see that the water inside the pipe vaporizes completely within a half second after the burst of the glass disc. The agreement between simulation and measurement is acceptable, especially at the later stage. In contrast, considerable deviations are present at the early stage of depressurization. The lowest pressure in the simulation, which corresponds to the maximum superheat degree and is required to trigger the onset of flashing, is obviously lower than that in the measurement. Furthermore, the rate of vapor production is under-predicted at the beginning, while in the period from 0.1 s to 0.2 s, it is over-predicted. It is acknowledged that the simplifying assumption of a constant mean bubble size is responsible for the deviations. It is inconsistent with the physical picture of the process, although it is widely used. The size of pre-existing nuclei in the sub-cooled liquid is substantially smaller than the prescribed value of 1 mm. This indicates a significant under-prediction of interfacial area density available for the initiation of flashing. On the other hand, steam bubbles grow rapidly during the flashing process, leading to a mean size much larger than 1 mm in a short time. Nevertheless, the effect of this uncertainty is weakened with the increase of the void fraction. The effect of prescribed bubble size was investigated in [27].

The discharged mass flow from the blowdown pipe is shown in Figure 2c below. Due to lack of experimental information, the results are compared with those obtained by two classical system codes, i.e., RELAP5 and TRACE, from the recent work [28]. One can see that the maximum value is reached shortly after the burst, where the difference between the codes is largest. According to RELAP5, the mass flow increases rapidly from 0 to 100 kg/s at the moment of the burst, while TRACE and CFX give a significantly lower maximum value with a delay around 0.1 s. In addition, the change of mass the flow rate with time delivered by CFX is much more gentle and smooth in comparison to that by RELAP5 and CFX.

4.2. Flashing-Induced Instability Problem in Natural Circulation

4.2.1. Description of the Problem

Since the middle of the 1980s, it has been recognized that the application of passive safety systems can contribute to improving the economics and reliability of nuclear power plants. In some advanced designs, natural circulation is used as a means for residual heat removal. For example, in a German BWR (boiling water reactor) concept, namely, the KERENA™ reactor [29], the containment cooling condenser (CCC) is the key component of a natural cooling circuit; see the red shaded region in Figure 3. In the case of overpressure, surplus steam in the containment condenses on the outside wall of CCC. Heat released by condensation is transferred to the cooling water inside the CCC tubes and finally to the shielding/storage pool vessel (SSPV) through natural circulation.

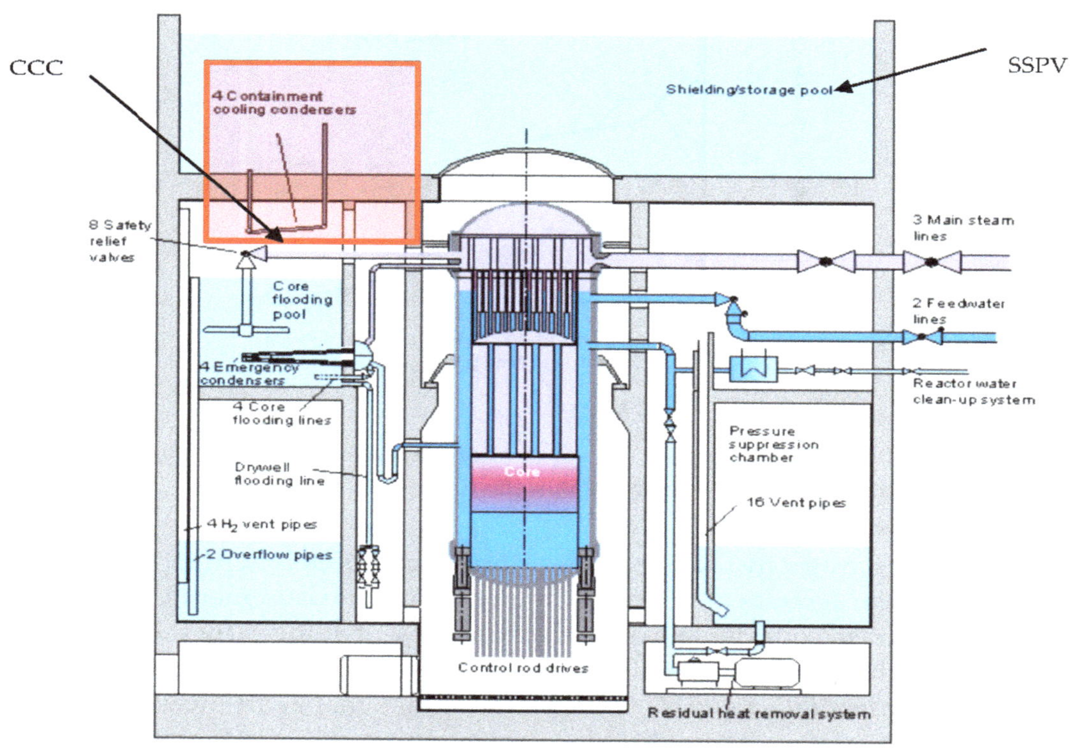

Figure 3. Section picture of KERENA™ containment [30]. CCC: containment cooling condenser; SSPV: shielding/storage pool vessel.

In the experimental investigation performed by AREVA [30], a severe water hammer was experienced. The flow instability is believed to come from the rapid formation and subsequent destruction of steam bubbles inside the circuit. Natural circulation systems are characterized by a downcomer and a riser, which connect to a heated and a cooled section at the ends, respectively. Nearly saturated warm circulating water can flash to vapor inside the riser if no boiling occurs in the heated section. Liquid temperature remains constant till flashing begins if the heat loss through the riser is negligible (adiabatic), while saturation temperature decreases with increased altitude.

The production of vapor in a flashing process taking place in natural circulation has a periodic oscillation character [31]. The circulation flow rate increases as a result of vapor production, which further leads to a low liquid temperature entering the riser. It can be lower than the saturation temperature at the exit of the riser. As a consequence, the flashing is suppressed, which leads to a decrease in flow rate and subsequently an increase in liquid temperature. Therefore, flashing can take place again in the adiabatic riser and generates a self-sustained oscillating flow. The feature of high frequency oscillation and rapid phase change requires high resolution and can cause numerical instability and convergence problems in high-resolved CFD simulations.

4.2.2. Simulation Results

For the simulation, a 3D pipe section is constructed on the basis of the riser in the AREVA test facility; see Figure 4. Two inlets are connected to two CCC condensers. For details on the experiments please, refer to [27,29]. Boundary conditions, such as inlet temperature, mass flow rate and outlet pressure level are provided by the experimental data. Furthermore, the single-phase condition at the inlet is assured by choosing a proper time segment according to the measurement. The start time for simulation is chosen as 4600 s. Furthermore, a mono-disperse approach is utilized, which distinguishes itself from the last case by the prescription of bubble number density instead of bubble size. A constant value of 5×10^4 m^{-3} is assumed in the following simulation. The settings are believed to be closer to the physical process of flashing since the bubbles are allowed to grow [4].

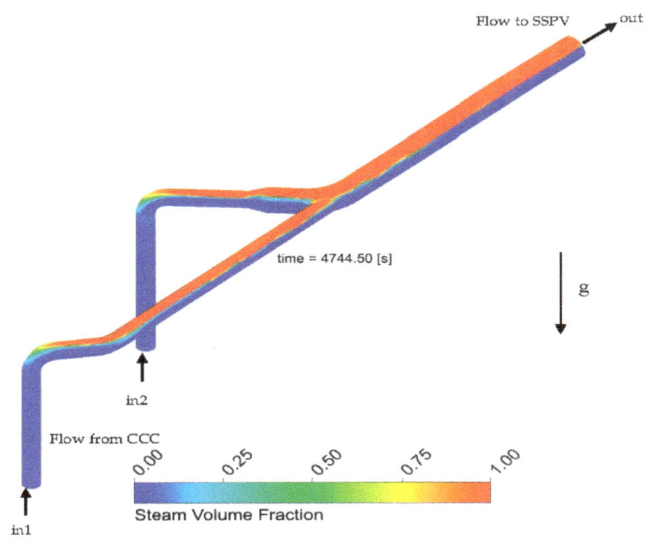

Figure 4. Segment of the riser pipe applied in the simulation colored by steam volume fraction at t = 4744.50 s.

Figure 4 shows the distribution of steam inside the domain at the moment of 4744.50 s. The red color symbolizes steam, while the blue is water. As discussed above, the liquid at the exit becomes first superheated and initiates the flashing process. This gives rise to a high steam volume fraction at the upper part, while single phase flow exists still in the two straight legs near the inlets. Since the gravitational force is considered, steam is accumulated at the top side of the inclined pipe leading to a stratified flow pattern. A broad range of flow patterns and regime transition represents challenges for current two-phase CFD technology.

The temporal course of the inlet pressure and liquid temperature (the measured pressure at the outlet and temperature at the inlet are given as boundary conditions) is depicted in Figure 5. During the time segment from 4600 s to 5200 s, the production of vapor goes periodically through onset, intensification and ceasing, and thermo-hydraulic parameters oscillate correspondingly. The maximum amplitude of pressure and temperature oscillation can reach 0.8 bar and 50 K, respectively. The predicted period agrees well with the measured one, and the amplitude of temperature profile is also satisfying. Nevertheless, clear deviation is observed at pressure valleys, where the onset of flashing occurs. Similar to the limitation of the last method with constant bubble size, the disagreement results from the prescription of a constant bubble number density and neglecting bubble dynamics, such as coalescence and break-up. The sensitivity study on the prescribed values for bubble number density was performed in the previous work of [32].

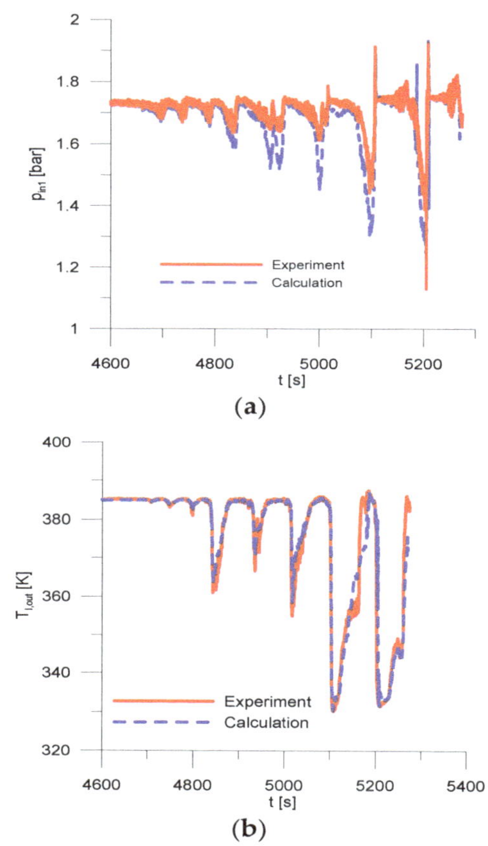

Figure 5. Comparison between calculation and experiment: (**a**) inlet pressure; (**b**) outlet temperature.

4.3. Abuaf [33] Nozzle Flashing Flow Test

4.3.1. Description of the Test

In nuclear safety analysis, the flashing of initially sub-cooled liquid through nozzles, orifices or other restrictions has been studied intensively for several decades due to the design basic accident of LOCA (loss-of-coolant accident). Much attention was paid to the critical flow problem, while a deep insight into the transient development of two-phase flow structure is still missing. In comparison with the single phase, the complexity of two-phase critical flow arises from both thermal and mechanical non-equilibrium effects at the interface, which prevent reliable prediction of the critical flow rate from using empirical relations.

The experimental investigation of water flashing through a converging-diverging nozzle presented in [33] has been widely used as a reference for numerical study or model development, e.g., [7,34]. In the frame of this work, several tests with different boundary conditions, such as inlet pressure, temperature and outlet pressure, are reproduced by means of CFD technology. The geometrical sketch of the circular nozzle is shown in Figure 6. Dynamic pressure decreases with the reduction of channel area in the converging part leading to a decrease in the saturation temperature. If it falls below the liquid temperature before reaching the neck position, the onset of flashing will occur near the neck and vaporization continues in the diverging part.

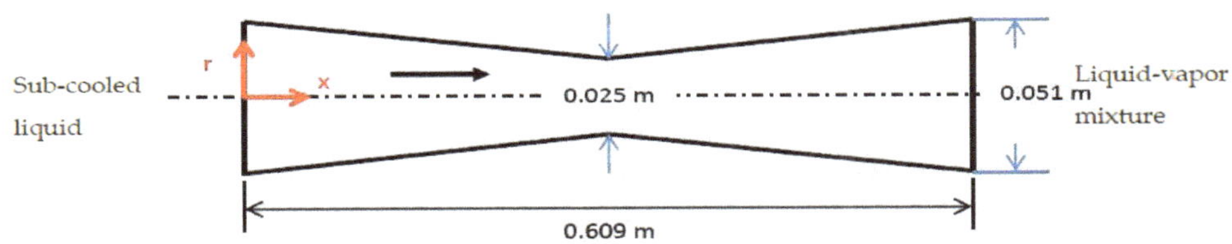

Figure 6. Circular converging-diverging nozzle used in the tests of [33].

4.3.2. Simulation Results

For this case, the generation and transportation of bubbles is traced by solving an additional transport equation for bubble number density. Similar setups were adopted in [6,7]; see Table 1. The formation of bubbles in superheated liquid is called "nucleation", which is deemed to take place under two different mechanisms, namely homogeneous and heterogeneous nucleation. Homogeneous nucleation occurs uniformly in the bulk caused by the thermodynamic state fluctuation of liquid molecules. In contrast, heterogeneous nucleation takes place at foreign "nuclei", which can be microscopic cavities on solid walls or dissolved gasses and impurities pre-existing in the sub-cooled liquid. The latter mechanism has been recognized to be dominant in the real processes, which is therefore taken into account in the current work. Since it allows both bubble size and number to vary, this method is obviously stepwise superior to the previous ones with prescribed bubble size or number density. However, it requires one more closure model, namely the nucleation model, which is still one weak spot in the numerical study of the boiling process. For investigations on the effect of nucleation models, please refer to [9].

From the spatial distribution of steam shown in Figure 7, one can see that the onset of vaporization occurs near the nozzle neck, and bubbles are formed overwhelmingly at the walls and then migrate to the center region. In other words, the mechanism of wall nucleation is dominant in comparison to bulk heterogeneous nucleation. As a result, the radial profile with a clear wall peak is built. The example presented below Figures 8–10 has the boundary conditions of inlet pressure 555.9 kPa, outlet pressure 402.5 kPa and inlet temperature 422.25 K.

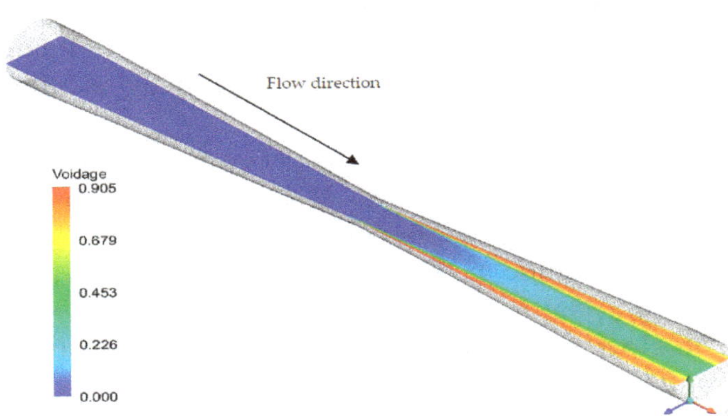

Figure 7. Distribution of the steam volume fraction inside the nozzle.

Furthermore, the nucleation process is found to take place in a narrow region (a few centimeters); see the red shaded region in Figure 8a. The cross-section averaged bubble number density increases steeply in the nucleation region and then remains almost constant in the stable two-phase region. The models described in [16,21] are chosen for the computation of the heterogeneous nucleation rate at the walls and in the bulk. The contribution of bulk nucleation is found to be trivial. In addition, the assumed small initial value is proven to have no influence on the final results [9].

The radial profile of bubble distribution in Figure 8b shows that the majority of steam bubbles remain in the neighborhood of nucleation spots (here, wall cavities), and the concentration in the center region is low and flat. The rate of bubbles migrating from the near-wall region to the nozzle center increases with the bubble size.

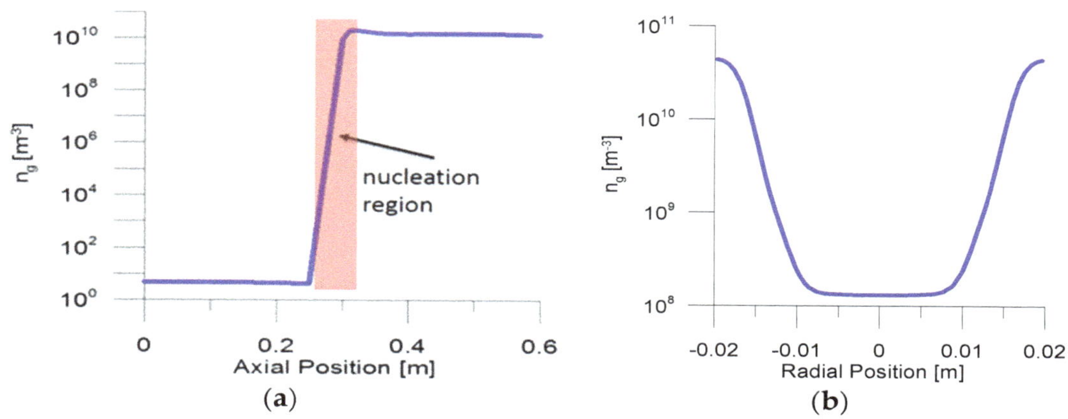

Figure 8. Distribution of bubble number density in (**a**) the axial direction and (**b**) the radial direction at $x = 0.458$ m.

In Figure 9, the simulation results are compared with the data. On the left side is the axial profile of absolute pressure (left y-axis) and steam volume fraction (right y-axis), and on the right side is the radial profile of the steam volume fraction at the position $x = 0.458$ m. Symbols and lines represent measurement and calculation, respectively.

The cross-section averaged steam volume fraction increases in the latter part of flow path, which is in an acceptable agreement with the data. On the other hand, the pressure profile exhibits obvious deviations in the diverging part of the nozzle. The predicted value is lower than the measured one. It suggests that too high a superheat (or pressure undershoot) is required for triggering the onset of flashing in comparison to the experiment. In other words, with the same superheat degree as observed in the experiment, the predicted inter-phase heat transfer rate is insufficient to overcome the latent heat and keep bubbles stable. The interfacial area density or inter-phase heat transfer coefficient is under-predicted at the moment of flashing onset. After triggering the flashing process, the pressure increases under the constraint of the outlet pressure, which is given as the boundary condition. Furthermore, the assumption of pressure equilibrium across the interface might introduce large errors in high pressure undershoot cases.

The transverse distribution of bubbles in the continuous medium is determined by non-drag forces, which depend sensitively on the mean bubble size. It is determined by the bubble number density and nucleation rate, since bubble coalescence and break-up are not considered. As shown in Figures 8b and 9b, bubbles are formed and accumulated in the near wall region and create a void fraction profile with a high wall-peak while a low value in the center region. A similar profile is obtained in the measurement. However, it displays a more gradual transition from the wall peak to the center valley. That means that more bubbles have migrated to the center due to the effect of non-drag forces, i.e., lift and turbulence dispersion. This may imply that bubble growth is under-estimated, and coalescence effects should be taken into account, since lift force pushes large bubbles to the center, while small bubbles to the wall. However, to refine the results, reliable data of local bubble size distributions are indispensable, which are unfortunately missing in most experimental investigations. Furthermore, sufficient turbulent dispersion would also help to improve the agreement. According to the FAD (Favre averaged drag) model, the dispersion force is proportional to eddy viscosity, which is probably under-predicted by the SST model with BIT source terms.

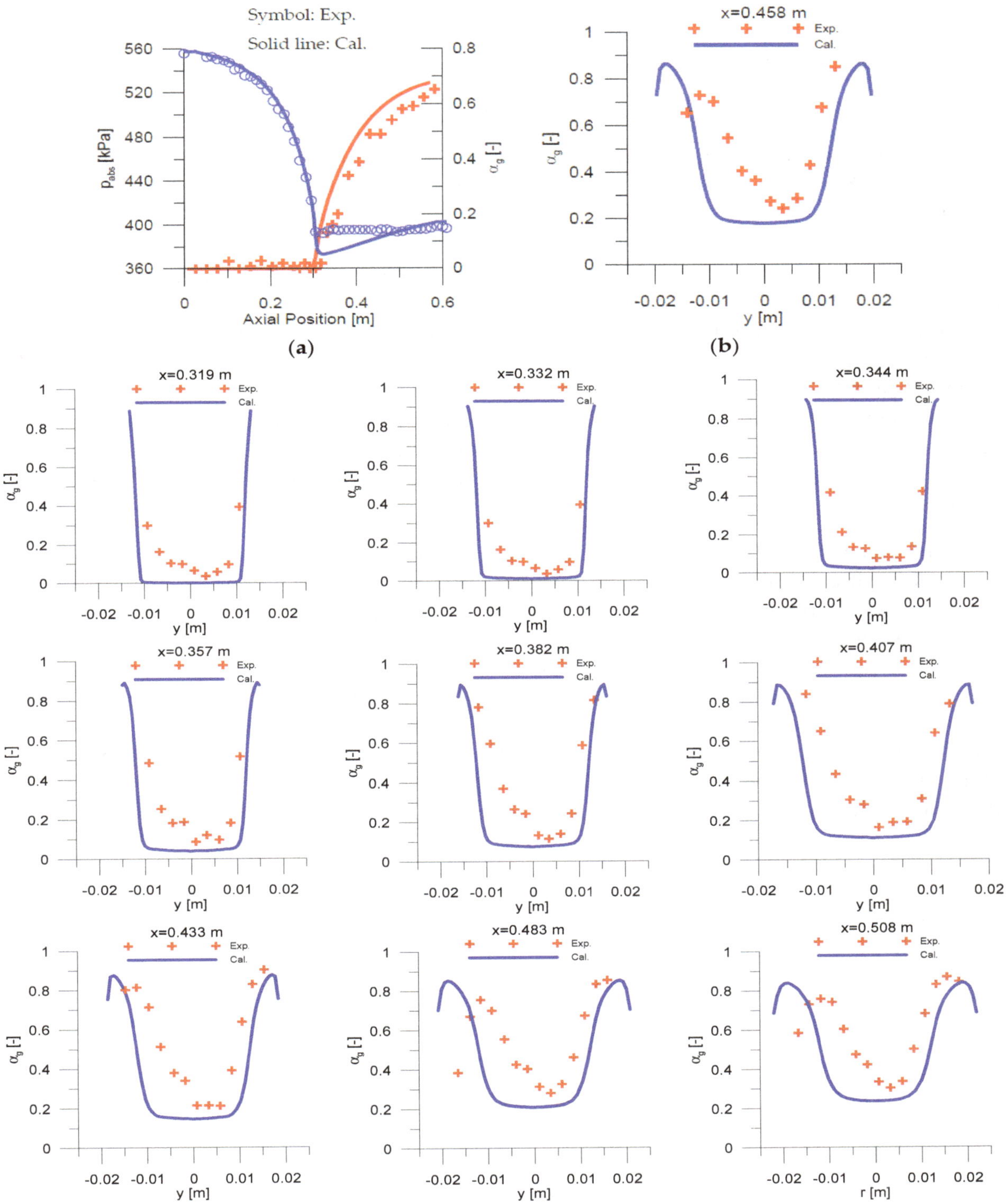

Figure 9. *Cont.*

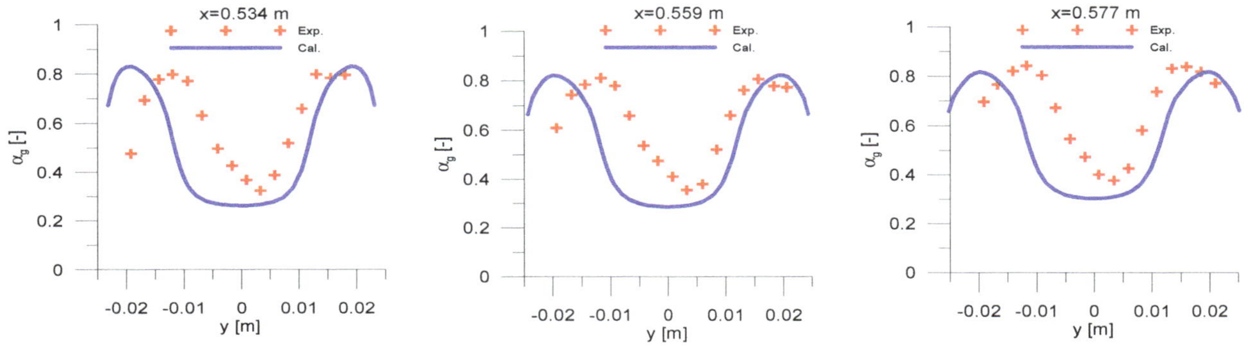

(c)

Figure 9. Comparison between calculation and experiment: (**a**) axial profile of the pressure and void fraction; (**b**) radial profile of the void fraction; (**c**) radial profile of the void fraction at different axial positions.

To give more details about the flow structure, the radial profile of the void fraction is depicted in Figure 9c for several axial positions other than $x = 0.458$ m shown above. The formation of bubbles starts around the throat position ($x = 0.305$ m; see Figure 6) and first leads to a monotonous peak near the wall. It shifts slowly to the nozzle center due to the effect of non-drag forces along the axis, and an "M-shape" profile is developed over the cross-section. The simulated results agree well with the measured ones except for an over-prediction of the wall-peak at the beginning. Another important observation is that in the experiment, the profile is not symmetrical with respect to the axis. According to the experimenter, the asymmetry might be due to the presence of pressure taps on one side of the nozzle [33], which act as nucleation centers.

Several other tests with different inlet and outlet pressures, as well as inlet temperatures are simulated. The critical mass flow rates are summarized in Figure 10 below. The critical rate of flow through the nozzle depends on a variety of parameters, inlet pressure, quality, pressure-undershoot, and so on. In general, increasing the inlet stagnation pressure or decreasing outlet pressure leads to an increase in the critical mass flow rate. Furthermore, the presence of vapor reduces the average density of the two-phase mixture and, thus, favors the deceleration process in the diverging part of the nozzle.

It is shown that the simulated results are in an overall good agreement with the measured ones. The maximum deviation is within 7%. Nevertheless, the simulation tends to under-predict the critical mass flow rate with the increasing of the inlet pressure and decreasing of the outlet pressure.

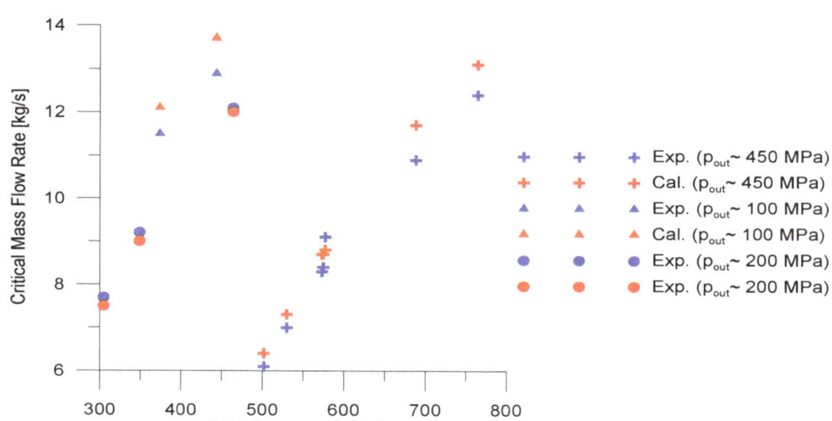

Figure 10. Dependence of the critical mass flow rate on inlet and outlet pressure in the nozzle.

5. Poly-Disperse Simulations

Although it is commonly used, the mono-disperse approach presented above is self-evident limited to situations with a constant bubble size or a narrow distribution. On the other hand, flashing processes are always accompanied with rapid vaporization and a high void fraction. This suggests that a broad spectrum of bubble sizes can be present, which asks for a poly-disperse simulation method. However, to the author's knowledge, no publications in this aspect are so far available for flashing situations, although there is a wide range of poly-disperse simulations for isothermal flows and a few for sub-cooled boiling.

In this work, the first attempts are made to apply the inhomogeneous MUSIG (multiple size group) model [35,36] to controlled pressure release transients, which were carried out on the TOPFLOW facility (see below). MUSIG is a poly-disperse method available in CFX for computing the mean bubble size of a spectrum. It is one kind of class method approximating the size spectrum with several discrete classes; see Figure 11.

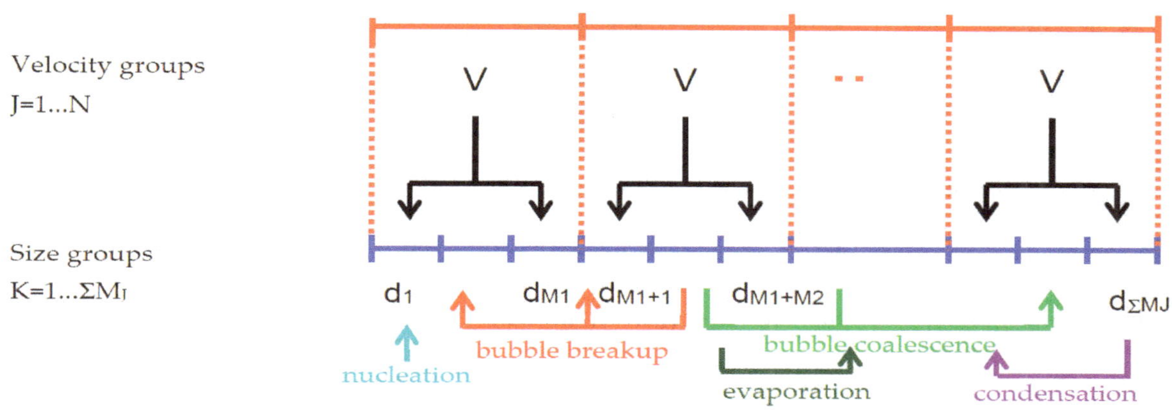

Figure 11. Schema of the inhomogeneous multiple size group (MUSIG) model.

The division of size groups is finer than that of velocity groups due to the limitation of computational speed. Separate momentum equations are solved for each velocity group while all phenomena that change the size distribution, such as nucleation, coalescence and breakup, are considered within the sub-size groups. The exchange between these size groups due to the above-mentioned phenomena is reproduced by solving additional transport equations with corresponding source/sink terms for the fraction of each group f_i. As a result, the mean Sauter diameter of bubbles belonging to a velocity group can be obtained from these size fractions. The MUSIG approach has been expounded in detail elsewhere [24,36].

5.1. TOPFLOW Pressure Release Experiment

One major test section equipped in the TOPFLOW facility is a vertical pipe with a nominal diameter of 200 mm; see Figure 12. The pressure release experiment was carried out at this section. For details about the multipurpose thermal-hydraulic test facility TOPFLOW at Helmholtz-Zentrum Dresden-Rossendorf, please refer to the work of [37].

During the pressure release experiment, water was circulated with a velocity of about 1 m/s and flows upwards through the test section; see Figure 12. The transient pressure course is controlled by the blow-off valve located above the steam drum, where saturation conditions are always guaranteed. As a result, cavitation in the circulation pump below is avoided. At the same time, the maximum evaporation rate in the test section is limited [27].

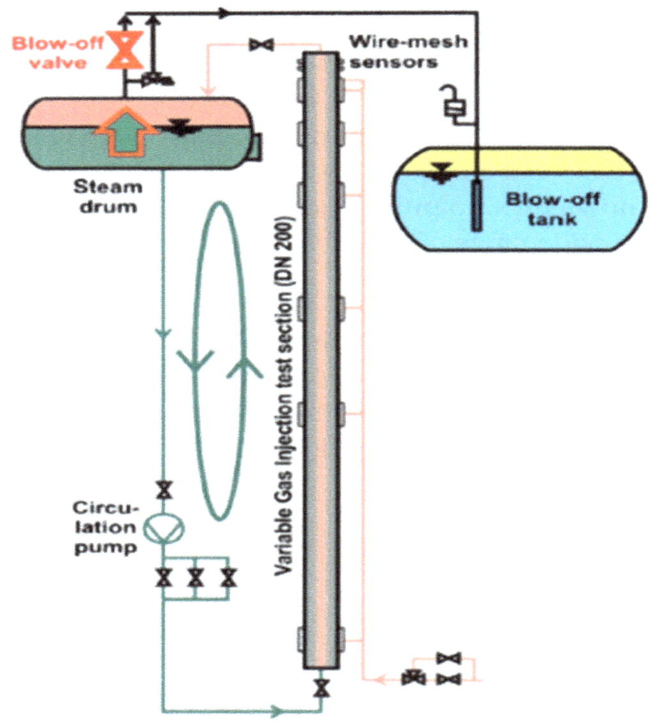

Figure 12. Schema of the experimental procedure.

The blow-off valve was opened to a maximum level and closed again according to the ramp shape as shown in Figure 13.

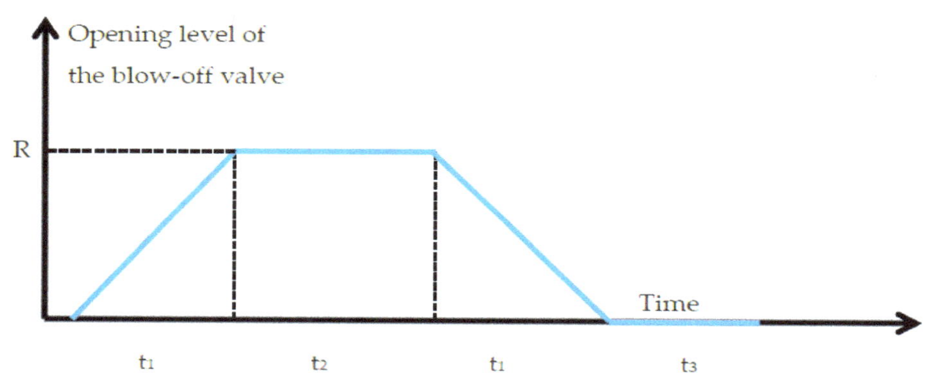

Figure 13. Operation of the blow-off valve.

The operation speed of the valve is controlled by time t_1, while the maximum opening degree and its duration are represented by R and t_2. Correspondingly, the generation and disappearance of steam inside the pipe are determined by the pressure release rate and the operation of the valve. Measurements of the volume fraction, gas velocity and bubble size distribution were realized with the aid of wire-mesh sensors (WMS) at the top; see Figure 12. The highly-resolved data of bubble size distributions and radial void fractions are optimal for the validation of simulation results obtained by the poly-disperse method. Rapid expansion of bubbles and a broad size spectrum were observed in the experiment, and for reliable simulation results, it is necessary to use a poly-disperse method. A detailed introduction to the experimental procedure and discussion on the influence of pressure level on the data were given by Lucas [38].

Two test cases under different pressure levels are investigated in the current work, whose experimental conditions are summarized in Table 2 below.

Table 2. Experimental conditions of the investigated test cases.

Case No.	Pressure (bar)	R (%)	t_1 (s)	t_2 (s)	t_3 (s)
1	10	60	21	30	30
2	65	20	7	56	30

Measured inlet temperature, mass flow rate and outlet pressure are used as boundary conditions in the simulation.

5.2. Simulation Results

The aim of poly-disperse simulations is to obtain a realistic mean bubble size by tracing the dynamic change of bubble size distributions, which is important in the case of a broad bubble size spectrum. For this purpose, besides highly-resolved data, reliable closure models for all physical processes that affect the bubble size are indispensable. In this work not only nucleation and phase change, but also bubble coalescence and break-up have been taken into account (see Figure 11). For heterogeneous nucleation, the same models as for the nozzle flow discussed above in Section 4.3.2 were employed, while the models presented in [24] were used for bubble coalescence and break-up. The change resulting from these phenomena is implemented as source or sink in the MUSIG size fraction transport equations mentioned above.

The evolution of the normalized bubble size distribution is displayed in Figures 14 and 15 for the two cases listed in Table 2, respectively. In the vertical axis label, $\Delta\alpha_i$ represents the void fraction portion belonging to the size group i, while $\sum\Delta\alpha i$ is the total void fraction. The red line represents experimental data, while the blue one simulation results.

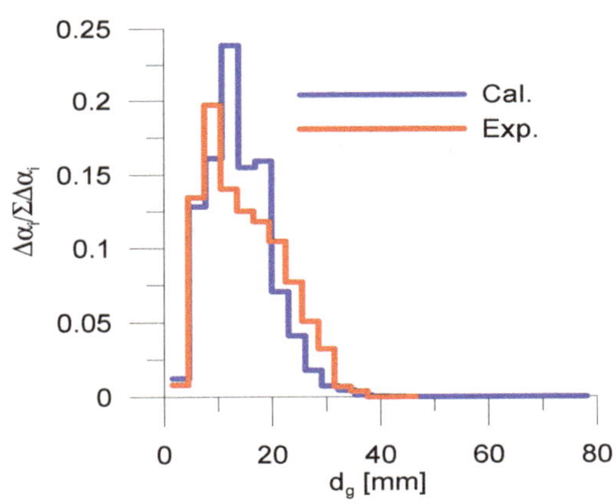

(**a**)

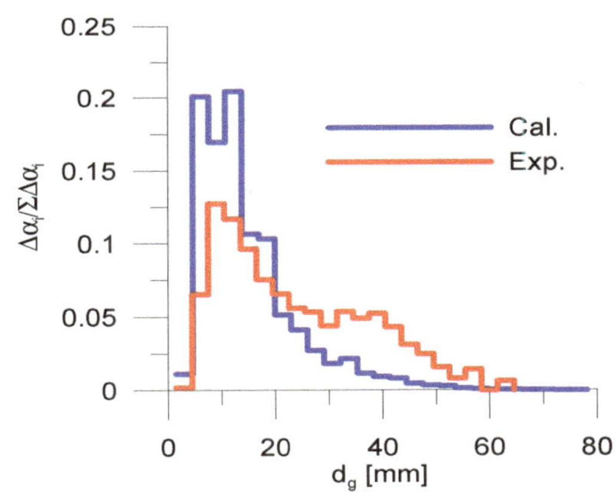

(**b**)

Figure 14. Evolution of normalized bubble size distribution at the wire-mesh sensor (WMS) plane (Case 1): (**a**) t = 39 s; (**b**) t = 45 s.

In Case 1, whose pressure level is 10 bar, bubble growth within the period from t = 39 s to 45 s is significant. The agreement between calculation and experiment at t = 39 s is satisfying, whereas six seconds later, the fraction of large bubbles is obviously under-predicted. As a result, the mean bubble size is smaller than the measured one, although both of them increase. One possible reason for this discrepancy is that the mechanism for inter-phase mass transfer is not completely reproduced by the "thermal phase change model". Furthermore, the break-up rate of large bubbles may be overestimated corresponding to the coalescence rate of small bubbles. The predictability of coalescence and break-up models depends on a number of input parameters, such as turbulence intensity and interfacial shear stress. The evaluation of these parameters in such complex two-phase flows is often difficult. In addition, the superposition of various mechanisms, as well as of coalescence and break-up effects increases the difficulty in further development and improvement of these closures.

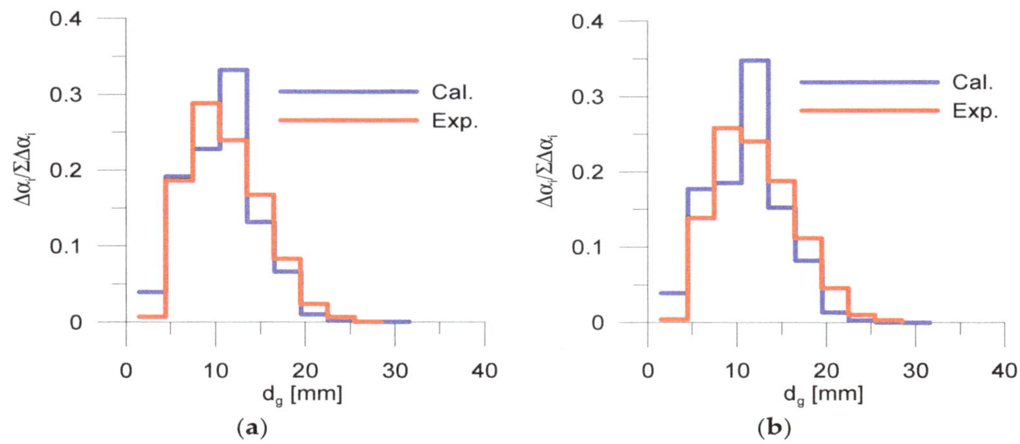

Figure 15. Evolution of normalized bubble size distribution at the WMS plane (Case 2): (**a**) t = 29 s; (**b**) t = 49 s.

The growth of bubbles is substantially slowed down with the increase of pressure level. As shown in Figure 15, in Case 2, the change of bubble size distribution within 20 s is trivial. Consequently, the agreement at both time points is acceptable.

The different performance of the applied model setup at two pressure levels suggests that the contribution of pressure non-equilibrium at the interface to inter-phase mass transfer can be significant in Case 1. An approach that takes both thermal and mechanical non-equilibrium into account is necessary to ensure the reliability of predictions in both low and high pressures.

The pressure and velocity field shown in Figures 16 and 17 below is intended to provide further insight into the flashing flow. In the simulation, the pressure boundary condition was specified for the outlet (top), and the measured pressure was imposed on the boundary. The calculated pressure at the inlet (bottom) was found in good agreement with the data, as shown in Figure 16. This implies that the pressure field in the whole domain is well reproduced. Water flashes into steam during the depressurization, whose rate is controlled by the opening of the blow-off valve (see Figure 13).

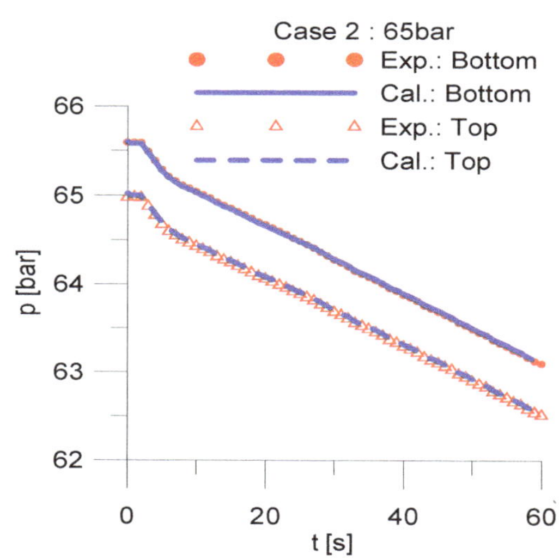

Figure 16. Evolution of the cross-section averaged absolute pressure at the top and bottom of the test section (Case 2).

The comparison of the gas vertical velocity component at the WMS plane, where data are available, is shown in Figure 17. Instead of a core peak as observed in the measurement, the calculated radial profiles show a transit peak. The velocity profile is similar to the void fraction ones, which depends on the magnitude of non-drag forces. Since the bubble size is well captured (see Figure 15), the discrepancy is caused either by an under-prediction of turbulence dispersion force or an over-prediction of lift force. Under the role of lift force, large bubbles move away from the wall and accumulate in the center, while turbulence dispersion tends to counteract this effect. However, for further evaluation of these models, reliable reference information, e.g., turbulence, is still missing.

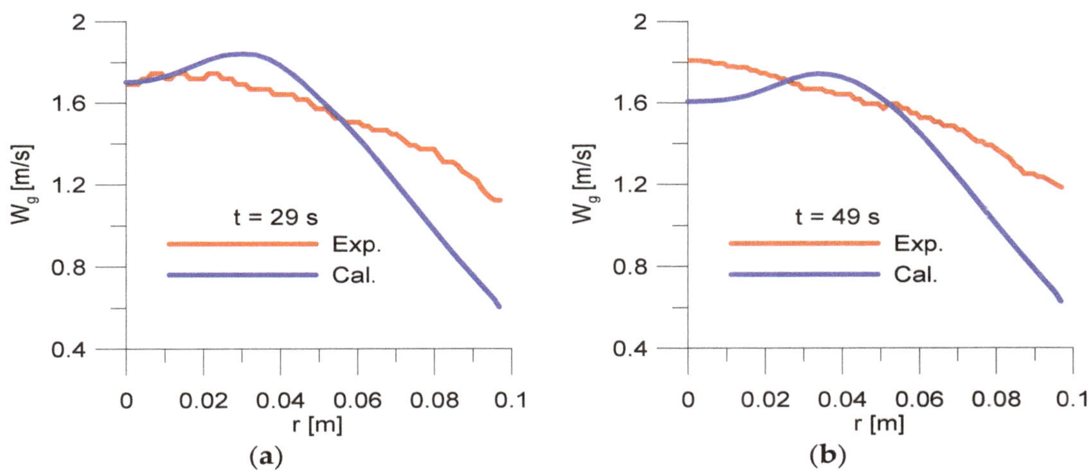

Figure 17. Evolution of radial gas velocity profiles at the WMS plane (Case 2): (**a**) t = 29 s; (**b**) t = 49 s.

Improving the two-phase turbulence, as well as dispersion and lift force modelling will be one of the emphases of future work. Related theoretical and experimental work has been planned and begun.

6. Conclusions

Flashing of liquid to vapor due to pressure drop represents highly complex two-phase situations. Due to its relevance to nuclear safety analysis, the flashing phenomenon has been extensively studied for several decades. Nevertheless, there exists a need to update the analytical approach to a sophistication level that matches available computational technologies, e.g., from system codes to CFD codes. CFD simulations with the simplified mono-disperse approaches, which are commonly used and often deliver satisfying results, are however limited for cases with nearly constant bubble size or number density. Since these conditions are normally not fulfilled, they are shown to have difficulty in capturing the onset of flashing, which consequently affects the agreement in the subsequent phase change stage. In addition, the results are sensitively dependent on the prescribed values and the spatial distribution of phases on the mean bubble size. The poly-disperse approach is promising for removing these restrictions, since it is free of any open parameters and reproduces the realistic change of bubble size distributions. However, as suggested in [12], more elaborate models require more empirical correlations or assumptions for phase interactions which are so far insufficiently tested. As a result, the prediction accuracy of the sophisticated method is greatly affected.

Therefore, before we can exploit the benefits of CFD simulations for flashing flows, we have to understand the physical sub-phenomena, such as nucleation characteristics and inter-phase transfer laws sufficiently, and be able to specify them precisely with closure models. For this purpose, highly-resolved and comprehensive data achieved by experiments or direct numerical simulations are indispensable, especially local bubble size, phase distribution, turbulence, velocity and pressure fields, which are often unavailable, e.g., in Tests 4.1, 4.2 and 4.3. The wire-mesh senor technique applied in the TOPFLOW pressure release experiment can provide the above measurements. Nevertheless, the necessary information on liquid velocity and turbulence parameters is still missing. The development of new measurement techniques is undergoing [39].

The presented results show that the Eulerian CFD technology reproduces global parameters, such as pressure and flow rates, reliably, even in complex practical situations. Nevertheless, the prediction of local phenomena, e.g., phase distribution and velocity fields, is still insufficient. More efforts are required in the assessment and improvement of closures.

Acknowledgments: This work was carried out at Helmholtz-Zentrum Dresden-Rossendorf in the frame of a research project funded by EON Kernkraft GmbH.

Author Contributions: Dirk Lucas and Yixiang Liao conceived of the simulations and evaluated the results. Yixiang Liao performed the simulations and wrote the paper.

Abbreviations

ANSYS	an American computer-aided engineering software developer headquartered south of Pittsburgh in Cecil Township, Pennsylvania, United States
AREVA	a French multinational group specializing in nuclear power and renewable energy headquartered in Paris La Défense
CCC	containment cooling condenser, a passive nuclear safety system
CFD	computational fluid dynamics
CFX	a commercial CFD code developed by ANSYS company
CSNI	committee on the Safety of Nuclear Installations, a committee of the OECD/NEA
FLUENT	a commercial CFD code developed by ANSYS company
IAPWS-IF97	International Association for the Properties of Water and Steam – Industrial Formulation 1997
KERENA™	a mid-power boiling water reactor developed jointly by the AREVA company and the German energy supply company E.ON
NEA	Nuclear Energy Agency, an intergovernmental agency organized under OECD
NEPTUNE-CFD	a French code created by EDF (Électricité de France) and CEA (Commissariat à I'Energie Atomique) for nuclear reactor thermal-hydraulics simulation and analyses
OECD	Organization for Economic Co-operation and Development
RELAP5	abbreviation of "Reactor Excursion and Leak Analysis Program", A component-oriented reactor systems thermal-hydraulics analysis code developed at Idaho National Laboratory
RPI	Rensselaer Polytechnic Institute
SSPV	Shielding/Storage Pool Vessel in KENERA reactor
TRACE	TRAC/RELAP Advanced Computational Engine, a component-oriented reactor systems thermal-hydraulics analysis code
TRAC	transient reactor analysis code, another reactor system code
WMS	wire mesh sensor, a measurement technique

References

1. Giese, T.; Laurien, E. A thermal based model for cavitation in saturated liquids. *Z. Angew. Math. Mech.* **2001**, *81*, 957–958.
2. Giese, T.; Laurien, E. Experimental and numerical investigation of gravity-driven pipe flow with cavitation. In Proceedings of the 10th International Conference on Nuclear Engineering (ICONE10), Arlington, VA, USA, 14–18 April 2002.
3. Laurien, E.; Giese, T. Exploration of the two fluid model of two-phase flow towards boiling, cavitation and stratification. In Proceedings of the 3rd International Conference on Computational Heat and Mass Transfer, Banff, AB, Canada, 26–30 May 2003.
4. Laurien, E. Influence of the model bubble diameter on three-dimensional numerical simulations of thermal cavitation in pipe elbows. In Proceedings of the 3rd International Symposium on Two-Phase Flow Modelling

and Experimentation, Pisa, Italy, 22–24 September 2004.
5. Frank, T. Simulation of flashing and steam condensation in subcooled liquid using ANSYS CFX. In Proceedings of the 5th Joint FZR & ANSYS Workshop "Multiphase Flows: Simulation, Experiment and Application", Dresden, Germany, 26–27 April 2007.
6. Maksic, S.; Mewes, D. CFD-Calculation of the flashing flow in pipes and nozzles. In Proceedings of the Joint U.S.—European Fluids Engineering Division Conference (ASME FEDSM'02), Montreal, QC, Canada, 14–18 July 2002.
7. Marsh, C.A.; O'Mahony, A.P. Three-dimensional modelling of industrial flashing flows. In Proceedings of the Computational Fluid Dynamics (CFD 2008), Trondheim, Norway, 10–12 June 2008.
8. Mimouni, S.; Boucker, M.; Laviéville, J.; Guelfi, A.; Bestion, D. Modelling and computation of cavitation and boiling bubbly flows with the NEPTUNE_CFD code. *Nucl. Eng. Des.* **2008**, *238*, 680–692. [CrossRef]
9. Janet, J.P.; Liao, Y.; Lucas, D. Heterogeneous nucleation in CFD simulation of flashing flows in converging-diverging nozzles. *Int. J. Multiph. Flow* **2015**, *74*, 106–117. [CrossRef]
10. Edwards, A.R.; O'Brien, T.P. Studies of phenomena connected with the depressurization of water reactors. *J. Br. Nucl. Enery Soc.* **1970**, *9*, 125–135.
11. Liao, Y.; Lucas, D.; Krepper, E.; Rzehak, R. Assessment of CFD predictive capacity for flash boiling. In Proceedings of the Computational Fluid Dynamics for Nuclear Reactor Safety-5 (CFD4NRS-5), Zurich, Switzerland, 9–11 September 2014.
12. Wallis, G.B. Critical two-phase flow. *Int. J. Multiph. Flow* **1980**, *6*, 97–112. [CrossRef]
13. Jones, O.C., Jr.; Zuber, N. Bubble growth in variable pressure fields. *J. Heat Transf.* **1978**, *100*, 453–459. [CrossRef]
14. Shin, T.S.; Jones, O.C., Jr. An active cavity model for flashing. *Nucl. Eng. Des.* **1986**, *95*, 185–196. [CrossRef]
15. Shin, T.S.; Jones, O.C. Nucleation and flashing in nozzles—1. *Int. J. Multiph. Flow* **1993**, *19*, 943–964. [CrossRef]
16. Blinkov, V.N.; Jones, O.C.; Nigmatulin, B.I. Nucleation and flashing in nozzles—2. *Int. J. Multiph. Flow* **1993**, *19*, 965–986. [CrossRef]
17. Blander, M.; Katz, J.L. Bubble nucleation in liquids. *AIChE J.* **1975**, *21*, 833–848. [CrossRef]
18. *Release 14.5. ANSYS CFX-Solver Theory Guide*; ANSYS: Canonsburg, PA, USA, 2012.
19. Riznic, J.; Ishii, M. Bubble number density in vapor generation and flashing flow. *Int. J. Heat Mass Transf.* **1989**, *32*, 1821–1833. [CrossRef]
20. Ranz, W.E.; Marshall, W.R. Evaporation from drops—Part I. *Chem. Eng. Prog.* **1952**, *48*, 141–146.
21. Ranz, W.E.; Marshall, W.R. Evaporation from drops—Part II. *Chem. Eng. Prog.* **1952**, *48*, 173–180.
22. Rohatgi, U.; Reshotko, E. Non-equilibrium one-dimensional two-phase flow in variable area channels. In *Non-Equilibrium Two-Phase Flows, Proceedings of the Winter Annual Meeting, Houston, TX, USA, 30 November–5 December 1975*; Meeting Sponsored by the American Society of Mechanical Engineers; American Society of Mechanical Engineers: New York, NY, USA, 1975; Volume 1, pp. 47–54.
23. Rzehak, R.; Krepper, E. CFD modeling of bubble-induced turbulence. *Int. J. Multiph. Flow* **2013**, *55*, 138–155. [CrossRef]
24. Liao, Y.; Rzehak, R.; Lucas, D.; Krepper, E. Baseline closure model for dispersed bubbly flow: Bubble coalescence and breakup. *Chem. Eng. Sci.* **2015**, *122*, 336–349. [CrossRef]
25. Garner, R.W. *Comparative Analyses of Standard Problems, Standard Problem 1 (Straight Pipe Depressurization Experiments)*; Interim Report I-212-74-5.1; Aerojet Nuclear Company: Rancho Cordova, CA, USA, 1973.
26. Stello, V. *Summary of All Participants Results in Comparison to the Experimental Data for the CSNI Standard Problem 1*; For the CSNI-Ad Hoc Working Group on Emergency Core Cooling; OECD/NEA: Paris, Franch, 1975.
27. Liao, Y.; Lucas, D.; Krepper, E.; Rzehak, R. Flashing evaporation under different pressure levels. *Nucl. Eng. Des.* **2013**, *265*, 801–813. [CrossRef]
28. Gurgacz, S.; Pawluczyk, M.; Niewiński, G.; Mazgaj, P. Simulation and analysis of pipe and vessel blowdown phenomena using RELAP5 and TRACE. *J. Power Technol.* **2014**, *94*, 61–68.
29. Dardy, A. *The AREVA Reactor Range—The Nuclear Safety Reference for Europe*; AREVA Day: Warsaw, Poland, 2010.
30. Leyer, S.; Wich, M. The Integral test facility Karlstein. *Sci. Technol. Nucl. Install.* **2012**, *2012*, 439374. [CrossRef]
31. Manera, A. Experimental and Analytical Investigations on Flashing-Induced Instabilities in Natural Circulation Two-Phase Systems. Ph.D. Thesis, Delft University of Technology, Delft, The Netherlands, 2003.

32. Liao, Y.; Lucas, D. 3D CFD simulation of flashing flows in a converging-diverging nozzle. *Nucl. Eng. Des.* **2015**, *292*, 149–163. [CrossRef]
33. Abuaf, N.; Wu, B.J.C.; Zimmer, G.A.; Saha, P. *A Study of Nonequilibrium Flashing of Water in a Converging-Diverging Nozzle: Volume 1—Experimental*; U.S. Nuclear Regulatory Commission: Washington, DC, USA, 1981.
34. Riznic, J.; Ishii, M.; Afgan, N. Mechanistic model for void distribution in flashing flow. In Proceedings of the Transient Phenomena in Multiphase Flow (ICHMT Seminar), Dubrovnik, Yugoslavia, 24–30 May 1987.
35. Krepper, E.; Lucas, D.; Frank, T.; Prasser, H.-M.; Zwart, P.J. The inhomogeneous MUSIG model for the simulation of polydispersed flows. *Nucl. Eng. Des.* **2008**, *238*, 1690–1702. [CrossRef]
36. Liao, Y.; Lucas, D. Poly-disperse simulation of condensing steam-water flow inside a large vertical pipe. *Int. J. Therm. Sci.* **2016**, *104*, 194–207. [CrossRef]
37. Schaffrath, A.; Kruessenberg, A.K.; Weiss, F.P.; Beyer, M.; Carl, H.; Prasser, H.M.; Schuster, J.; Schuetz, P.; Tamme, M.; Zimmermann, W. TOPFLOW—A new multipurpose thermalhydraulic test facility for the investigation of steady state and transient two-phase flow phenomena. *Kerntechnik* **2001**, *66*, 209–212.
38. Lucas, D.; Beyer, M.; Szalinski, L. Experiments on evaporating pipe flow. In Proceedings of the 14th International Topical Meeting on Nuclear Reactor Thermalhydraulics (NURETH-14), Toronto, ON, Canada, 25–30 September 2011.
39. Ziegenhein, T.; Zalucky, J.; Rzehak, R.; Lucas, D. On the hydrodynamics of airlift reactors, Part I: Experiments. *Chem. Eng. Sci.* **2016**, *150*, 54–65. [CrossRef]

5

Experimental Research on Hydraulic Collecting Spherical Particles in Deep Sea Mining

Guocheng Zhao [1,2], Longfei Xiao [1,2,*], Tao Peng [1,2] and Mingyuan Zhang [1]

1 State Key Laboratory of Ocean Engineering, Shanghai Jiao Tong University, Shanghai 200240, China; guocheng.zhao@sjtu.edu.cn (G.Z.); Pengtao@sjtu.edu.cn (T.P.); mzbzmy@sjtu.edu.cn (M.Z.)
2 Collaborative Innovation Center for Advanced Ship and Deep-Sea Exploration, Shanghai 200240, China
* Correspondence: xiaolf@sjtu.edu.cn.

Abstract: Hydraulic collecting is the key technology in deep sea mining and dredging engineering. It determines economic benefits of the project and environmental issues. However, mechanistic studies of hydraulic collecting are rarely described. In this study, the mechanism of collecting spherical particles is researched by dimensional analysis and experimental study. The experimental system is established to carry out three kinds of tests including 253 different test cases. The empirical model of collecting performance prediction is established by the tests of vertical force characteristics and vertical incipient motion characteristics of particles in suction flow field. The results show that the vertical suction force coefficient (C_{vs}) decreases exponentially with the ratio of bottom clearance to diameter of the particle (h/d), increases linearly with the ratio of diameter of the suction pipe to diameter of the particle (D/d), and is nearly independent of Reynolds number (Re). The empirical formula of vertical force and criterion-formula of vertical incipient motion of particles are obtained with the maximum tolerance less than 15%. The phenomenon that the vortex could help strengthen the suction force was observed in the tests. In addition, the characteristics of suction flow field were obtained by flow visualization tests, and applied to explain the force characteristics of particles in the suction flow field.

Keywords: deep sea mining; manganese nodules exploitation; hydraulic collecting; suction flow field; dimensional analysis

1. Introduction

Enormous mineral resources are deposited in the international seabed, of which polymetallic nodules have attracted more and more attention around the world for their high economic value and easy prospecting. With increasing demand for various mineral resources and gradual exhaustion of land resources in modern society and industry, research and development of seabed mineral resources is becoming increasingly urgent [1]. Therefore, deep sea mining is thought to be a key approach to the sustainable development of human beings and many countries around the world are stepping up exploration and exploitation of seabed mineral resources [2,3]. Most of the deep sea mining technologies and systems for exploration, recovery and processing were originally been done in the 1970s. Four subsystems are usually included in deep sea mining systems: surface systems, pipe systems, buffer systems and collector systems [4]. Chung [5,6] presented a brief summary of the developments in mining systems and technologies in the past years, and pointed out that the mobility, safety, collection efficiency and sweep efficiency of the collector system are the most important parameters in deep sea mining system design. The collecting part in the miner system is also considered as the key technology, because commercial production must achieve high sweep efficiency [7], and profitable deep sea mining exploitation is only feasible on the premise that there is a nodule collector with maximum collecting capacity of 140 kg of wet nodules per second [8]. Except for the high collecting efficiency, eco-friendly mining processes are also in demand because deep sea mining could be a new environmental challenge related to ocean biology [9,10]. To pick up nodules, a variety of collecting methods such as hydraulic methods, mechanical methods and hybrid collection methods have been developed. Ocean Management Incorporated (OMI's) sea trial results in 1978 showed that the hydraulic method has higher collection efficiency than the mechanical method [11]. Therefore, the problem of hydraulic collecting will be studied in this paper.

The mechanism of hydraulic collecting is fairly complicated because the collecting efficiency is influenced by a lot of factors such as particle size, bottom clearance, collector structure and flow rate. Several studies have been conducted to examine the process of hydraulic collecting. Hong et al. [12] dealt with experimental approaches for enhanced understanding of the hydraulic performance of a hybrid pick-up device. The experiments were conducted in a 2-D flume tank. By parametric experiments they found position and shape of baffle plates are significant factors for effective design of hydraulic nodule lifter. Yang et al. [13] discussed major parameters and their influences on the performance of the pick-up device by tests. The results showed that the hydraulic pick-up device with proper dimensions and parameters could get high pick-up rate and low content sediments. Tsutsui [14] had an experimental study of the flow field and the aerodynamic force on a sphere above a plane. The surface pressures on the sphere and the plane were measured by inclined multi-pipe manometers and the results were compared with photographs showing the flow visualization of the sphere. Lim et al. [15] analyzed flow field characteristics with outflow discharge from a collecting device in deep ocean by using software FLUENT. The study revealed seawater velocity and streamline distributions along with complicated flow characteristics downstream including nodule particles behavior. Zhao et al. [16] carried out studies of hydraulic collection of single spherical particle at various collecting bottom clearances and flow rates by both experiments and numerical simulations. It was verified that the vertical force prediction is feasible for single ore particle in collecting condition based on numerical simulation. The study revealed that the variation of the wake vortex is the dominant factor of force vibration.

The problems of hydraulic lifting have been studied, which could be helpful to understand the process of hydraulic collecting. Chung et al. [17] conducted experimental investigations to study the shape effect of solids on pressure drop in a 2-phase vertically upward transport. Sobota et al. [18] had an experimental study on nodule and water velocities at upward flow in a pipeline of $D = 150$ mm. The slip velocity values of 10-, 30- and 50-mm-dimeter nodules at the volume concentration 10% were measured. The results illustrated that the slip velocity values are of the same order as the fall velocity of these solids. Jiang et al. [19] had an investigation into the concentration characteristics of large size particles in vertical pipes for hydraulic lifting. Particles of different diameters (5–30 mm) were used for experimental studies. A formula for the radial distribution of particle concentration was developed and validated. Jiang et al. [20] did the experiments on single particle settling velocity in the static fluid, floating velocity for a single particle and critical velocity in a vertical hydraulic lifting system. They had found that the larger the size and concentration of particles were, the more possibly the particles moved to the pipe wall. Yang et al. [21] analyzed the influence of factors such as particle concentration, diameter, gradation and boundary conditions on single and group suspension velocity by doing numerous experiments. A relation between particle group and single particle suspension velocity ratio and volumetric concentration was given. Behavior of single particle in vertical lifting pipe is the foundation of the group particles in vertical lifting pipe. In addition, the studies of deep sea air-lift have been researched. Fan et al. [22] investigated the performance of an air-lift pump for artificial upwelling of deep water theoretically and experimentally with a vertical pipe of 0.4 m diameter and 28.3 m height. An empirical model of the air-lift artificial upwelling is presented. The performance of the model has been confirmed by the experimental findings. Pougatch et al. [23] presented a numerical model of the three-phase flow in the upward air-lift pipe. And the influence of the pipe diameter on the lifting efficiency has been investigated numerically. Ma et al. [24] assessed the technological feasibility and profitability analyses in terms of solid production rate, energy consumption per tonnage of mineral, and profitability per tonnage of mineral. The effects of submergence ratio, pipe diameter, particle diameter, mining depth, and gas flux rate are investigated.

The research of flow around a sphere is the foundation of study on hydraulic collecting. Achenbach [25] had experiments on the flow past spheres in the Reynolds number range $5 \times 10^4 \leq Re \leq 6 \times 10^4$. He compared his results with other available data and pointed out the dependence of friction forces on Reynolds number. Johnson and Patel [26] investigated the flow of an incompressible viscous fluid past a sphere at different Reynolds numbers by Detached Eddy Simulation (DES). Constantinescu and Squires [27] applied Large Eddy Simulation (LES) and DES to investigate the flow around a sphere at a Reynolds number of 10,000 in the subcritical regime.

Collecting efficiency and environmental issues from the collecting process are the biggest challenges for deep sea mining. Therefore, providing sufficient suction force and avoiding unnecessary disturbance by predicting the performance of collecting accurately is very important. In view of the previous studies including experiments and numerical simulations, however, there are few studies treated the mechanism study of hydraulic collecting and the research of collecting performance prediction.

In this study, mechanism of hydraulic collecting is researched by dimensional analysis and experiments. The characteristics of vertical force, vertical incipient motion and wake flow of spherical particles in suction flow field have been investigated. The empirical formula of vertical force of particles is obtained from measurements and further verified by the vertical incipient motion tests. Furthermore, the criterion-formula of vertical incipient motion of particles is derived to predict the collecting performance. The characteristics of flow field were observed by flow visualization tests. The results of this study will serve to further understand the hydrodynamic characteristics of the hydraulic collecting in deep sea mining and provide reference for the design of the deep sea miner collector.

2. Mechanism of Hydraulic Collecting

Firstly, the simplified models of manganese nodules and hydraulic collecting are founded. Then the method of dimensional analysis is used to study the mechanism of hydraulic collecting and instruct the design of the test program. At last, the empirical formula of vertical force and criterion-formula of vertical incipient motion of particles are derived to predict the hydraulic collecting performance.

2.1. Model of Manganese Nodules

Manganese nodules occur as potato-shaped concretions on the seafloor of abyssal plains in about 4000–6000 m water depth in all major oceans. They form two-dimensional deposits on the surface or within the first 10 cm of the deep-sea sediments (Figure 1).

Figure 1. High abundance of large nodules on the seafloor of the Peru Basin (ca. 4000 m water depth) [28].

Nodules in the eastern tropical Pacific and the central Indian Ocean are of special economic interest due to their high enrichment of metals such as Ni, Cu, Co, Mo, Li, rare earth elements (REEs), and Ga [28]. The dry bulk density of nodules ranges between 1.00 and 2.40 g/cm^3 [29,30]. They can have different sizes, shapes, and surface textures and can reach a size of up to 15 cm in diameter (general range from 2 cm to 10 cm). Some extremely large specimens of 21 cm have been found in the Peru Basin [31]. Moritani [32] did a study about the description and classification of manganese nodules. It is found that nodule occurrence is generally confined to the sediment-water interface of the sea bottom, in which nodules are buried in the sediments in different degrees. The relative position of occurrence seems to result in difference in surface texture, namely rough texture on the buried parts and smooth ones on the exposed parts. And the most common shapes of nodules are either ideal individual forms of spheroidal, ellipsoidal, discoidal, etc. or polylobate or intergrown aggregate form.

In terms of the studies, the assumptions are made to simplify the physical problems in this study:

(1) The shape of particles is spheroidal as a common shape of nodules.

(2) The surface texture of particles is smooth because the roughness of the exposed parts of nodules is usually small.

(3) The particles are on a flat surface because some parts of the mining area are flat and they are usually where we start deep sea mining; in addition, the sediments near the particles are so soft that the particles are always rolling out from sediments before being lifted by the collecting pipe.

(4) No deposit sediments surround particles because sediments contents are very low in the process of hydraulic collecting in engineering test.

2.2. Model of a Kind of Hydraulic Collecting

An underwater mining system was developed for operations and the flexible riser concept was validated in the Indian seas at 410 m water depth in 2000 jointly by National Institute of Ocean Technology (NIOT) and Institut für Konstruktion (IKS) of the University of Siegen (Germany). The underwater mining systems are shown in Figure 2 [33]. The collecting principle of these underwater mining systems is to suck the particles through a suction pipe. The suction flow field was formed near the bottom of suction pipe when water pump was running. The particles are lifted and sucked into the pipe by vertical suction force induced by pressure difference near the particles.

(a) (b)

Figure 2. (**a**) Underwater mining system with manipulator and cutter for 500 m water depth; (**b**) Enhanced underwater mining system for 500 m water depth [33].

According to the collecting principle of these underwater mining systems, the physical model in this study is simplified as Figure 3.

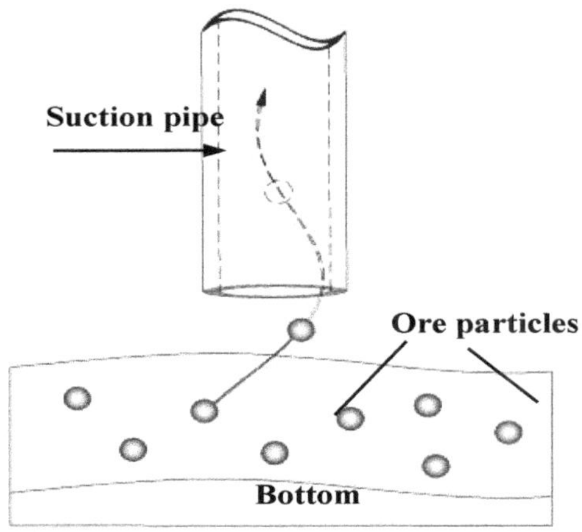

Figure 3. A Kind of hydraulic collecting model.

The main parameters are given in Figure 4, which represent various cases and determine the characteristics of flow field. The main parameters in relation with the vertical force of the particle in this study are listed in Table 1.

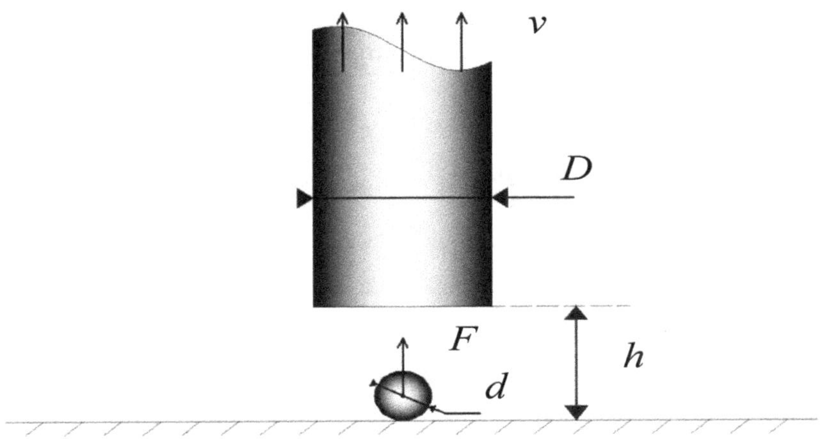

Figure 4. Main parameters of the model.

Table 1. Definition of main parameters.

Physical Parameter	Unit	Definition
d	m	Diameter of particle
h	m	Bottom clearance [1]
D	m	Diameter of suction pipe
F	N	Vertical suction force of particle
F_B	N	Buoyancy in water
G	N	Gravity of particle
ρ	kg/m^3	Density of particle
ρ_w	kg/m^3	Density of water
v	m/s	Flow velocity in pipe

[1] Bottom clearance h: Indicate the vertical height of the suction pipe; use the vertical distance from the pipe inlet to the bottom boundary.

The vertical suction force of the particle is similar to the drag force because both of them are induced by the pressure difference near the sphere or the velocity slippage between liquids and solids. The differences between them are: (1) different definitions, a fluid flowing past the particle exerts a force on it, the vertical force is the component of this force that is normal to ground (parallel to the gravity direction) while the drag force is the in-line component of the force parallel to the flow direction; (2) the drag force is usually defined in an oncoming flow with one direction while the vertical force is defined in the suction flow field which is clearly different with an oncoming flow. In this research, the flow velocity in pipe is used for 2 reasons: (1) the average velocity in pipe is stable along the whole pipe; (2) the flow velocity in pipe is one of the key parameters for engineering application because we could change the vertical force on particles by changing pump flow (the average velocity in pipe) to ensure collection efficiency in different cases.

2.3. Dimensional Analysis

2.3.1. Vertical Force F of Particles in Suction Flow Field

To indicate the vertical force of particles in suction flow field, vertical suction force coefficient C_{vs} is introduced. And the characteristics of flow field can be described by Reynolds number Re. Because of the complexity of the suction flow field, it is difficult to define Re and C_{vs} by using normal definition. In this study, the definition of Re and C_{vs} is proposed to be based on the flow velocity v in the collecting pipe as characteristic velocity.

The vertical force F of the particle fixed on the bottom boundary depends on the fluid density of water ρ_w, the viscosity coefficient of water μ, the velocity of the flow in pipe v, the diameter of the particle d, the diameter of the suction pipe D and bottom clearance h. As a result, the function of F is defined as:

$$F = f(\rho_w, \mu, v, D, d, h) \tag{1}$$

The dimensions of these parameters are as follows:

F	ρ_w	μ	v	d	D	h
MLT^{-2}	ML^{-3}	$ML^{-1}T^{-1}$	LT^{-1}	L	L	L

Therefore, we can get three basic dimensions of this problem with L, M and T, and then the theorem of dimensional analysis is used to obtain:

$$\Pi_1 = \frac{F}{\rho_w v^2 d^2} \tag{2}$$

Because d^2 of the spherical particle section has area dimension, the cross-section area A of the sphere is used instead:

$$\Pi_{1,\text{modified}} = \frac{F}{\frac{1}{2}\rho_w v^2 A} = C_{vs} \tag{3}$$

If the viscosity coefficient of fluid μ is used as dependent variable, then it can be achieved:

$$\Pi_2 = \frac{\mu}{\rho_w v d} \tag{4}$$

It can be derived by formula transformation:

$$\Pi_{2,\text{modified}} = \frac{\rho_w v d}{\mu} = Re \tag{5}$$

It is important to emphasize that the velocity v used in the Re definition is the flow velocity in the suction pipe, rather than the velocity around the particle.

In the same way, we can get the dimensionless variables:

$$\Pi_3 = \frac{D}{d} \tag{6}$$

$$\Pi_4 = \frac{h}{d} \tag{7}$$

The final dimensional relationships can be obtained:

$$\Pi_{1,\text{modified}} = f(\Pi_{2,\text{modified}}, \Pi_3, \Pi_4) \tag{8}$$

The vertical suction force coefficient is following:

$$C_{vs} = \frac{F}{\frac{1}{2}\rho_w v^2 A} = f(Re, \frac{D}{d}, \frac{h}{d}) \tag{9}$$

It is found that the vertical suction force coefficient C_{vs} could be related to the Reynolds number Re, D/d, and h/d.

2.3.2. Critical Bottom Clearance h for the Vertical Incipient Motion of Particles

According to the force balance analysis of the particle, we compare the theoretical wet weight mg-F_{buoy} of spherical particle to the vertical force calculated by C_{vs}. Therefore, the relationship between C_{vs} and Froude number Fr should be studied.

Based on the force analysis of the particles, the vertical incipient motion condition is as follows:

$$\vec{F} + \vec{F}_{buoy} = m\vec{g} \tag{10}$$

Using the functions of vertical force and volume of spherical particle:

$$F = \frac{1}{2}\rho_w A v^2 \times C_{vs} = \frac{1}{8}\pi\rho_w C_{vs} v^2 d^2 \tag{11}$$

$$mg - F_{buoy} = \frac{4}{3}g(\rho - \rho_w)\pi r^3 = \frac{1}{6}g(\rho - \rho_w)\pi d^3 \tag{12}$$

The following function can be obtained from Equation (10):

$$C_{vs} = \frac{4}{3}\frac{\rho - \rho_w}{\rho_w}\frac{gd}{v^2} \tag{13}$$

In terms of the definition of the Froude number Fr:

$$Fr = \frac{v}{\sqrt{gd}}\sqrt{\frac{v^2}{gd}} \tag{14}$$

The relationship between C_{vs} and Fr could be obtained:

$$C_{vs} = \frac{4}{3}\frac{\rho - \rho_w}{\rho_w}\frac{1}{Fr^2} \tag{15}$$

In consequence, if the vertical suction force coefficient determined by D/d and h/d is greater than the vertical suction force coefficient calculated by Fr, the particles will be lifted. Therefore, the criterion-formula is established to predict the vertical incipient motion of particles.

3. Experimental Description

In order to investigate the characteristics of ore particles in the suction flow field, the test system has been designed and constructed in the laboratory and three kinds of experiments are carried out: (1) vertical force tests: to measure the vertical force of particles in various cases; (2) vertical incipient motion tests: to measure the critical bottom clearances of particles and further validate the accuracy of the vertical force tests; (3) flow visualization tests: to observe the characteristics of flow field around the particle.

3.1. Test Set Up

Figure 5 is the sketch of vertical force tests. In Figure 5a, the spherical particle is fixed on the three-component force sensor, separated by an intermediate acrylic board. Tests aimed to measure the vertical force of the particle without relative velocity with the pipe. In Figure 5b, the spherical particle is free on the bottom to measure the vertical incipient motion clearance in suction flow field.

The particle is also still relative with the pipe before starting to lift in most tests. But in some exception, the particle will rotate around a vertical axis on the bottom and the phenomenon will be discussed later.

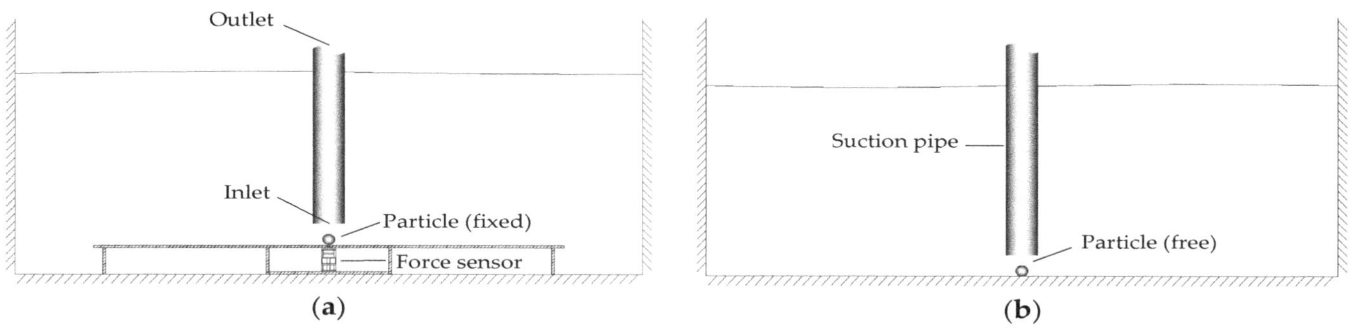

Figure 5. (**a**) Sketch of vertical force tests and (**b**) vertical incipient motion tests.

The whole test system consists of three parts: suction, movement and measurement as shown in Figure 6. The suction part is mainly composed of glass flume and pump circulation device. The size of glass flume is 2.5 m × 1.5 m × 1.0 m. The water in the flume is connected to the pump through a pipe, in which the rated power is 22 kW and the maximum flow rate is 100 t/h. The LDG-SIN-DN100 electromagnetic flow meter (made by MEACON, Hangzhou, Zhejiang, China) is installed at the outlet of the pump to measure the flow rate through the pump. Considering the fact that the flow rate at the inlet of the pipe is almost the same as the measured one, the average flow velocity is calculated and called flow velocity in pipe. In addition, a 0.7 m baffle is installed and two parallel honeycomb plates are placed to weaken the impact of water wave on force measurement. The movement part is composed of an ER50-C10 six-degree-of-freedom robot (made by Effort, Wuhu, Anhui, China), whose lower arm is fixed to the inlet of the pipe. By adjusting the motion of the mechanical arm to change the position of the pipe entrance, the precision can reach 0.01 mm and 0.01°. The measurement part adopts three-component force sensor. The top of the sensor is rigidly connected with the measured particle, and its bottom is rigidly connected to the stainless-steel plate placed on the bottom of the flume. In addition, an acrylic plate is installed under the particle, so that the acrylic plate can be considered as a boundary, and the force sensor has no effect on the flow field above the plate. In the experiment, the vertical distance between the inlet and the particle is measured as the vertical distance between the bottom of pipe and the acrylic plate, and it is defined as the bottom clearance h.

The force sensor is connected with the data acquisition system, and it can measure the data of the vertical suction force of the particles within a period of time. To ensure the accuracy of the measurement results, the layout structure of force sensor in the experiment has been improved many times, and the system tolerance is successfully maintained within 1%. In addition, in previous tests, it was found that if the measured particle is too heavy or too light, it will cause an oscillation of the measured value. Therefore, the particles used in the vertical force tests are designed to balance their buoyancy with the proper gravity, so that the measurement results can be more accurate.

In each of the test cases, the critical bottom clearance h was determined for the incipient motion of particles by accurately controlling the position of the ER50-C10 robot. The flow velocity is measured by the electromagnetic flow meter and shows the value of quantity of flow. Figure 7 shows the experimental value in the recorder of the flow in the pipe for a period of 200 s. The red line in the figure means that the target value of the flow quantity is 56.55 t/h in terms of the specified flow velocity. In comparison, the tolerance between the average of the experimental value and the target value is within 0.3%. The standard deviation is 0.1323 t/h, which is 0.2% of the target value of flow quantity. Thus the tolerance of flow velocity is small enough and can be acceptable in this study.

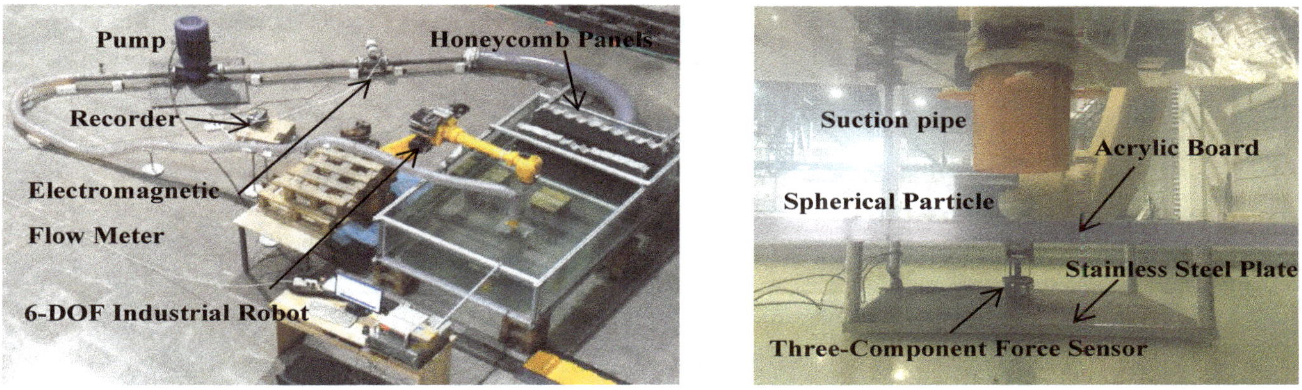

Figure 6. (**a**) Overlook of the test system; (**b**) Front view of the suction part.

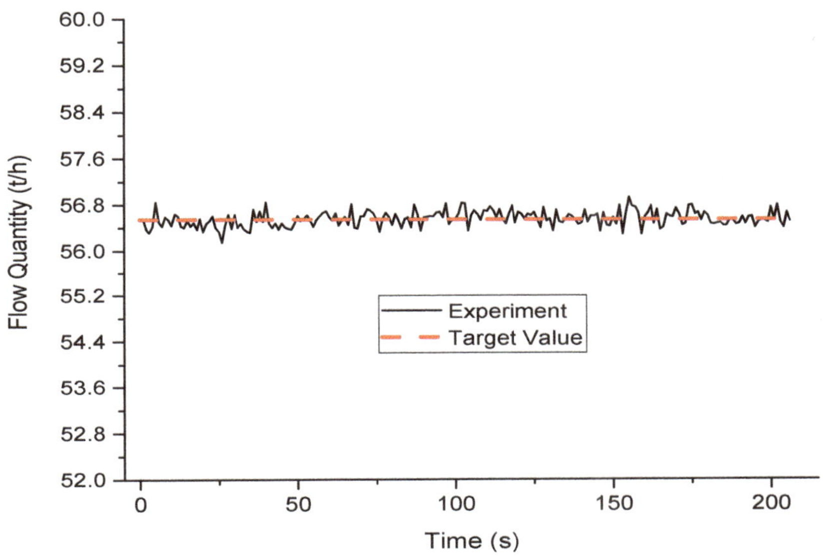

Figure 7. Target value (56.55 t/h) and experimental value of the flow quantity.

3.2. Test Program

To examine the effects on C_{vs} of Re, h/d, D/d and other parameters, a series of values of each factor has been chosen. The diameters of suction pipe are 0.075 m, 0.1 m and 0.125 m, and the diameter of spherical particles are 0.032 m, 0.036 m and 0.040 m. In total, 234 cases were carried out with 3~5 flow velocities in pipe ranging from 1 m/s to 2 m/s. In this range, the vertical forces of particles can be bigger and smaller than its weights in water. Moreover, each case was repeated three times with 3 min each time. The water temperature in this study is 10 °C and the dynamic viscosity coefficient μ equals 1.308×10^{-3} Pa·s. Therefore the range of dimensionless parameters can be determined. The range of Re is from 30,581 to 61,162, the range of D/d is from 1.875 to 3.906, and the range of h/d is from 1.389 to 2.375.

To obtain the incipient motion characteristics of spherical particles in suction flow field and further validate the accuracy of the vertical force tests, a total of 18 cases of vertical incipient motion tests were carried out with different parameters, as shown in Table 2. Each test was repeated ten times to measure the bottom clearance of vertical incipient motion of the particles when they were lifted into the suction pipe. The single particle is free on the bottom and right below the suction pipe.

Table 2. Parameters for vertical incipient motion tests of single particle.

Physical Parameter	Unit	Value
Particle density ρ	kg/m^3	2164.5, 2495.7
Particle diameter d	m	0.040, 0.036, 0.035
Pipe diameter D	m	0.100, 0.075
Ratio of diameter D/d		1.87, 2.50, 2.78, 2.89
Flow velocity in pipe v	m/s	2, 1.8, 1.6
Dynamic viscosity coefficient μ	Pa·s	1.308×10^{-3}

In addition, more vertical incipient motion tests of single and multiple particles were carried out to study the relations between the vertical incipient motion characteristics of single and multiple particles, such as the critical bottom clearance of each particle and the sequence of each particle to be lifted. Each test was also repeated ten times to measure the bottom clearance and the sequence of lifted particles. The parameters are listed in Table 3.

Table 3. Parameters for vertical incipient motion tests of multiple particles.

Physical Parameter	Unit	Value
Particle diameter d	m	0.019, 0.025, 0.030, 0.039
Particle weight G	N	0.092, 0.194, 0.352, 0.792
Pipe diameter D	m	0.1
Ratio of diameter D/d		5.21, 4.06, 3.33, 2.54
Flow velocity in Pipe v	m/s	2

Flow visualization tests were carried out to observe the characteristics of suction flow field. To compare with the results in references [16,34], the test cases are set as Table 4.

Table 4. Experimental parameters for flow visualization tests.

Physical Parameter	Unit	Case1	Case2
Bottom clearance h	m	0.07	0.07
Flow velocity in Pipe v	m/s	2.000	1.635
Reynolds number Re		61,162	50,000

4. Results and Discussion

Based on the measurements, the empirical formula of vertical force and criterion-formula of vertical incipient motion of particles are established separately. By comparing experiments of single and multiple particles, the correspondence of both incipient motion characteristics is obtained. The phenomenon that the vortex could help strengthen the suction force was found in both tests. In addition, the characteristics of flow field were observed by flow visualization tests, and applied to explain the force characteristics of particles in suction flow field.

4.1. Vertical Force Characteristics of Particles in Suction Flow Field

4.1.1. Influence of Re on C_{vs}

Figure 8 shows the test results of C_{vs} for different Re and h/d. It can be seen that the vertical suction force coefficient is nearly independent of Reynolds number when Re is in the range from 30,581 to 61,162. This range is subjected to the subcritical flow regime for flow around a sphere in a uniform flow, which also has the feature that the vertical suction force coefficient is nearly independent of Reynolds number [22]. Therefore, Equation (9) for C_{vs} can be simplified to:

$$C_{vs} = f(\frac{D}{d}, \frac{h}{d}) \tag{16}$$

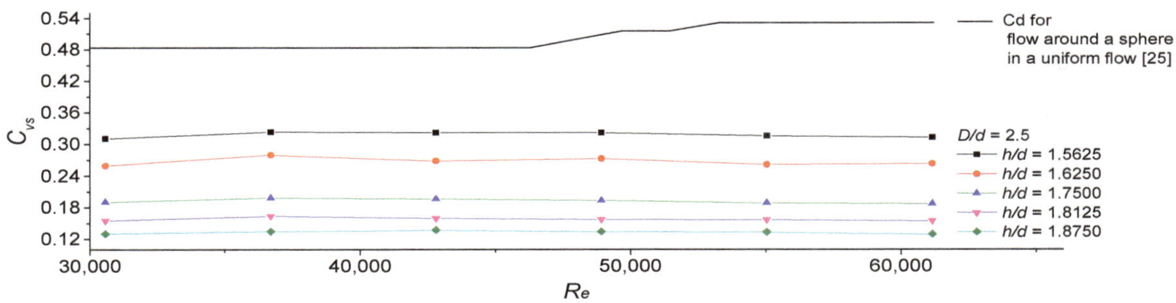

Figure 8. Relations between vertical suction force coefficients and Reynolds numbers.

In addition, the vertical suction force coefficients of spheres in suction flow are smaller than those in a uniform flow. One reason for this phenomenon is that the velocity close to the solid surface of particle is much smaller than that in the pipe. The velocity used to define C_{vs} is the velocity in pipe, however, the velocity used to define C_d is the oncoming flow velocity which is similar to that close to the solid surface of particles.

4.1.2. Influences of D/d and h/d on C_{vs}

The test results of C_{vs} for different D/d and h/d are shown in Figures 9 and 10. As shown in Figure 9, with the increase of D/d, the C_{vs} almost increases linearly. The coefficients of determination (R^2) of linear fitting functions for seven different fitting curves are over 0.9804.

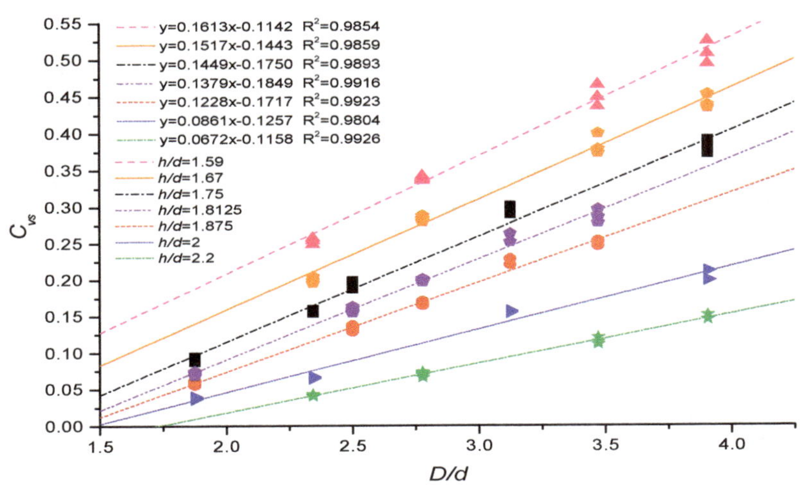

Figure 9. Relations between vertical suction force coefficients and ratios of D to d.

Unlike the influence of D/d, Figure 10 shows that the C_{vs} decreases exponentially with h/d in the hydraulic collecting process. The coefficients of determination (R^2) of exponential fitting functions are over 0.9996 and thus the fitting quality is extremely good. In consequence, h/d is a dominant factor to influence the C_{vs}. If the deep sea miner collector could work with a small h/d, large amount of collecting energy consumption will be saved and the disturbance to the sea floor will be minimized.

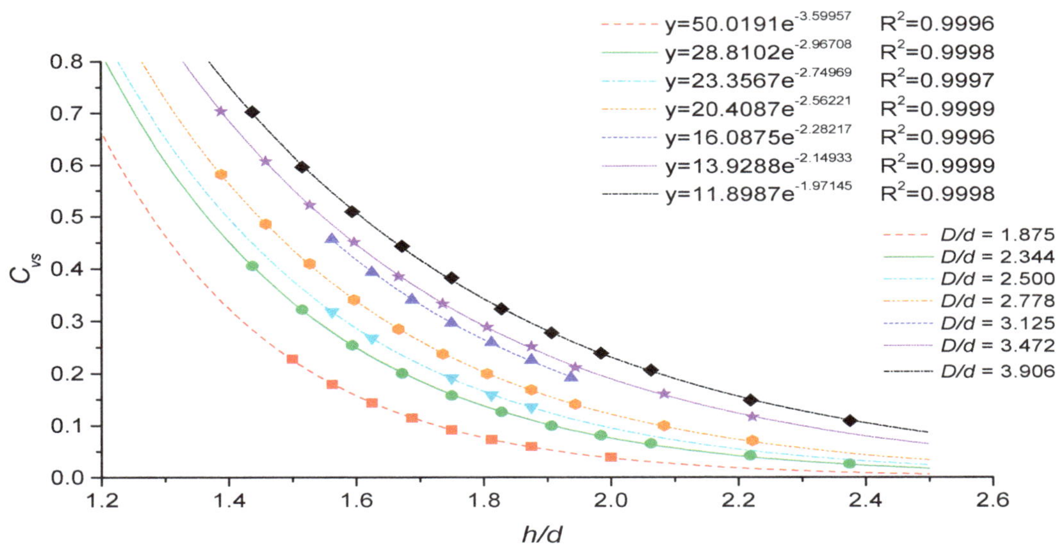

Figure 10. Relations between vertical suction force coefficients and ratios of h to d.

4.1.3. Empirical Formula among D/d, h/d and C_{vs}

The empirical formula among D/d, h/d and C_{vs} is required to predict the collecting performance under different cases. As a result, the functions between the natural logarithm of C_{vs} ($\ln C_{vs}$) and h/d for different D/d are drawn in Figure 11. An interesting phenomenon is observed that all the linear fitting lines tend to meet at one point.

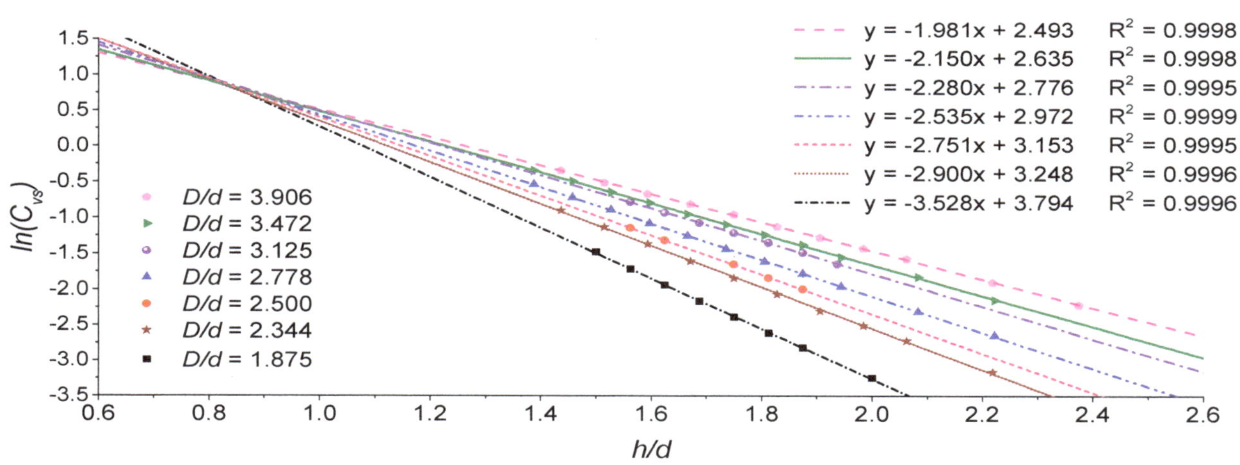

Figure 11. Relations between $\ln(C_{vs})$ and h/d.

By further analyzing the linear fitting parameters listed in Table 5 and making data fittings between the D/d and the two parameters, other two exponential fitting formulas for those parameters can be obtained respectively, as shown in Table 6.

Table 5. Linear fitting parameters of $\ln(C_{vs})$~h/d.

D/d	Slope k	Intercept b	R^2
3.906	−1.981	2.493	0.9998
3.472	−2.15	2.635	0.9998
3.125	−2.28	2.776	0.9995
2.778	−2.535	2.972	0.9999
2.5	−2.751	3.153	0.9995
2.344	−2.9	3.248	0.9996
1.875	−3.528	3.794	0.9996

Table 6. Exponential fitting relations between parameters and D/d.

Fitting Formula	R^2
$k = -5.7005(D/d)^{-0.7876}$	0.9958
$b = 5.3305(D/d)^{-0.5669}$	0.9944

As a result, the empirical formula among D/d, h/d and $\ln C_{vs}$ is established:

$$\ln(C_{vs}) = -5.70\left(\frac{D}{d}\right)^{-0.788}\left(\frac{h}{d}\right) + 5.33\left(\frac{D}{d}\right)^{-0.567} \quad (17)$$

The graph of the empirical formula can be drawn as in Figure 12.

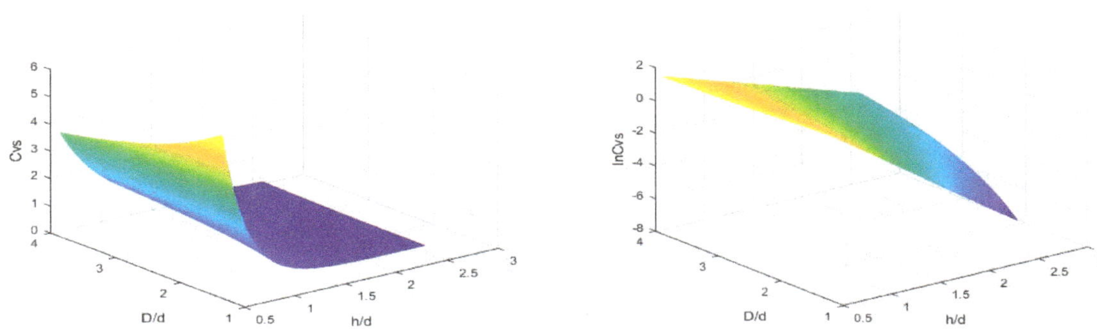

Figure 12. Relations among C_{vs}, h/d and D/d.

Figure 13 shows the tolerance between calculated value C_{vs}' by the empirical formula Equation (17) and measured value C_{vs} by experiments for 234 cases.

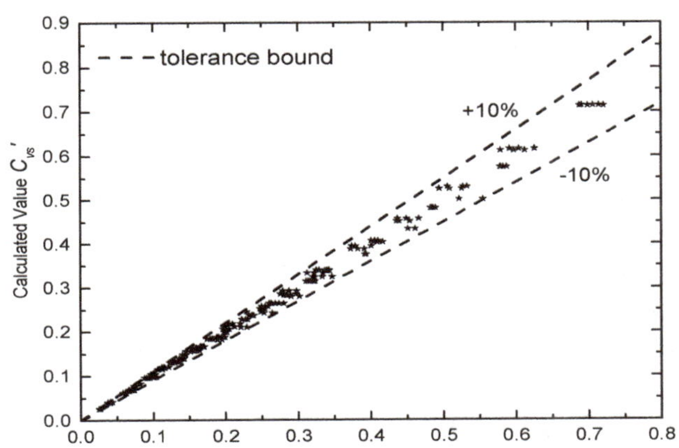

Figure 13. Tolerance between calculated C_{vs}' and measured C_{vs} for 234 cases.

The maximum tolerance is less than 10% and this indicates that the suction force prediction formula is sufficiently accurate to predict the vertical force of particles in different cases.

4.2. Vertical Incipient Motion Characteristics of Particles in Suction Flow Field

4.2.1. Criterion-Formula of Vertical Incipient Motion of Particles

In this study, the vertical suction force coefficient C_{vs} is used to represent the characteristics of vertical force of the particle in suction flow field. Because the equation ($F + F_{buoy} = mg$) is the critical condition to trigger the vertical incipient motion of particles, therefore, if C_{vs} determined by D/d and h/d is bigger than that calculated by Equation (15), the particles will be lifted theoretically.

Figure 14 shows the tolerance of C_{vs} between the theoretical values by Equation (15) and the predicted values by Equation (17). The dash dotted line and solid line in the graph represent the theoretical values based on two kinds of densities: $\rho = 2164.5$ kg/m^3 and $\rho = 2495.7$ kg/m^3. Four dotted lines represent the upper and lower bounds with 15% tolerance of the two lines, respectively. The two kinds of data points are obtained based on the measurements of the cases. It can be seen that the measured data are distributed on both sides of the theoretical lines within the range of 15% tolerance. Therefore, the vertical force tests are further verified by the vertical incipient motion tests. Furthermore, the criterion-formula Equation (15) of vertical incipient motion of particles is obtained with the maximum tolerance less than 15% and can be used to predict the collecting performance in various cases. As a result, it is feasible to provide sufficient suction force and avoid unnecessary disturbance by predicting the performance of collecting accurately.

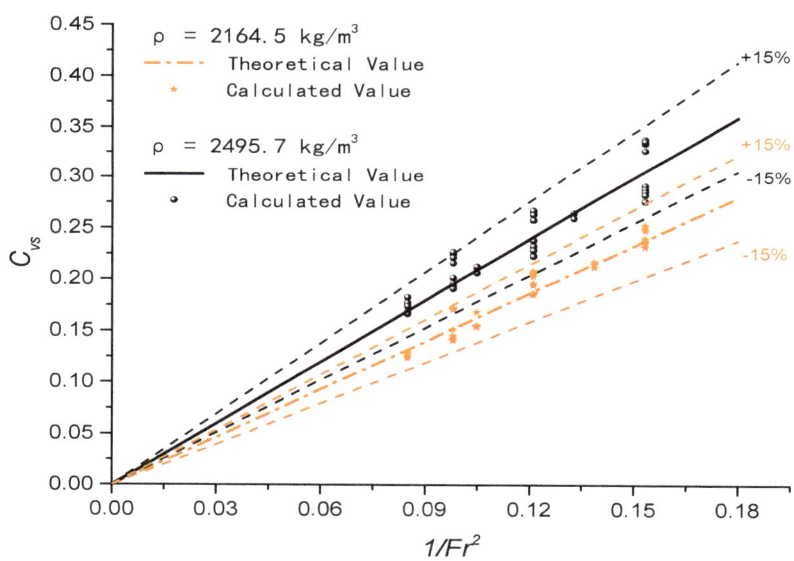

Figure 14. Relations between C_{vs} and $1/Fr^2$.

4.2.2. Vertical Incipient Motion Characteristics of Single Particles

Table 7 shows the test results of critical bottom clearance h in relation with D/d for the incipient motion of single particle. Each of the test cases for a single particle was repeated ten times and the recorded critical bottom clearance was represented by h_i ($i = 1, 2, \ldots, 10$). Then, the mean value $\overline{h}$ and standard deviation of the measurements were calculated. The test results show that the smaller particle has a higher standard deviation of h, which means the h of smaller particle exhibits more significant fluctuation. By observing the video taken by a high-speed camera, it is found that the particle d19.2 is always in circular motion with a large h. When particles remain almost still, however, the h is small. The reason for high standard deviation of particle d19.2 could be explained like this: the particle d19.2 tends to rotate around a vertical axis on the bottom before being sucked into the pipe because of the significant impact of vortex. The irregularity of vortex devotes to the high standard deviation of h and the vortex could help strengthen the suction force.

Table 7. Results of single particle tests.

D/d	5.21	4.05	3.32	2.54
h_i (i = 1, 2, . . . , 10)	d19.2	d24.7	d30.1	d39.4
h_1 (mm)	59.81	60.76	65.00	69.32
h_2 (mm)	62.8	61.16	66.44	69.46
h_3 (mm)	69.18	63.00	67.45	70.45
h_4 (mm)	84.53	63.83	68.51	71.36
h_5 (mm)	61.18	60.94	66.23	69.39
h_6 (mm)	63.57	61.82	66.52	69.57
h_7 (mm)	65.28	61.86	66.86	69.71
h_8 (mm)	73.77	63.58	67.63	70.79
h_9 (mm)	65.77	62.23	66.92	69.81
h_{10} (mm)	67.81	62.6	66.98	69.89
$\overline{h}$ (mm)	67.37	62.18	66.85	70.03
Standard deviation (mm)	6.90	1.02	0.93	0.80

4.2.3. Vertical Incipient Motion Characteristics of Multiple Particles

Table 8 lists the results of the sequence of lifting particles and the critical bottom clearances of test cases for four particles, which are right below the fixed pipe. Similarly, each of the test cases for multiple particles was repeated 10 times. It is clear that for most cases the particles d39.39 and d19.2 would be lifted firstly, and this is consistent with the results in Table 7, where the $\overline{h}$ of particles d39.39 and d19.2 are larger than the other particles.

Table 8. Results of multiple particle tests.

D/d	5.21	4.05	3.32	2.54
No.i, h_i (i = 1, 2 . . . ,10)	d19.2	d24.7	d30.1	d39.4
No.1	2nd	3rd	4th	1st
h_1 (mm)	70.32	63.90	60.33	70.32
No.2	1st	2nd	4th	3rd
h_2 (mm)	88.16	88.16	61.54	68.23
No.3	2nd	4th	3rd	1st
h_3 (mm)	65.79	62.12	62.12	67.84
No.4	2nd	3rd	3rd	1st
h_4 (mm)	64.24	64.24	64.24	68.17
No.5	4th	2nd	3rd	1st
h_5 (mm)	63.82	66.25	66.25	67.50
No.6	1st	3rd	4th	2nd
h_6 (mm)	83.38	64.29	61.06	68.47
No.7	1st	3rd	3rd	4th
h_7 (mm)	69.60	61.22	61.22	59.78
No.8	2nd	4th	3rd	1st
h_8 (mm)	68.92	63.82	63.82	68.92
No.9	2nd	4th	3rd	1st
h_9 (mm)	67.59	59.63	61.03	67.59
No.10	1st	3rd	4th	2nd
h_{10} (mm)	91.85	60.72	60.72	70.65

Especially, in Nos.1, 3, 4, 8 and 9, the particle d39.39 is lifted firstly and the particle d19.2 is lifted secondly. Moreover, the relevant experiment video in Figure 15 shows that particles remain almost still before lifted. In Nos.2, 6, 7 and 10, on the other hand, the particle d19.2 is lifted firstly and the video shows that particles are in circular motion before lifted, as shown in Figure 16. As a result, the smallest particle is influenced by the circular motion significantly. The instability of the fluid field generally causes the smallest particle to be sucked firstly, and this supports the supposition in Section 4.2.2 that the vortex could help strengthen the suction force. Therefore, the characteristics of collecting single particle would be meaningful for further research on collecting multiple particles.

Figure 15. A snapshot of multiple particles without circular motion.

Figure 16. A snapshot of multiple particles with circular motion.

4.3. Flow Visualization Results

Flow visualization experiments were carried out to observe the characteristics of suction flow field. The visualizations in Figure 17b are similar to the numerical results [16]. The streamlines gather in the middle and the flow field exhibit cohesive characteristics.

(a) (b)

Figure 17. (**a**) Velocity vector distribution from numerical solution (h = 0.07 m, v = 2 m/s, Re = 61162) [16]; (**b**) Flow visualization results (h = 0.07 m, v = 2 m/s, Re = 61162).

Furthermore, by comparing the flow visualization test results in Figure 18, it is evident that the wake vortex separation point in suction flow is much more near the top of the particle than that in uniform flow. The closer the separation point of the wake vortex near the top of the particle, the smaller area on the particle is influenced by the wake vortex, leading to much smaller vertical suction force coefficient. Therefore, this phenomenon also contributes to the result that the vertical suction force coefficients of spheres in suction flow are smaller than those in a uniform flow in Figure 8. In addition, this physical phenomenon could explain why tests could have good repeatability and the prediction for both vertical force and incipient motion of particles in suction flow field could be accurate.

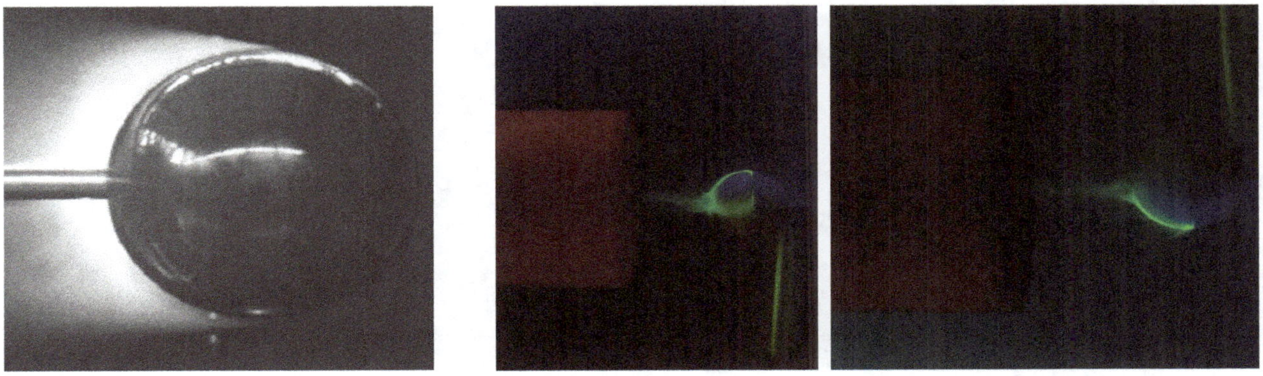

(a) (b)

Figure 18. Flow visualization results for flow around a sphere in the uniform flow (Re = 50,000) [34] (**a**); and the suction flow (Re = 50,000) (**b**).

5. Conclusions

The mechanism of hydraulic collecting is researched by dimensional analysis and experimental studies. The experimental system is established in laboratory to carry out three kinds of tests including 253 different test cases. The results show that the vertical suction force coefficient (C_{vs}) decreases exponentially with the ratio of bottom clearance to diameter of the particle (h/d, range from 1.875 to 3.472), increases linearly with the ratio of diameter of the suction pipe to diameter of the particle (D/d, range from 1.875 to 3.472), and is nearly independent of Reynolds number (Re, range from 30,581 to 61,162). The empirical formula of vertical force of particles is obtained with respect to h/d and D/d, and the maximum tolerance is less than 10%.

The empirical formula of vertical force of particles in suction flow field is further verified by the vertical incipient motion tests. The criterion-formula of vertical incipient motion of particles is obtained to predict the collecting performance with the maximum tolerance less than 15%. As a result, it is feasible to provide sufficient suction force and avoid unnecessary disturbance by predicting the performance of collecting accurately. By comparing experiments on single and multiple particles, the correspondence of both incipient motion characteristics is obtained, and thus the research on a single particle could reveal some principles of multiple particles. In addition, the phenomenon that the vortex could help strengthen the suction force is observed in the tests.

The characteristics of suction flow field were observed by flow visualization tests. It is found that the wake vortex separation point is near the top of the particle and very small area on the particle is influenced by the unstable wake vortex. This physical phenomenon could contribute to the result that the C_{vs} in a suction flow is much smaller than the C_d in a uniform flow and explain why the flow field around the particle is stable enough to guarantee the good repeatability in tests.

Author Contributions: Conceptualization, G.Z. and L.X.; Data curation, G.Z. and M.Z.; Formal analysis, G.Z. and M.Z.; Investigation, G.Z., L.X., T.P. and M.Z.; Methodology, G.Z., L.X., T.P. and M.Z.; Resources, L.X. and T.P.; Supervision, L.X. and T.P.; Validation, G.Z., L.X., T.P. and M.Z.; Visualization, G.Z.; Writing—original draft, G.Z., L.X. and M.Z.

Acknowledgments: The present work is supported by the National Key Research and Development Program of China (2016YFC0304103). The support is gratefully acknowledged.

Abbreviations

The following abbreviations are used in this article:

C_{vs}	vertical suction force coefficient
D	diameter of suction pipe
DES	Detached Eddy Simulation
d	diameter of particle
F	vertical suction force of particle
F_B	buoyancy in water

Fr	Froude number
G	gravity of particle
h	bottom clearance
IKS	Institut für Konstruktion
LES	Large Eddy Simulation
NIOT	National Institute of Ocean Technology
OMI	Ocean Management Incorporated
Re	Reynolds number
REE	rare earth element
v	flow velocity in pipe
ρ	density of particle
ρ_w	density of water
μ	dynamic viscosity coefficient

References

1. Bury, J. Mining mountains: neoliberalism, land tenure, livelihoods, and the new Peruvian mining industry in Cajamarca. *Environ. Plan. A* **2005**, *37*, 221–239. [CrossRef]
2. Mero, J.L. The mineral resources of the sea. *Elsevier Oceanogr. Ser.* **1965**, *1*, 1–5.
3. Willums, J.O.; Bradley, A. MIT's deep sea mining project. In Proceedings of the 6th Offshore Technology Conference, Houston, TX, USA, 6–8 May 1974; pp. 1072–1076.
4. Chung, J.S. Deep-Ocean Mining Issues and Ocean Mining Working Group (OMWG). In Proceedings of the 3rd ISOPE Ocean Mining Symposium, Goa, India, 8–10 November 1999.
5. Chung, J.S. Deep-ocean Mining Technology: Development II. In Proceedings of the 6th ISOPE Ocean Mining Symposium, Changsha, China, 9–13 October 2005; pp. 1–6.
6. Chung, J.S. Deep-ocean mining technology III: Developments. In Proceedings of the 8th ISOPE Ocean Mining Symposiumth, Osaka, Japan, 20–24 September 2009.
7. Chung, J.S. Advances in manganese nodule mining technology. *Mar. Technol. Soc. J.* **1985**, *19*, 39–44.
8. Herrouin, G.; Lenoble, J.P.; Charles, C.; Mauviel, F.; Bernard, J.; Taine, B. A Manganese Nodule Industrial Venture Would Be Profitable: Summary of a 4-Year Study in France. In Proceedings of the Offshore Technology Conference, Houston, TX, USA, 1–4 May 1989.
9. Boetius, A.; Haeckel, M. Mind the seafloor. *Science* **2018**, *359*, 34–36. [CrossRef] [PubMed]
10. Ma, W.; Schott, D.; Lodewijks, G. A new procedure for deep sea mining tailings disposal. *Minerals* **2017**, *7*, 47. [CrossRef]
11. McFarlane, J.; Brockett, T.; Huizingh, J.P. *Analysis of Mining Technologies Developed in the 1970s and 1980s*; International Seabed Authority: Kingston, Jamaica, 2008.
12. Hong, S.; Choi, J.S.; Kim, J.H.; Yang, C.K. Experimental study on hydraulic performance of hybrid pick-up device of manganese nodule collector. In Proceedings of the Third ISOPE Ocean Mining Symposium, Goa, India, 8–10 November 1999; pp. 69–77.
13. Yang, N.; Tang, H. Several considerations of the design of the hydraulic pick-up device. In Proceedings of the Fifth ISOPE Ocean Mining Symposium, Tsukuba, Japan, 15–19 September 2002; pp. 119–122.
14. Tsutsui, T. Flow around a sphere in a plane turbulent boundary layer. *J. Wind Eng. Ind. Aerodyn.* **2008**, *96*, 779–792. [CrossRef]
15. Lim, S.J.; Kim, J.W.; Jung, S.T.; Cho, H.Y.; Lee, S.H. Deep Seawater flow Characteristics around the Manganese Nodule Collecting Device. *Procedia Eng.* **2015**, *16*, 544–551. [CrossRef]
16. Zhao, G.; Xiao, L.; Lu, H.; Chen, Z. A Case Study of Hydraulic Collecting a Single Spherical Particle. In Proceedings of the 27th International Ocean and Polar Engineering Conference, San Francisco, CA, USA, 25–30 June 2017; pp. 30–38.
17. Chung, J.S.; Yarim, G.; Savasci, H. Shape effect of solids on pressure drop in a 2-phase vertically upward transport: silica sands and spherical beads. In Proceedings of the Eighth International Offshore and Polar Engineering Conference, Montreal, PQ, Canada, 24–29 May 1998; Volume 1, pp. 58–65.
18. Sobota, J.; Boczarski, S.; Petryka, L.; Kotlinski, R.; Stoyanova, V. Slip Velocity in Nodules Vertical Flow-Experimental Results. In Proceedings of the Fourth ISOPE Ocean Mining Symposium, Szczecin, Poland, 23–27 September 2001; pp. 127–131.

19. Jiang, L.; Li, P.; Tian, L.; Han, W. Concentration Characteristics of Large-Size Particles in Vertical Pipes for Hydraulic Lifting. In Proceedings of the Sixth ISOPE Ocean Mining Symposium, Changsha, China, 9–13 October 2005; pp. 106–109.
20. Jiang, L.; Li, P.; Tian, L.; Han, W. Experiment Study on Critical Velocity in Vertical Pipes for Hydraulic Lifting. In Proceedings of the Sixth ISOPE Ocean Mining Symposium, Changsha, Hunan, China, 9–13 October 2005; pp. 115–118.
21. Yang, N.; Chen, G.; Tang, D.; Jin, X.; Xiao, H. Behavior of single particle and group particles in vertical lifting pipe in China. In Proceedings of the Ninth ISOPE Ocean Mining Symposium, Maui, HI, USA, 19–24 June 2011; pp. 153–157.
22. Fan, W.; Chen, J.; Pan, Y.; Huang, H.; Chen, C.T.A.; Chen, Y. Experimental study on the performance of an air-lift pump for artificial upwelling. *Ocean Eng.* **2013**, *59*, 47–57. [CrossRef]
23. Pougatch, K.; Salcudean, M. Numerical modelling of deep sea air-lift. *Ocean Eng.* **2008**, *35*, 1173–1182. [CrossRef]
24. Ma, W.; van Rhee, C.; Schott, D. Technological and Profitable Analysis of Airlifting in Deep Sea Mining Systems. *Minerals* **2017**, *7*, 143.
25. Achenbach, E. Experiments on the flow past spheres at very high Reynolds numbers. *J. Fluid Mech.* **1972**, *54*, 565–575. [CrossRef]
26. Johnson, T.A.; Patel, V.C. Flow past a sphere up to a Reynolds number of 300. *J. Fluid Mech.* **1999**, *378*, 19–70. [CrossRef]
27. Constantinescu, G.S.; Squires, K.D. LES and DES investigations of turbulent flow over a sphere at *Re* = 10,000. *Flow Turbul. Combust.* **2003**, *70*, 267–298. [CrossRef]
28. Hein, J.R.; Koschinsky, A. Deep-ocean ferromanganese crust and nodules. In *Earth Systems and Environmental Sciences, Treatise on Geochemistry*, 2nd ed.; Holland, H., Turekian, K., Eds.; Elsevier: Amsterdam, The Netherlands, 2013; pp. 273–291.
29. Hein, J.R.; Mizell, K.; Koschinsky, A.; Conrad, T.A. Deep-ocean mineral deposits as a source of critical metals for high- and green-technology applications: comparison with land-based resources. *Ore Geol. Rev.* **2013**, *51*, 1–14. [CrossRef]
30. Blöthe, M.; Wegorzewski, A.V.; Müller, C.; Simon, F.; Kuhn, T.; Schippers, A. Manganese-cycling microbial communuties inside deep-sea manganese nodules. *Environ. Sci. Technol.* **2015**, *49*, 7692–7700. [CrossRef] [PubMed]
31. Von Stackelberg, U. Growth history of manganese nodules and crusts of the Peru Basin. In: Manganese mineralization: Geochemistry and mineralogy of terrestrial and marine deposits. *Geol. Soc. Spec. Publ.* **1997**, *119*, 153–176. [CrossRef]
32. Moritani, T. Description, classification, and distribution of manganese nodules. *Geol. Surv. Japan Cruise Rep.* **1977**, *8*, 136–158.
33. Atmanand, M.A.; Shajahan, M.A.; Deepak, C.R.; Jeyamani, R.; Ravindran, M.; Schulte, E.; Panthel, J.; Grebe, H.; Schwarz, W. Instrumentation for underwater crawler for mining in shallow waters. In Proceedings of the International Symposium of Autonomous Robots and Agents, Singapore City, Singapore, 26 May 2000.
34. Bakić, V.; Perić, M. Vizualization of Flow around a Sphere for Reynolds Numbers between 22,000 and 400,000. *Therm. AeroMechanism* **2005**, *12*, 307–315.

6

Study the Flow behind a Semi-Circular Step Cylinder (Laser Doppler Velocimetry (LDV) and Computational Fluid Dynamics (CFD))

S. M. Sayeed-Bin-Asad *, Tord Staffan Lundström and Anders Gustav Andersson

Division of Fluid and Experimental Mechanics, Luleå University of Technology, SE-971 87 Luleå, Sweden; staffan.lundstrom@ltu.se (T.S.L.); anders.g.andersson@ltu.se (A.G.A.)

* Correspondence: sayeed.asad@ltu.se.

Academic Editor: Bjørn H. Hjertager

Abstract: Laser Doppler Velocimetry (LDV) measurements, flow visualizations and unsteady Reynolds-Averaged Navier-Stokes (RANS) Computational Fluid Dynamics (CFD) simulations have been carried out to study the turbulent wake that is formed behind a semi-circular step cylinder at a constant flow rate. The semi-circular cylinder has two diameters, a so-called step cylinder. The results from the LDV measurements indicate that wake length and vortex shedding frequency varies with the cylinder diameter. This implies that a step cylinder can be used to attract fish of different size. By visualizations of the formation of a recirculation region and the well-known von Kármán vortex street behind the cylinder are disclosed. The simulation results predict the wake length and shedding frequency well for the flow behind the large cylinder but fail to capture the dynamics of the flow near the step in diameter to some extent and the flow behind the small cylinder to a larger extent when compared with measurements.

Keywords: laser doppler velocimetry (LDV); computational fluid dynamics (CFD); fish migration; step cylinder; shedding; vortex street

---

1. Introduction

Hydro and nuclear power are the major sources of electricity in Sweden. Swedish Energy [1] has recently published some statistics regarding power production in Sweden showing that hydro and nuclear power stand for 47% and 34% of the total electricity in Sweden, respectively. In order to increase the percentage renewable energy overall, to keep the target of greenhouse gas emission [2], systems are constantly developed to harvest wind, wave, tidal, and solar energy. These sources of energy are typically irregular and continuous power production is required to cover for this irregularity. The availability, storage capacity, and flexibility of hydropower make it a perfect contender for the power regulation role. However, hydropower has some local impacts on the environment, such as creating dams in rivers and changing water flow directions, which lead to some problems for fauna passage, in general, and migrating fishes, in particular. Fish migration issues are mostly studied from a biological point of view and more detailed studies from a fundamental fluid mechanics point of view could lead to new insights and solutions.

Some hydropower dams have fishways or fish ladders to allow fish to migrate past dams, as well as past rotating turbines and, during swimming or passing this fishway or fish ladder, fish have to tackle different flow obstructions or disturbances, like turbine intakes, stones, and concrete structures. Fluid flow characteristics in fish ladders or fishways during fish migration are crucial when designing effective fishways so that fishes migrate in an efficient way. Studies [3–7] have shown that cylindrical obstructions can be used to create favorable conditions for upstream migrating fishes of appropriate size. The concept is to create periodic vortices that the fish uses in an energy efficient way (less energy than in a free stream) to maintain a position in the flow. Liao [8] describes the structure of the flow around a half cylinder and the positions of fish swimming within the wake behind and around the cylinder. It was found that the fishes exploit the flow conditions created by the semicircular cylinder when swimming upstream around it. Apart from this, Liao discovered the altered shedding frequencies and wakes for various cylinder diameters. Different sizes of fish exploit vortices with different shedding frequencies. Therefore, it is important to have proper shedding for appropriate sizes of fish to swim past obstructions.

There are many applications for step full cylinders, including airport control towers, street lamp posts, antenna members and radio communication towers. Hence, there are several driving forces for engineers to examine this type of flow phenomena, both numerically and experimentally.

Dunn et al. [9] carried out velocity measurement with laser doppler velocimetry (LDV) and flow visualization around a step cylinder. One result is that the recirculation near a step has significant effects on the shedding frequencies as compared to locations away from the step. Similarly, another study from Dunn et al. [10] also found significant effects of free-stream shearing on vortex-shedding characteristics from a step-cylinder. Chris Morton et al. [11–16] carried out several studies using both numerical and experimental with 2D and 3D particle image velocimetry (PIV) investigations on step cylinders, such as single step dual step, etc. Detailed studies of the wake vortex dynamics behind step-cylinder are provided. He [17] has recently used tomographic-PIV (3D-PIV) behind a step cylinder for three-dimensional (3D) information of vortex shedding. Rafati [18] conducted a series of experimental studies with various diameters of step cylinders using tomographic-PIV measurement. One result is that for high flow (higher velocity than free stream velocity), streamwise vortices are introduced into the wake that complicates the vortex dynamic, and vortex interactions. Rafati [18] also produced a table of studies on step cylinders available (Table 1).

Table 1. Previous studies on flow behind step cylinders. Reproduced from [18].

Study	Re_D	D/d
N.Ko et al. [19]	8×10^4	2
Yagita et al. [20]	8×10^2–10^4	1.25–5
Lewis et al. [21]	67–200	1.14–1.76
Dunn et al. [10]	62–1230	2
Norberg [22]	3×10^3–13×10^3	1.25–2
N.Ko et al. (1984, 1990, 1992) [23–25]	8×10^4	1.33–2.78
Chua et al. [26]	4.72×10^3	3
Morton et al. [11,27,28]	150–2100	2
Dunn et al. [9] (span-wise sheared flow)	152–764	1.92

Despite having some investigations of flow behavior around stepped cylinders, little or no scientific works, especially experimental, on half cylinders with steps has been performed. Therefore, in the current study the flow around such an object is studied with LDV and computational fluid dynamics (CFD). The flow dynamics of the step and diameter change is interesting from a fluid mechanics point of view but as the half cylinders also have shown to be interesting for fish migration, the difference in wake formation and shedding of vortices behind the different cylinder diameters could also be designed to fit different sizes and species of fish.

2. Experimental Arrangement

Experiments were carried out in a water flume at the John Field Laboratory at Luleå University of Technology, Sweden. The flume is 7.5 m long with a cross-section of 0.295 m × 0.310 m. The semi-circular step cylinder employed in this investigation is made of acetal (POM) and the diameter of the small (d) and large (D) cylinders are 50 mm and 100 mm, respectively. Figure 1 describes the configuration and measuring zones of the cylinder centerline.

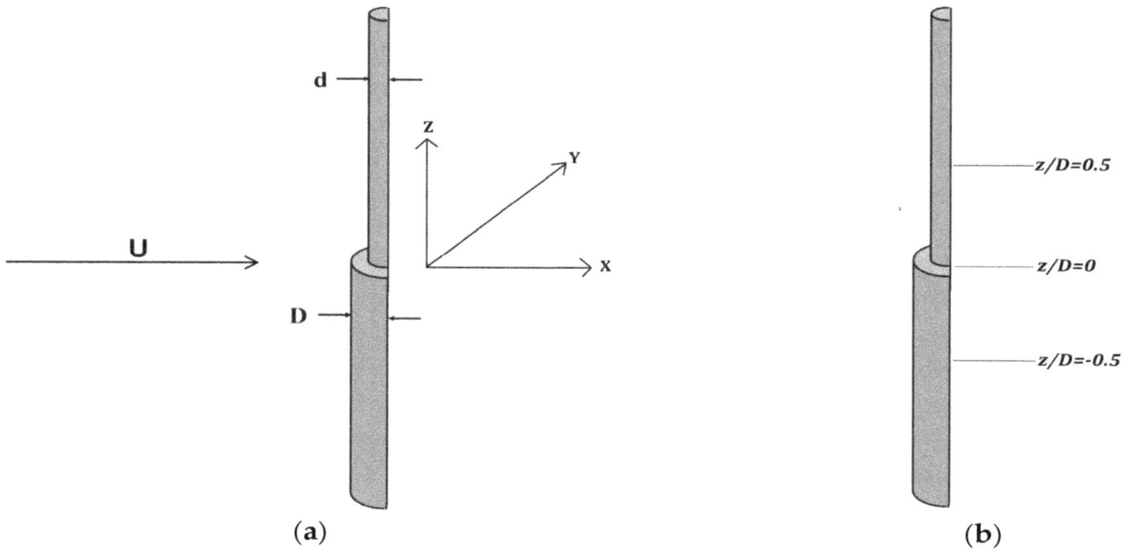

Figure 1. Cylinder configuration; (**a**) flow direction and coordinate description; and (**b**) measuring zones.

A submersible pump was used to pump the water through the channel from a rectangle water reservoir tank with 2 m^3 storage capacity. The water reservoir is placed at the outlet of the channel and the temperature of the water was controlled using an automatic cooling arrangement. This is essential since variation in water temperature changes its viscosity and refractive index properties. Changes in water temperature leads to laser beam displacements from the desired point of measurements. A Danfoss MassFlo Corolios flow meter was used to control the water flow rate.

To obtain a uniform velocity distribution through the channel there is a steel net and a honeycomb at the inlet of the channel. The thickness of the honeycomb is 75 mm and the diameter of its wholes is 7.6 mm. The steel net geometry consists of 2.5 mm × 2.5 mm square holes and it is 0.8 mm thick. A schematic arrangement of the experimental set-up is presented Figure 2.

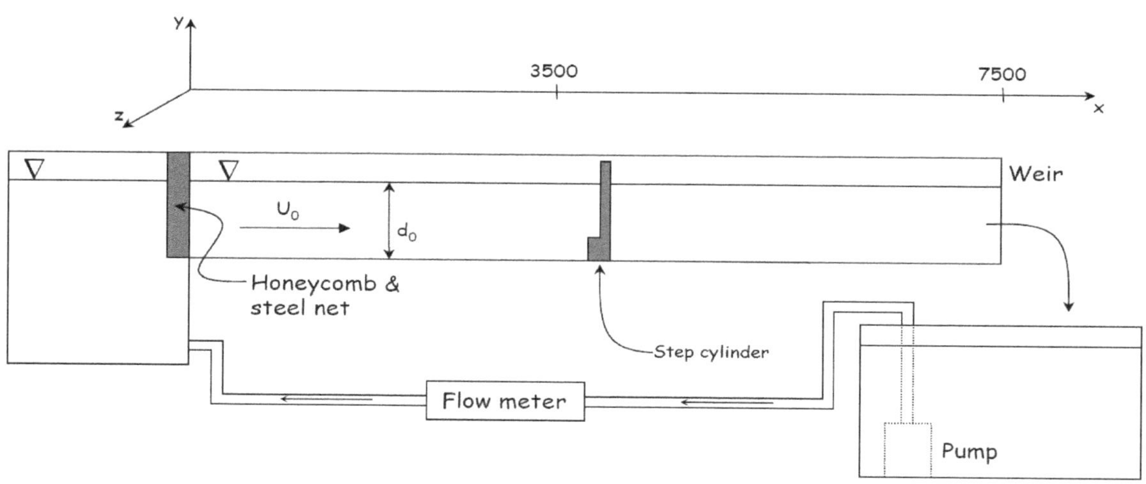

Figure 2. Schematic presentation of water circulation in the flume.

This LDV system, used in this investigation, is a two component configuration with an 83 mm diameter optical fiber probe and a front lens with 360 mm focal length which consists of a 300 mW diode-pumped solid-state laser, and transmitting optics, including a beam splitter Bragg-cell, photo detector, and signal processor. The system is used in backscatter mode so that the probe is both transmitting and receiving signals. The laser has two channels with a wavelength of 532 and 561 nm, respectively. The laser power is 300 mW for each channel. The smallest dimensions of the measurement volume are approximately 0.123×0.117 mm^2 for both colors, and the system is operated in coincidence mode, which ensures correlated velocity components. The water was seeded with polyamide particles with a diameter of 5 μm (Dantec's PSP-5). TSI's *FlowSizer 64* software (Shoreview, MN, USA) was used for the data acquisition and data acquisition rates were tried to maintain greater than 50 Hz. The 2D-LDV probe was mounted on a 3D traversing mechanism (can travel maximum 600 mm in each direction) controlled by the software. The sampling time was maintained as 121 seconds at each measurement point, corresponding to at least 10,000 samples, which are essential for statistical precision. All of the measurements were carried out in coincidence mode [29] to obtain an equal number of velocity counts for each direction, and the free stream velocity (U_∞) was determined from repeated measurements at a position 155 mm upstream of the cylinder at $z/D = 0$ as Norberg [30] determined. Free stream (U_∞) is also calculated theoretically using the flow rate (0.0077 m^3·s^{-1}) and it is found to be $U_\infty = 0.1451$ m·s^{-1}. Repeatability tests are completed to estimate the first-order variable uncertainty for the experiments [31]. Four different measurements for all experiments were completed to estimate the random error introduced by the experimental facility and changes in experimental conditions, such as the water temperature, position of the cylinder, etc.

Flow visualization was also conducted to explore the wake structure using a digital single-lens reflex (DSLR) camera (Nikon D90) and two tungsten light heads to illuminate the flow area. Two food dyes (red and green) were injected from both sides of the cylinder using a programmable syringe pump running at constant speed. Videos were recorded from both the side and top of the flume and the recordings were converted into images in equal time step using MATLAB (The MathWorks, Natick, MA, USA).

Except otherwise mentioned for scaling the results, all positions and distances are scaled with D and velocities with the free stream velocity U_∞.

3. Numerical Setup

The simulations were performed with the commercial software Ansys CFX 16 (ANSYS, Inc, Canonsburg, PA, USA). A structured grid with 10.1 million nodes and an average wall resolution of $y^+ = 0.7$ and maximum $y^+ = 2.8$ was created in ANSYS IcemCFD (275 Technology Drive, Canonsburg, PA, USA). The volume of fluids (VOF) method was used to account for both the water and air phase. The numerical domain did not include the full flume, the inlet was placed 1 m upstream the cylinder and the outlet was placed 1.5 m downstream to reduce computational time. To generate a fully developed inlet profile an additional simulation model was created for the upstream part of the flume with a constant velocity inlet assumed at the honeycomb and the velocity profile from the outlet of that simulation was used as inflow for the domain with the cylinder. A second-order accurate advection scheme was used to solve the flow equations and a second-order backward Euler transient scheme was used for the temporal discretization. A timestep of 0.025 s was selected which corresponds to a RMS Courant number of ~5 and the relevant transient results were saved every 10 time steps. The streamwise velocity component was monitored at 16 points in the central plane on three different heights corresponding to the measurement positions from the LDV. The shear stress transport (SST) turbulence model was selected [32] as it utilizes the near-wall capabilities of the k-ω model and at the same time uses the bulk flow from the k-ε model where the k-ω is weaker. The flow was initialized with a constant velocity corresponding to the measured mass flow (0.0077 m^3·s^{-1}). After an initial settling period of ~1000 iterations the flow became more periodic and a developed vortex shedding was assumed. The simulation data was then saved for 2500 iterations or 62.5 s resulting in a total computational time of ~4 weeks on 64 CPU cores.

4. Results and Discussion

Migrating fishes swimming around cylindrical obstructions face vortices and, interestingly, these vortices provide extra energy to fishes to improve their swimming abilities. These vortices are influenced by wake and shedding frequency, but these wake and vortex shedding frequencies are greatly dependent on the cylinder diameter [8]. A cylinder with different diameters, such as a step cylinder, could generate an interesting flow behind it and variable wake and shedding frequencies. This warrants this study of wake and vortex shedding behind a semicircular step cylinder with LDV and CFD.

4.1. Wake

Time-averaged measurements of centerline velocity distribution behind the step cylinder at $z/D = -0.5$, $z/D = 0.0$, and $z/D = 0.5$ are shown in Figure 3a–c and notice that for all positions the negative values captured indicate a wake. A main result is that behind the step ($z/D = 0.0$), the wake is longer compared to the situation behind both the large and small cylinders (see Figure 3a–c and Table 2). The simulations, on the other hand, yield that the length of the wake behind the large cylinder ($z/D = -0.5$) is shorter than for the other cases and the time averaged local velocity $\bar{u}$ swiftly goes back toward free stream velocity, U_∞. There is, therefore, good agreement between the simulation and experiments (see Figure 3a). The agreement is also fairly good between simulations and measurements at the step while the simulations severely over predict the length of the wake behind the small cylinder ($z/D = 0.5$) and the simulated results never approach the free stream velocity in the interval of measurement (see Figure 3c) and Table 2 also yields that although simulations compare well to experiments behind the large cylinder and at the step, the simulations always overpredict the length of the wake.

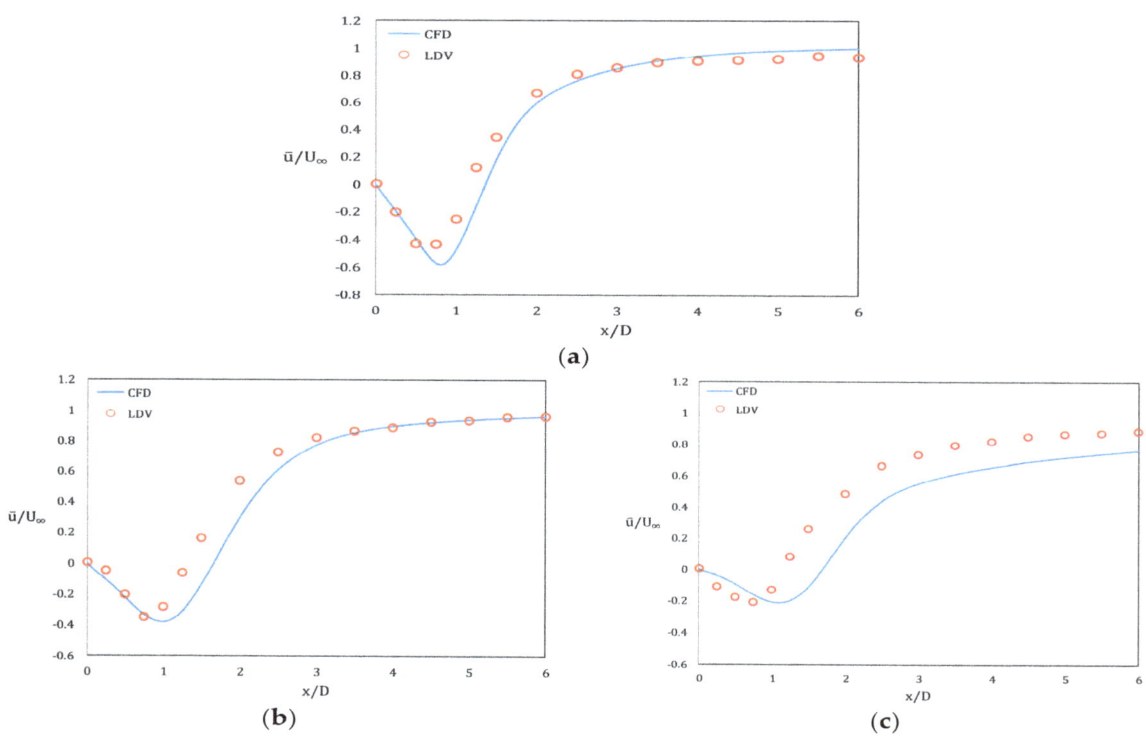

Figure 3. Wake centerline velocity at (**a**) $z/D = -0.5$; (**b**) $z/D = 0.0$; and (**c**) $z/D = 0.5$.

Table 2. Wake length behind the step cylinder.

Description	Position (z/D)	Case	Position (x/D)	Wake Length (mm)
Downward 50 mm	−0.5	CFD	1.4	139
		LDV	1.25	125
Along step	0	CFD	1.7	169
		LDV	1.5	150
Upward 50 mm	0.5	CFD	1.76	176
		LDV	1.25	125

The wake behind the cylinder can be qualitatively visualized by the average of streamwise velocity as derived in the simulations, as seen in Figure 4. The wake is symmetrical in the horizontal plane (Figure 4a) which suggests that the velocity field was averaged for a sufficiently long time; the bow wake [7] in front of the cylinder is also captured. When scrutinizing the vertical plane (Figure 4b) the wake is longer down-stream the smaller cylinder. It can also be seen that within the wake behind the step cylinder a recirculation region develops immediately after the cylinder where reverse flow ($\bar{u} < 0$) is observed (Figures 3 and 4), which extends roughly one diameter downstream of the cylinder. Mean velocity, $\bar{u}$ rapidly approaches the free stream velocity U_∞ after the recirculation region. The velocity after the wake is not capable of recovering completely as the drag-induced momentum deficit must remain constant behind the cylinder (see Figure 3). The continuum is, instead, reserved by a slight increase in the level of the surface.

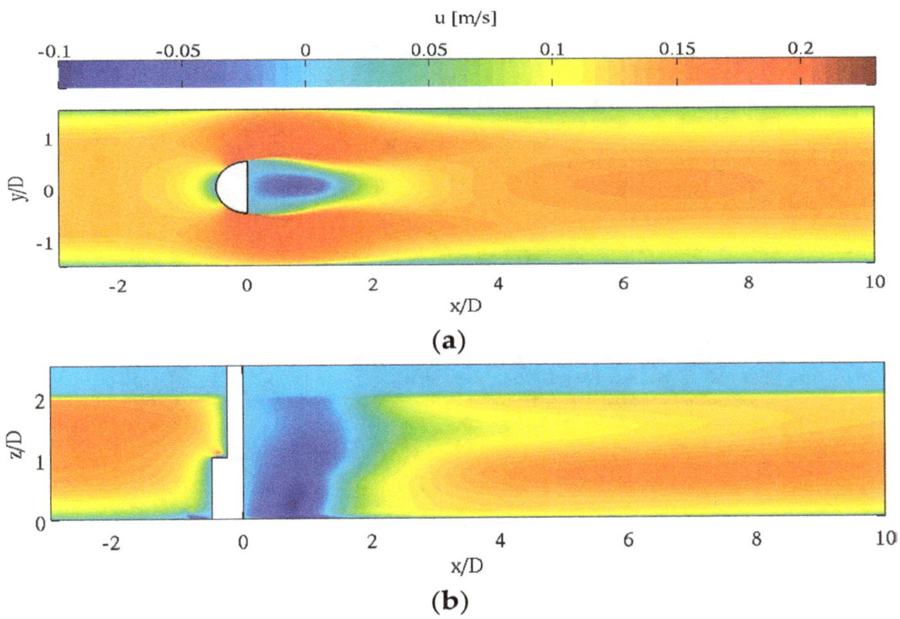

Figure 4. Averaged streamwise velocity at (**a**) $z/D = -0.5$; and (**b**) $y/D = 0.0$.

The wake structures have also been visually observed from the side of a semi-circular cylinder with a step. For instance, the structures of a wake were found behind the step cylinder at $z/D = -0.5$ with a 1 s interval (t, $t + 1$ and $t + 1 + 1$) are presented in Figure 5a–c, respectively. Dyes (red and green) were injected from the both sides of the step cylinder. It is seen that the dyes were swept into the well-known von-Karman vortex street right after downstream of the cylinder. Additionally, three-dimensional (3D) flow structures were also observed behind the cylinder, and a mainly downward flow was noticed.

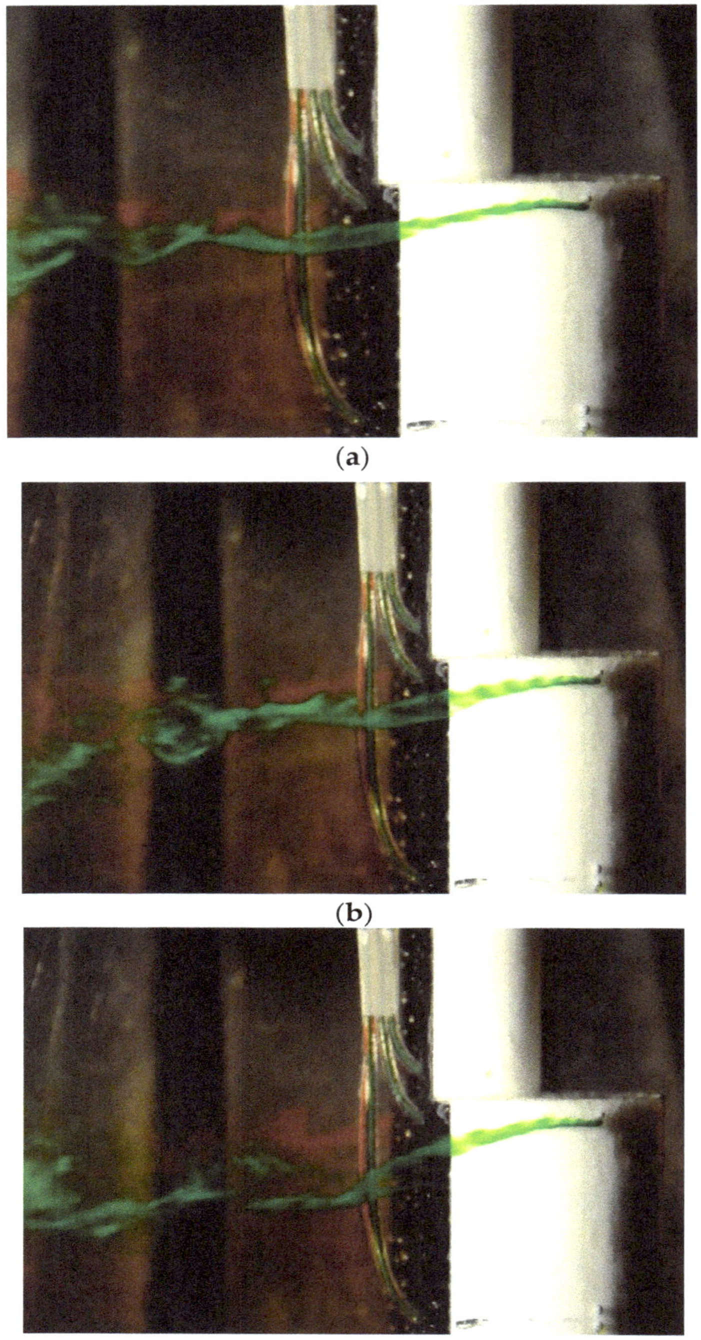

(**a**)

(**b**)

(**c**)

Figure 5. Wake structures around step cylinder at $z/D = -0.5$; (**a**) $[t = 0]$; (**b**) $[t + 1]$; (**c**) $[t + 1 + 1]$.

4.2. Vortex Shedding

The vortex shedding behind the cylinder cannot be described with averaged data. Therefore, Figure 6 shows an instant velocity field captured at the final time step of the simulation (t = 87.5 s) for the horizontal planes in the center of the large cylinder, small cylinder, and at the cylinder step, respectively. As seen there is a clear shedding for all planes. The impacts of channel blockage, i.e., the ratio of the cylinder diameter (D or d) to channel width (W) should be taken into consideration. The blockage ratio, D/W and d/W, are 0.34 and 0.17 for large and small cylinders, respectively, in this study.

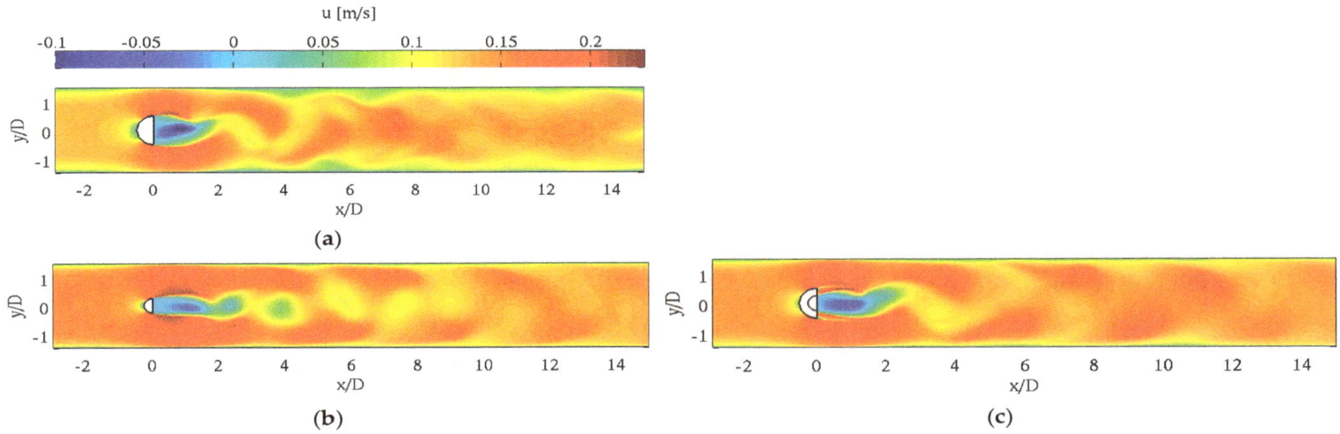

Figure 6. Streamwise velocity of the final time step (t = 87.5 s) at (**a**) $z/D = -0.5$; (**b**) $z/D = 0.5$; and (**c**) $z/D = 0.0$.

The shedding can also be studied in the vorticity field, for instance, Figure 7 shows the vertical vorticity component at one of the last time steps (t = 86.5 s). Counter-rotating vortices can be seen for all three planes and in the plane at the step there seems to be some interaction between the vortices shed from the small and large cylinder. Visualization experiments were also carried out for $z/D = -0.5$ and z/D = 0.5 to verify the vorticity obtained from CFD. Figure 7 depicts the comparison of these with CFD at a similar stage of vortex shedding and it is shown that there is a qualitative agreement between experiments and simulations.

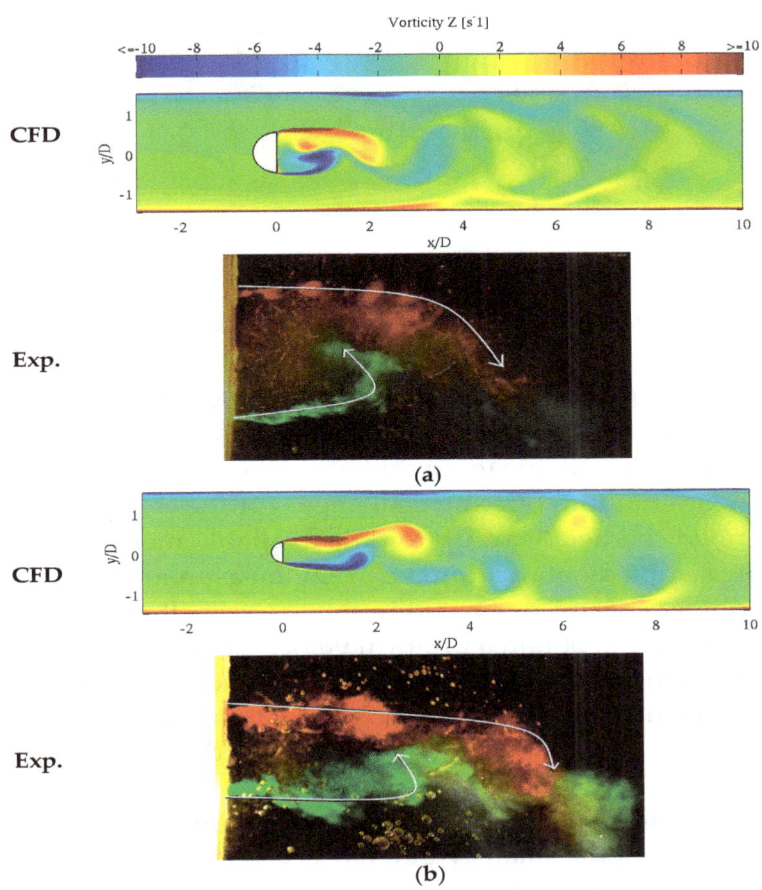

Figure 7. *Cont.*

CFD

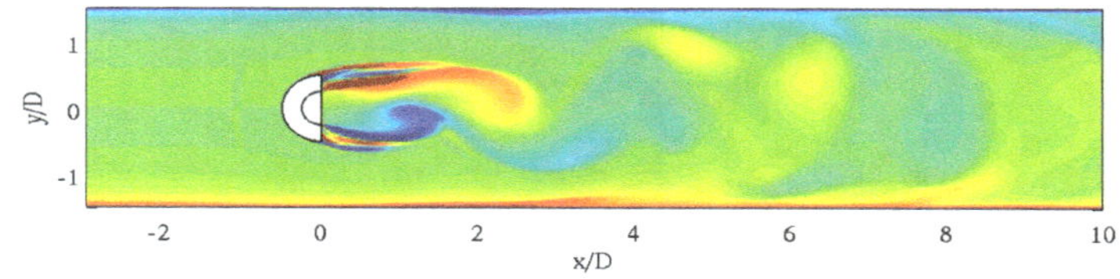

(c)

Figure 7. Vertical vorticity component (t = 86.5 s) at (**a**) $z/D = -0.5$; (**b**) $z/D = 0.5$; and (**c**) $z/D = 0.0$ along with the experimental visualization for $z/D = -0.5$ and $z/D = 0.5$.

Another interesting variable is the fluid pressure. Figure 8 shows the pressure in the final time step at $z/D = -0.5$ and $z/D = 0.5$. Zones where the pressure is lower than the surrounding pressure can be found throughout the vortex street, which could be of some importance when predicting and describing fish movement in this type of flow.

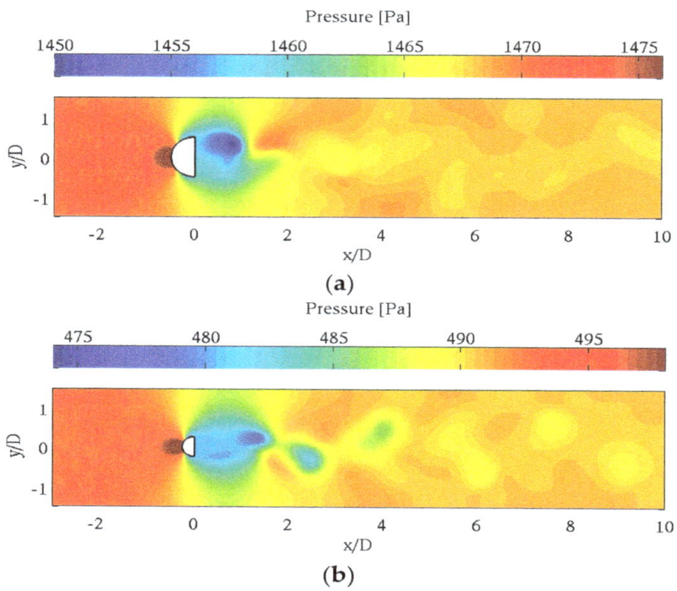

Figure 8. Fluid pressure at the final time step (t = 87.5 s) at (**a**) $z/D = -0.5$; and (**b**) $z/D = 0.5$.

The shedding frequency (f) was determined from the velocity spectra using the fast Fourier transform (FFT). The concept of LDV is that measurements are carried out if a particle travels through the volume of measurement, meaning that often there is, no signal available [33–35]. Thus, sampling data from LDV are random offering challenges for the estimation of the power spectrum. Nobach [29,31,36] described the details of up-to-date available methods to analyze the frequency spectrum from arbitrarily sampled data acquired by LDV measurements.

The estimation of power spectra presented in this paper is the interpolation of the arbitrarily or randomly sampled LDV data to obtain a uniform velocity over time, which is then re-sampled in equal steps with a known sampling frequency close to the average sampling frequency. The mean velocity for each signal was then subtracted as the fluctuations around the mean u' describing the periodicity. The dominant peak in each spectrum is then assumed to correspond to the vortex shedding frequency (f).

The shedding frequency for the large cylinder and step are evaluated at $x/D = 2.5$, $y/D = 0.75$ and the shedding frequency from the small cylinder at $x/d = 2.5$ and $y/d = 0.75$, see Figure 9. For the shedding behind the large cylinder the dominating peak gives f = 0.34 Hz for the experiments and f = 0.38 Hz for the simulations corresponding to Strouhal numbers $St_{LDV} = fD/U_\infty \approx 0.234$ and

$St_{CFD} \approx 0.262$ which can be considered to be in good agreement. For the shedding behind the smaller cylinder the simulation gives the same frequency as behind the large cylinder whereas the LDV gives a peak at $f = 0.6$ Hz which is more consistent with other studies [37–39] behind fully-circular stepped cylinders. For the shedding behind the step, the experiments yield two dominant peaks, one at $f = 0.36$ Hz, which should correspond to the shedding from the large cylinder and $f = 0.6$ Hz which corresponds to the shedding from the small one, whereas the simulation once again gives the same frequency as for the large cylinder, not capturing the dynamics of the small cylinder or the step. This may be traced to the SST turbulence model's inability to capture the shedding correctly; other possible contributors could be that the time-step is too large during the experiments so that some spatial or temporal details are neglected or that the flow at the smaller cylinder is not yet fully developed even after 3500 iterations. In addition, the effect of blockage on vortex shedding for the flow past a step cylinder can also be significant. Some previous studies [37,38,40–43] have shown that the effect of channel blockage on vortex shedding for the flow past a step cylinder can also be significant as the channel blockage affects f, consequently influencing St. The increase of shedding frequency with the increase of the blockage ratios is evidently observed in these spectral results (Figure 6). Therefore, channel blockage could also be the reason on varying the shedding frequency behind step cylinder (large and small cylinders).

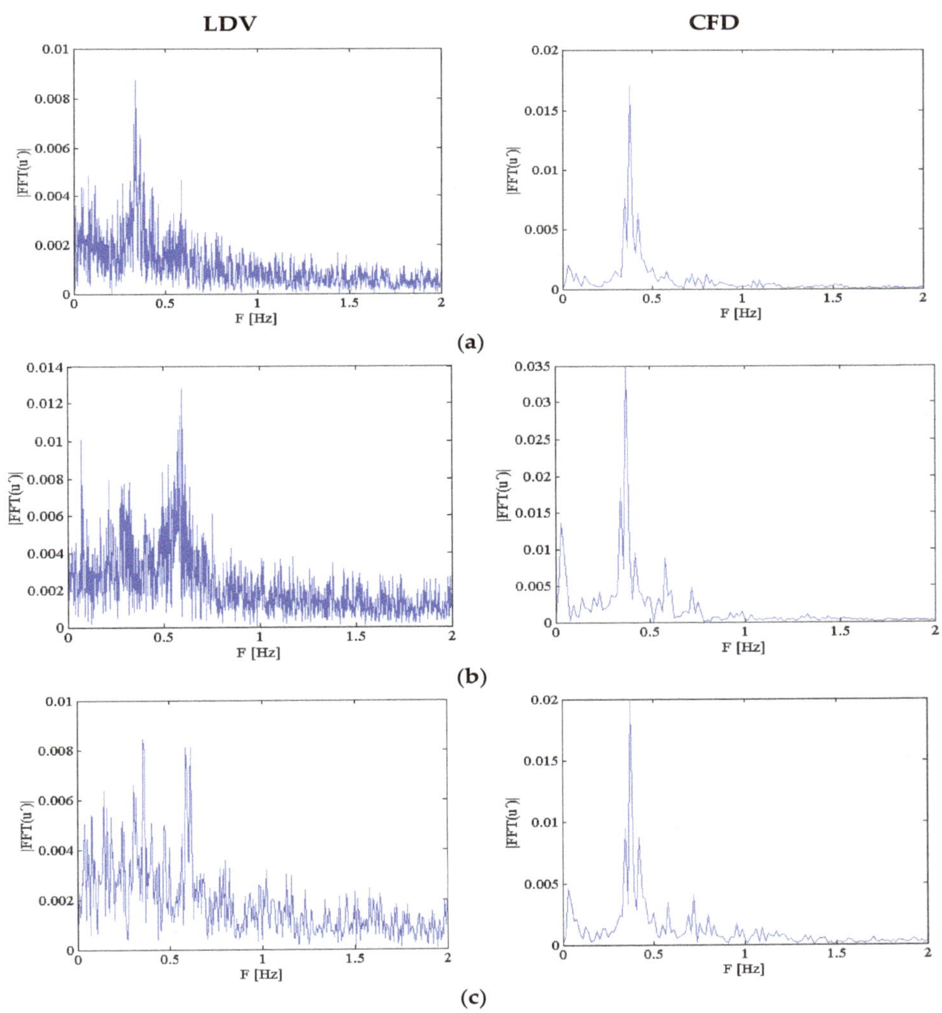

Figure 9. Single-sided FFT amplitude spectrum of the streamwise velocity fluctuations for LDV and CFD at (**a**) the large cylinder, $x/D = 2.5$, $z/D = -0.5$, $y/D = 0.75$ (**b**) small cylinder, $x/d = 2.5$, $z/D = 0.5$, $y/d = 0.75$; and (**c**) diameter step, $x/D = 2.5$, $z/D = 0.0$, $y/D = 0.75$.

5. Concluding Remarks

An investigation with LDV measurements and CFD simulations behind a semi-circular cylinder with a step has been conducted. Additional flow visualization was also carried out to explore the wake structure. A summary of key features of the wake flows obtained in this study is given below:

1. The main result is that the measured vortex shedding frequency is, $f = 0.34$ Hz behind the large cylinder and $f = 0.6$ Hz behind the small cylinder, while it has two dominating frequencies across the step, $f = 0.36$ Hz and $f = 0.6$ Hz. This implies that a step cylinder can be used to attract fish of different sizes and facilitate their motion up-stream the fishway.
2. Experiments yield that the wake length is longer along the step ($z/D = 0.0$) and it reduces for the large cylinder ($z/D = -0.5$), as well as the small cylinder ($z/D = 0.5$). The simulation over-predicts all of these wake lengths.
3. Flow visualization finds the development of three-dimensional (3D) flow structures along with well-known Von Karman vortex street.
4. Simulation results show acceptable agreement with the experimental measurements behind the large cylinder, but fail to capture the dynamics of the small cylinder diameter.

Future studies with tomographic-PIV would enable the detailed vortex dynamics of the flow behind the stepped half-cylinders to be investigated. Additional simulations with more advanced models, such as the Reynolds stress models and large eddy simulations, could also provide additional insights into this type of problem.

For the application of fish migration, it would also be interesting to test these configurations on fish in laboratory studies, as well as in the field, to see how different species and sizes of fish respond to different geometries and flow conditions.

Acknowledgments: This work has been funded by the collaboration initiative *StandUp for Energy*. The research program is a part of the Swedish government's commitment to high quality research in areas of strategic importance.

Author Contributions: S. M. Sayeed-Bin-Asad, Tord Staffan Lundström and Anders Gustav Andersson conceived and designed the experiments; S. M. Sayeed-Bin-Asad performed the experiments and analyzed the data; Anders Gustav Andersson performed the simulation; S. M. Sayeed-Bin-Asad wrote the paper; All authors read and approved the final manuscript.

References

1. Swedenergy. Svensk Energi—Swedenergy—AB. Available online: http://www.svenskenergi.se/ (accessed on 12 October 2016).
2. Kaveh, A. Experimental Investigation of a Kaplan Runner under Steady-State and Transient Operations. Ph.D. Thesis, Luleå University of Technology, Luleå, Sweden, 2016.
3. Liao, J.C.; Beal, D.N.; Lauder, G.V.; Triantafyllou, M.S. Fish exploiting vortices decrease muscle activity. *Science* **2003**, *302*, 1566–1569. [CrossRef] [PubMed]
4. Liao, J.C. A review of fish swimming mechanics and behaviour in altered flows. *Philos. Trans. R. Soc. B Biol. Sci.* **2007**, *362*, 1973–1993. [CrossRef] [PubMed]
5. Stewart, W.J.; Tian, F.B.; Akanyeti, O.; Walker, C.J.; Liao, J.C. Refuging rainbow trout selectively exploit flows behind tandem cylinders. *J. Exp. Biol.* **2016**, *219*, 2182–2191. [CrossRef] [PubMed]
6. Akanyeti, O.; Thornycroft, P.J.M.; Lauder, G.V.; Yanagitsuru, Y.R.; Peterson, A.N.; Liao, J.C. Fish optimize sensing and respiration during undulatory swimming. *Nat. Commun.* **2016**, *7*, 11044. [CrossRef] [PubMed]
7. Sayeed-Bin-Asad, S. Velocity Distribution Measurements in A Fishway Like Open Channel by Laser Doppler Anemometry (LDA). *EPJ Web. Conf.* **2016**. [CrossRef]
8. Liao, J.C.; Beal, D.N.; Lauder, G.V.; Triantafyllou, M.S. The Kármán gait: Novel body kinematics of rainbow trout swimming in a vortex street. *J. Exp. Biol.* **2003**, *206*, 1059–1073. [CrossRef] [PubMed]

9. Dunn, W.; Tavoularis, S. Vortex shedding from a step-cylinder in spanwise sheared flow. *Phys. Fluids* **2011**, *23*, 035109. [CrossRef]
10. Dunn, W.; Tavoularis, S. Experimental studies of vortices shed from cylinders with a step-change in diameter. *J. Fluid Mech.* **2006**, *555*, 409–437. [CrossRef]
11. Morton, C.; Yarusevych, S. Vortex shedding in the wake of a step cylinder. *Phys. Fluids* **2010**, *22*, 083602. [CrossRef]
12. Morton, C.; Yarusevych, S. Three-dimensional flow and surface visualization using hydrogen bubble technique. *J. Vis.* **2015**, *18*, 47–58.
13. Morton, C.; Yarusevych, S. On vortex shedding from low aspect ratio dual step cylinders. *J. Fluids Struct.* **2014**, *44*, 251–269. [CrossRef]
14. Morton, C.; Yarusevych, S. Vortex dynamics in the turbulent wake of a single step cylinder. *J. Fluids Eng.* **2014**, *136*, 031204. [CrossRef]
15. McClure, J.; Morton, C.; Yarusevych, S. Flow development and structural loading on dual step cylinders in laminar shedding regime. *Phys. Fluids* **2015**, *27*, 063602. [CrossRef]
16. Morton, C.R.; Yarusevych, S. A combined experimental and numerical study of flow past a single step cylinder. In Proceedings of the ASME 2010 3rd Joint US-European Fluids Engineering Summer Meeting Collocated with 8th International Conference on Nanochannels, Microchannels, and Minichannels, Montreal, QC, Canada, 1–5 August 2010; American Society of Mechanical Engineers: New York, NY, USA, 2010.
17. Morton, C.; Yarusevych, S.; Scarano, F. A tomographic particle image velocimetry investigation of the flow development over dual step cylinders. *Phys. Fluids* **2016**, *28*, 025104. [CrossRef]
18. Rafati, S. *Investigating Step Cylinder Wake Topology with Tomographic PIV*; Delft University of Technology: Delft, The Netherlands, 2014.
19. Ko, N.; Leung, W.; Au, H. Flow behind two coaxial circular cylinders. *J. Fluids Eng.* **1982**, *104*, 223–227. [CrossRef]
20. Yagita, M.; Kojima, Y.; Matsuzaki, K. On vortex shedding from circular cylinder with step. *Bull. JSME* **1984**, *27*, 426–431. [CrossRef]
21. Lewis, C.; Gharib, M. An exploration of the wake three dimensionalities caused by a local discontinuity in cylinder diameter. *Phys. Fluids A Fluid Dyn.* **1992**, *4*, 104–117. [CrossRef]
22. Norberg, C. An experimental study of the flow around cylinders joined with a step in diameter. In Proceedings of the 11th Australasian Fluid Mechanics Conference, Hobart, Australia, 14–18 December 1992.
23. Ko, W.-M.; Chan, S.-K. *Pressure Distributions on Circular Cylinders with Stepwise Change of the Diameter*; American Society of Mechanical Engineers: New York, NY, USA, 1984.
24. Ko, N.; Chan, A. In the intermixing region behind circular cylinders with stepwise change of the diameter. *Exp. Fluids* **1990**, *9*, 213–221. [CrossRef]
25. Ko, N.; Chan, A. Wakes behind circular cylinders with stepwise change of diameter. *Exp. Therm. Fluid Sci.* **1992**, *5*, 182–187. [CrossRef]
26. Chua, L.; Liu, C.; Chan, W. Measurements of a step cylinder. *Int. Commun. Heat Mass Transf.* **1998**, *25*, 205–215. [CrossRef]
27. Morton, C.; Yarusevych, S. Modeling Flow over a Circular Cylinder with a Stepwise Discontinuity. In Proceedings of the 39th AIAA Fluid Dynamics Conference, San Antonio, TX, USA, 22–25 June 2009.
28. Morton, C.; Yarusevych, S. Cross flow over cylinders with two stepwise discontinuities in diameter. In Proceedings of the Turbulence and Shear Flow Phenomena Conference, Ottawa, ON, Canada, 28–30 July 2011.
29. Nobach, H. Present Methods to Estimate the Cross-Correlation and Cross-Spectral Density for Two-Channel Laser Doppler Anemometry. In Proceedings of the 18th International Symposium on the Application of Laser and Imaging Techniques to Fluid Mechanics, Lisbon, Portugal, 4–7 July 2016.
30. Norberg, C. LDV-Measurements in the Near Wake of a Circular Cylinder. Available online: http://www.ht.energy.lth.se/fileadmin/ht/Norberg-FEDSM98-5202.pdf (accessed on 12 October 2016).
31. Hirt, C.W.; Nichols, B.D. Volume of fluid (VOF) method for the dynamics of free boundaries. *J. Comput. Phys.* **1981**, *39*, 201–225. [CrossRef]
32. Menter, F.R. Review of the shear-stress transport turbulence model experience from an industrial perspective. *Int. J. Comput. Fluid Dyn.* **2009**, *23*, 305–316. [CrossRef]

33. Morton, C.; Yarusevych, S. An experimental investigation of flow past a dual step cylinder. *Exp. Fluids* **2012**, *52*, 69–83. [CrossRef]
34. Velte, C.M.; George, W.K.; Buchhave, P. Estimation of burst-mode LDA power spectra. *Exp. Fluids* **2014**, *55*, 1–20. [CrossRef]
35. Müller, E.; Nobach, H.; Tropea, C. A refined reconstruction-based correlation estimator for two-channel, non-coincidence laser Doppler anemometry. *Meas. Sci. Technol.* **1998**, *9*, 442. [CrossRef]
36. Nobach, H. Laser Doppler Data Processing. Available online: http://ldvproc.nambis.de/programs/pyLDV.html (accessed on 10 October 2016).
37. Turki, S.; Abbassi, H.; Nasrallah, S.B. Effect of the blockage ratio on the flow in a channel with a built-in square cylinder. *Comput. Mech.* **2003**, *33*, 22–29. [CrossRef]
38. Zong, L.; Nepf, H. Vortex development behind a finite porous obstruction in a channel. *J. Fluid Mech.* **2012**, *691*, 368–391. [CrossRef]
39. Morton, C.R. Experimental and Numerical Investigations of the Flow Development over Circular Cylinders with Stepwise Discontinuities in Diameter. Master's Thesis, University of Waterloo, Waterloo, ON, Canada, 2010.
40. Sahin, M.; Owens, R.G. A numerical investigation of wall effects up to high blockage ratios on two-dimensional flow past a confined circular cylinder. *Phys. Fluids (1994–Present)* **2004**, *16*, 1305–1320. [CrossRef]
41. Mettu, S.; Verma, N.; Chhabra, R. Momentum and heat transfer from an asymmetrically confined circular cylinder in a plane channel. *Heat Mass Transf.* **2006**, *42*, 1037–1048. [CrossRef]
42. Okajima, A.; Yi, D.; Sakuda, A.; Nakano, T. Numerical study of blockage effects on aerodynamic characteristics of an oscillating rectangular cylinder. *J. Wind Eng. Ind. Aerodyn.* **1997**, *67*, 91–102. [CrossRef]
43. Park, C.-W.; Lee, S.-J. Free end effects on the near wake flow structure behind a finite circular cylinder. *J. Wind Eng. Ind. Aerodyn.* **2000**, *88*, 231–246. [CrossRef]

7

Numerical Study on the Transient Thermal Performance of a Two-Phase Closed Thermosyphon

Zhongchao Zhao *, Yong Zhang, Yanrui Zhang, Yimeng Zhou and Hao Hu

School of Energy and Power, Jiangsu University of Science and Technology, Zhenjiang 212000, China; 18896658850@163.com (Y.Z.); zyr1056187930@163.com (Y.Z.); 18851406053@163.com (Y.Z.); 00001111zy@sina.com (H.H.)

* Correspondence: zhongchaozhao@just.edu.cn.

Abstract: The transient thermal performance of phase change and heat and mass transfer in a two-phase closed thermosyphon are studied with computational fluid dynamics (CFD). A CFD model based on the volume of fluid technique is built. Deionized water is specified as the working fluid of this thermosyphon. The CFD model reproduces evaporation and condensation in the thermosyphon at different heating inputs. The average wall temperatures are also analyzed. Variations of average wall temperatures indicate that this thermosyphon reaches a steady state after 19 s, and starts to work in advance when the heating input increases. Moreover, thermal resistance is decreased until a minimum (0.552 K/W) by increasing the heating input, and the effective thermal conductivity is elevated to a maximum (2.07 × 106 W/m·K).

Keywords: thermosyphon; volume of fluid; multiphase flow; evaporation and condensation

1. Introduction

A two-phase closed thermosyphon, like a wicked heat pipe, transmits heat by the evaporation and condensation of a working fluid that is circulating in a sealed container. However, the thermosyphon relies on the force of gravity to return the working fluid from a condenser to an evaporator rather than relying on the capillary forces produced by the wick in the wicked heat pipe [1]. As a kind of high-efficiency heat transfer component, a thermosyphon, whose internal complex heat and mass transfer is the focus of this research about the heat transfer mechanism of a heat pipe, has already been widely used in the field of heat exchange. Thermosyphon technology is playing an important role in many industrial applications, particularly in the heat transfer of heat exchangers and in energy savings in applications. For the advantages of its light weight, wide operating temperature range, compact structure, flexibility, high capacity of heat transfer, and great isothermal performance, the thermosyphon has been widely used in many cooling fields that contain electronic components and products [2–4].

A thermosyphon consists of a condenser, an adiabatic section, and an evaporator section. The evaporator section is heated by a hot source, and the condenser section is cooled by a cold source. The adiabatic section placed between the condenser and evaporator sections is surrounded with thermal insulation layers. Heat is absorbed by the evaporator section with a liquid pool, which is situated at the bottom of the thermosyphon, following which liquid that absorbs the latent heat of evaporation is turned into vapor. Afterwards, vapor rises up to the condenser section at the top of the thermosyphon, where it is condensed and gives up latent heat. The condensed working fluid, which forms a liquid film along the walls, moves back to the evaporator under gravity.

A great variety of experiments have been done to understand the thermal characteristics of the thermosyphon [5–11], although with an unknown internal heat and mass transfer mechanism.

Annamalai and Ramalingam [12] built a model for a wicked heat pipe by using ANSYS CFX and performing an experimental study. They assumed that the section inside the heat pipe was filled by the vapor phase, and the liquid phase only occurred in the wick section. The wall temperature of the condenser and evaporator and the vapor temperature were compared. Legierski et al. [13] provided a model for simulating the temperature along the heat pipe walls at different moments and the vapor velocity inside the heat pipe. They considered the heat pipe as an open system, wherein the evaporator was treated as the inlet of mass flow and the condenser was treated as the outlet. The start-up time of the heat pipe was 20~30 s, and the evaporation coefficient of water was 1.58×10^{-3}. The range of effective thermal conductivity was also provided.

De Schepper et al. [14] employed the user-defined function (UDF) and volume of fluid (VOF) techniques to calculate the phase-change of a hydrocarbon feedstock. They proposed that UDFs were specified as source terms, which were used to simulate heating and boiling in the convection section of a steam cracker. Lin et al. [15] carried out a computational fluid dynamics (CFD) analysis of miniature oscillating heat pipes using VOF and mixture methods. Alizadehdakhel et al. [16] performed an experimental study under different operating conditions in a thermosyphon. They also created a CFD model to compare the predicted temperature profile with experimental results that employed the VOF model.

Fadhl et al. [17] carried out an experimental study and CFD analysis of a two-phase closed thermosyphon using the VOF technique. The experimental investigation was performed with the input heat powers of 100.41, 172.87, 225.25, 275.6, 299.52, and 376.14 W. The temperature profile predicted by CFD was consistent with the experimental data at the same input heat powers. They then [18] established a two-dimensional model with three working fluids, i.e., distilled water, R134a, and R404a. The refrigerants (R134a and R404a) and distilled water showed different boiling phenomena in the thermosyphon.

Most currently available studies only focused on the steady state thermal performance of the thermosyphon, but numerical investigation of the thermal performance during start-up remains limited. Therefore, the purpose of this study is to simulate a vapor–liquid two-phase flow, evaporation and condensation, and other complex physical processes during the operation of a closed thermosyphon. The VOF technique was employed to develop a CFD model to track the liquid–vapor interface. In order to calculate the phase change, the mass and energy sources compiled by UDFs were used to complete the FLUENT code.

2. CFD Modeling

The objective of this study was to analyze the mechanism of phase change and the thermal performance of working fluids in a two-phase closed thermosyphon. There are three main methods for the numerical calculations of Euler–Euler multiphase flows, which are the VOF method, the mixture method, and the Eulerian method. The mixture method and the Eulerian method are applicable when the volume fractions of dispersed-phase exceed 10%, wherein the phase is mixed or separated in the flows. As the phase is immiscible with multiphase flows in the thermosyphon, the VOF method was adopted as it can simulate either stratified flow or free surface flow.

2.1. VOF Model

A single set of Navier–Stokes equations in the VOF model were applied to immiscible phases, and the volume fraction of each phase was recorded into each cell throughout the computation domain. Thus, the summation of volume fraction of all phases is unity:

$$\alpha_l + \alpha_v = 1 \tag{1}$$

The cell is occupied by liquid phase if $\alpha_l = 1$ and by vapor phase if $\alpha_v = 0$. The relationship of $0 < \alpha_l < 1$ represents that the cell is at the interface between liquid and vapor phases. The continuity equations for the VOF model have the following forms:

$$\nabla \cdot \left(\alpha_l \rho_l \vec{u}\right) = -\frac{\partial}{\partial t}(\alpha_l \rho_l) + M_l \tag{2}$$

$$\nabla \cdot \left(\alpha_v \rho_v \vec{u}\right) = -\frac{\partial}{\partial t}(\alpha_v \rho_v) + M_v \tag{3}$$

where ρ_l is the density of liquid phase and ρ_v is the density of vapor phase.

The continuity equations treat velocity u as the mass-averaged velocity. Meanwhile, the rates of mass transfer passing the two-phase interface are represented as M_l and M_v.

Considering the effect of gravity and volumetric surface tension, the momentum equation for the VOF model is described as follows:

$$\frac{\partial}{\partial t}\left(\rho \vec{u}\right) + \nabla \cdot \left(\rho \vec{u} \vec{u}\right) = \rho \vec{g} - \nabla p + \nabla \cdot \left[\mu\left(\nabla \vec{u} + \nabla \vec{u}^{T}\right) - \frac{2}{3}\mu \nabla \cdot uI\right] + F_{CSF} \tag{4}$$

where g represents the acceleration of gravity, p represents the pressure, and I represents the unit tensor.

In Equation (4), the volumetric surface tension force is calculated by the continuum surface force (CSF) model proposed by Brackbill et al. [19]. Therefore, the surface tension, as a source term, has the following form:

$$F_{CSF} = 2\sigma_{vl} \frac{a_l \rho_l k_v \nabla a_v + a_v \rho_v k_l \nabla a_l}{(\rho_l + \rho_v)} \tag{5}$$

where σ_{vl} is the surface tension coefficient and k is the surface curvature.

In the momentum equation, density and viscosity rely on the volume fraction of the phases. Therefore, density ρ and the dynamic viscosity μ are expressed as follows:

$$\rho = a_l \rho_l + a_v \rho_v \tag{6}$$

$$\mu = a_l \mu_l + a_v \mu_v \tag{7}$$

where μ_l and μ_v are the dynamic viscosities of liquid and vapor, respectively.

The energy equation for the VOF model has the following form:

$$\frac{\partial}{\partial t}(\rho e) + \nabla \cdot \left(\vec{v}(\rho e + p)\right) = \nabla \cdot (k \nabla T) + E \tag{8}$$

where $k = \alpha_l k_l + \alpha_v k_v$. The energy source term used to compute the rates of heat transfer through the interface is represented as E.

In the VOF model, T is the mixture temperature rather than the temperature of a specific phase for the energy equation. Meanwhile, the internal energy e treated as a mass-averaged variable is given by the following:

$$e = \frac{\alpha_l \rho_l e_l + \alpha_v \rho_v e_v}{\alpha_l \rho_l + \alpha_v \rho_v} \tag{9}$$

where e_l and e_v are calculated on the basis of the specific heat c_p of the phase and the mixture temperature:

$$e_l = c_{p,l}(T - T_{sat}) \tag{10}$$

$$e_v = c_{p,v}(T - T_{sat}) \tag{11}$$

2.2. Mass and Heat Transfer Model

FLUENT 6.3 [20] was used to solve the governing equations of the VOF model. UDF was linked to the governing equations, which simulated the evaporation and condensation in the thermosyphon. In UDF, the mass and heat transfer rates are defined by source terms proposed by De Schepper et al. [14]. Therefore, all source terms are required to calculate the mass and heat transfer rates in the following forms:

For mass source terms in the evaporation process

$$M_l = -C\rho_l\alpha_l\frac{T_{mix} - T_{sat}}{T_{sat}} \tag{12}$$

$$M_v = -(S_{Ml}) \tag{13}$$

For mass source terms in the condensation process:

$$M_l = C\rho_v\alpha_v\frac{T_{sat} - T_{mix}}{T_{sat}} \tag{14}$$

$$M_v = -(S_{Ml}) \tag{15}$$

For energy source terms in the evaporation and condensation processes, respectively:

$$E_e = (-C\rho_l\alpha_l\frac{T_{mix} - T_{sat}}{T_{sat}}) \cdot LH \tag{16}$$

$$E_c = (C\rho_v\alpha_v\frac{T_{sat} - T_{mix}}{T_{sat}}) \cdot LH \tag{17}$$

where T_{mix} is the mixture temperature, T_{sat} is the saturation temperature, and LH is the latent heat. Coefficient C, used to compute the condensation/evaporation rate, is normally specified as 0.1 [17].

3. Geometry and Boundary Condition

A two-dimensional numerical model was built for simulating the two-phase flow, taking the axisymmetric structure of a thermosyphon into consideration. The geometry of the computational domain (Figure 1) is divided into the condenser, adiabatic, and evaporator sections. Furthermore, the lengths of evaporator and condenser sections are 100 mm, while the adiabatic section is fixed at 50 mm. The inner diameter and wall thickness are 8.32 and 0.6 mm, respectively. The grids were created with GAMBIT 2.4.6, and the type of grid was Quad. The grid independence was tested by different mesh sizes, and the average temperature of the evaporator (Te.av) and condenser (Tc.av) sections for different mesh sizes were monitored. For the water-charged thermosyphon and the heating power of 40 W, it was found that both Te.av and Tc.av were almost the same for a different number of grids, such as 156,900, 80,000, and 58,045.

Thus, the grid quantity of 80,000 was selected for the simulation analysis, and it was adopted in order to ensure the grid independence. Eight layers of cells were meshed, giving the initial size and growth factor of 0.035 and 1.2, respectively. A very thin liquid film was developed near the inner walls of the thermosyphon, and the boundary layer technique was applied on the left and right inner walls. In fluid dynamics, there is zero velocity between the fluid and the solid boundary, relatively. Therefore, a non-slip boundary condition was applied on the inner walls. A constant heat flux was specified as the boundary condition for the evaporator section, and a convection heat transfer coefficient was specified as the boundary condition for the condenser section. The heat flux, as the boundary condition for the adiabatic section, was zero, considering this section was insulated. Furthermore, water was adopted as the working fluid and fixed at 60% of the evaporator volume, and the density depending on the temperature in this model is defined by the steam table in the following form:

$$\rho_l = 249.46 + 6.625 \times T - 0.0184 \times T^2 + 1.532 \times 10^{-5} \times T^3 \tag{18}$$

where T represents the mixture temperature.

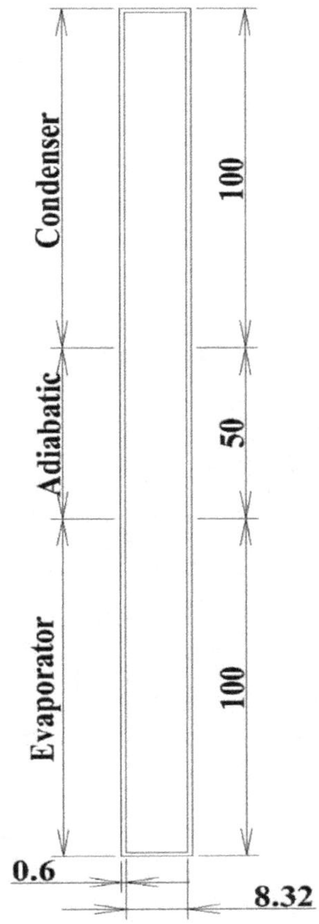

Figure 1. Model geometry and dimensions.

4. CFD Solution Method and Model Validation

Taking the start-up and steady-state operation into consideration, the whole simulation process of the thermosyphon is transient with a time step of 0.0001 s. In order to solve the pressure–velocity coupling, the SIMPLE algorithm was selected from the segregated algorithms provided by FLUENT. The first order upwind was adopted to discretize the momentum and energy equations, which reduced the difficulty of convergence. Meanwhile, Geo-Reconstruct for the volume fraction combined with PRESTO for the pressure was included in this model. In the present simulation, the scaled residual of mass and velocity components should be 10^{-4} for the convergence of numerical computation. Water vapor was taken as the primary (vapor) phase, and the saturation temperature was 308.55 K.

For validating the reliability of the numerical model and method, the experimental data should be compared with the results of the simulation. The thermosyphon was made of copper with the internal and external diameters of 8.32 mm and 9.52 mm, respectively (Figure 2a). The lengths of evaporator and condenser sections were designed to be 100 mm, and the length of the adiabatic section was designed to be 50 mm. The whole experimental apparatus consisted of a thermosyphon, heating system, cooling system, vacuum pumping system, and data acquisition system (Figure 2b). The filling ratio, which means the ratio of volume of liquid to that of thermosyphon, was 24%, and the working fluid was 3.26 g deionized water. The experiments were performed at the heating input powers of 40 W, 60 W, and 80 W. Three of the thermocouples were sited on the evaporator section, one on the adiabatic section, and two on the condenser section (one at the inlet and one at the outlet of the cooling water jacket, respectively).

Figure 3 shows the comparisons of the simulation and experimental results of the wall temperature of the evaporator and condenser during the start-up process at the heating input power of 40 W. The experimental data showed the same trend as the simulation results of the VOF model, and the start-up time is nearly equal. The maximum error of temperature difference of the evaporator and condenser between the simulation and the experiment are 0.53% and 1.02%, respectively. Considering the influence of the experimental environment, the errors are acceptable. Therefore, the numerical data were in good fit to the experimental data, and the numerical model and method were reliable.

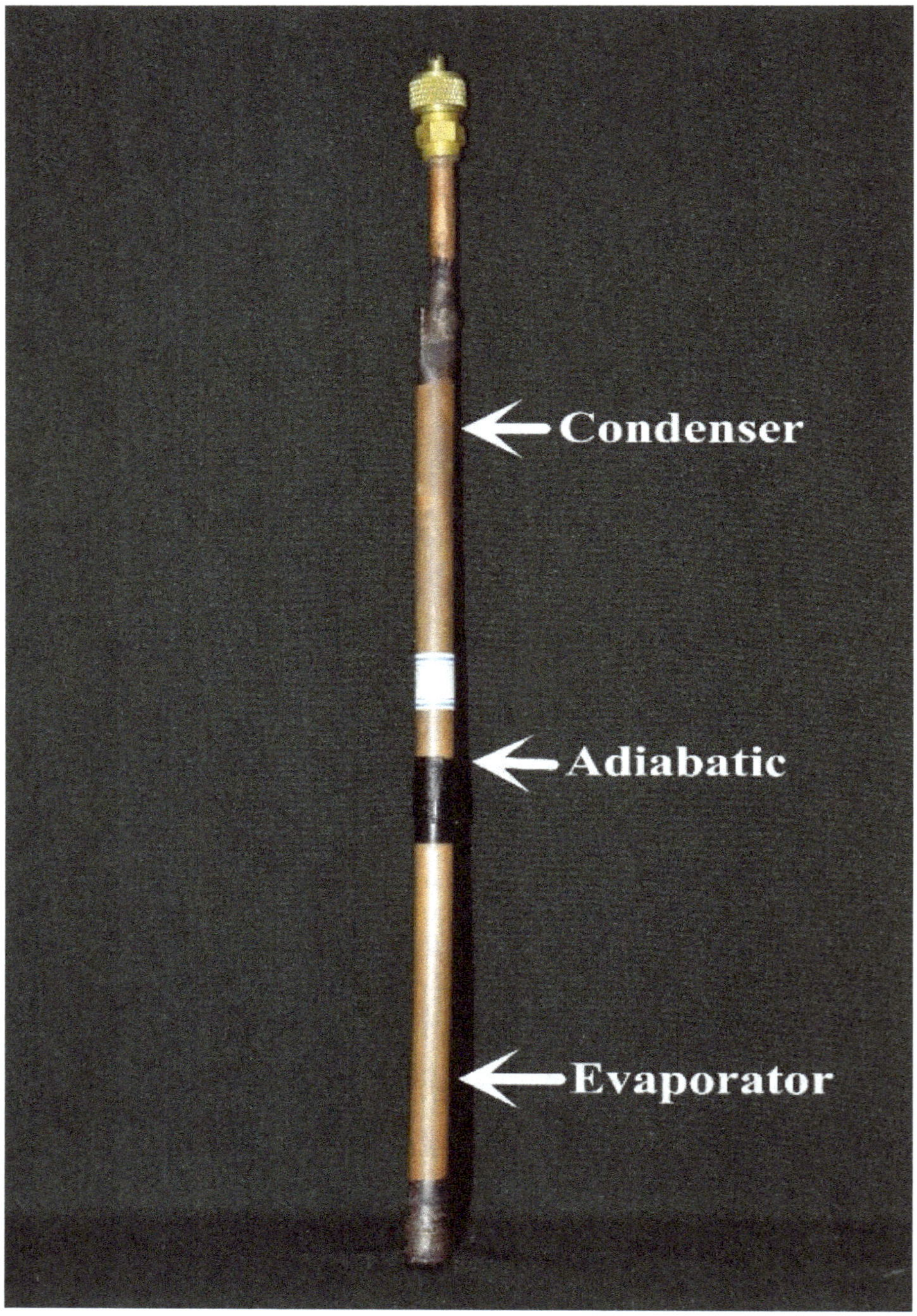

(a)

Figure 2. *Cont.*

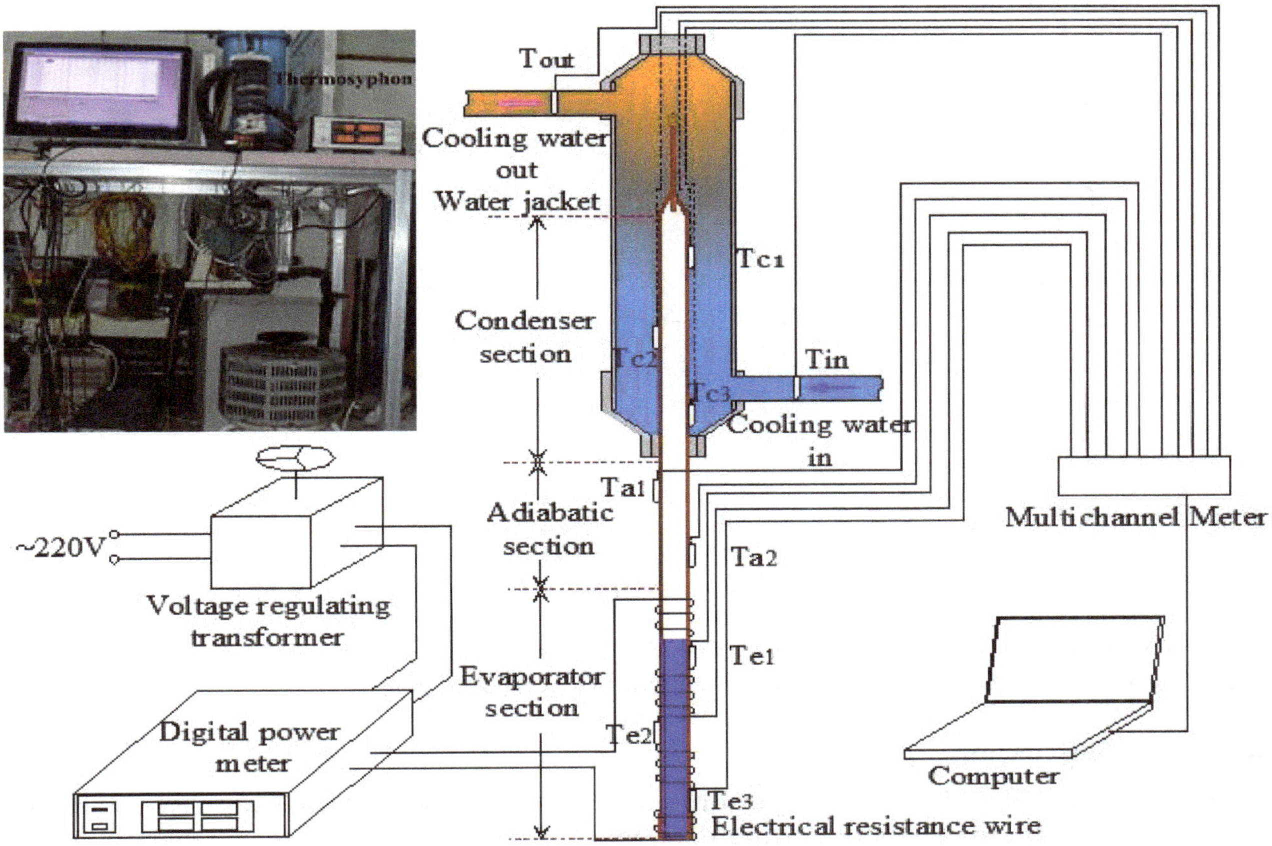

(**b**)

Figure 2. (**a**) The real thermosyphon and (**b**) Schematic diagram of experimental apparatus.

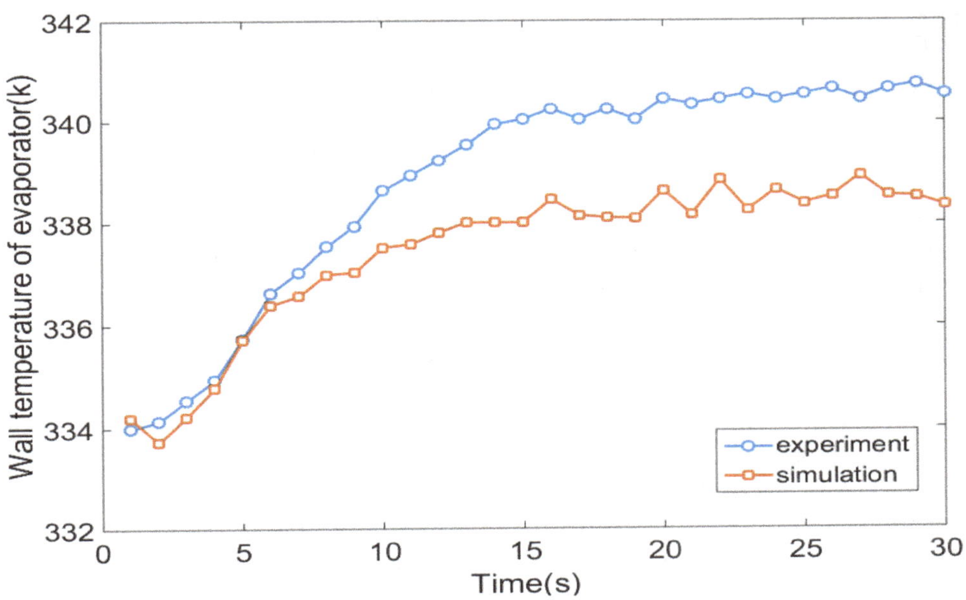

(**a**)

Figure 3. *Cont.*

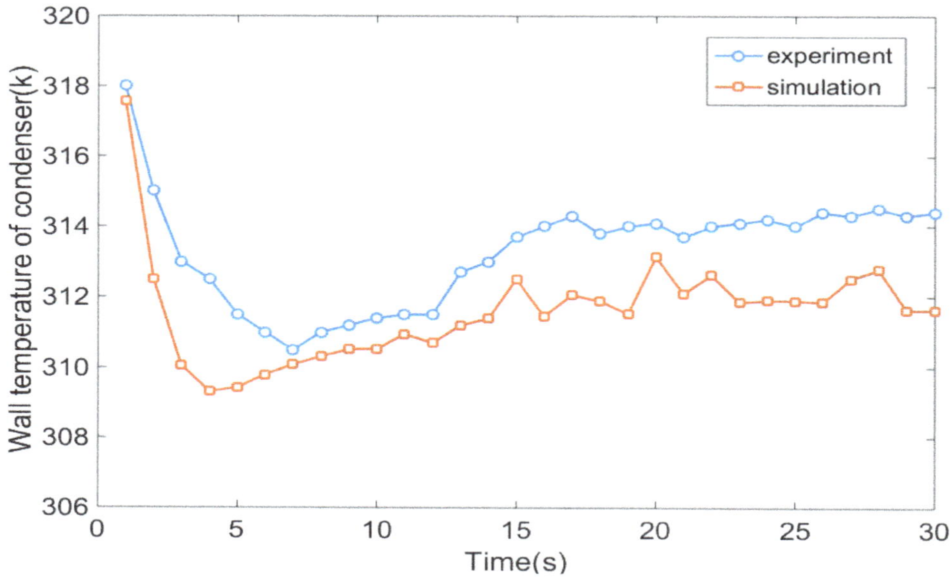

(**b**)

Figure 3. Start-up characteristics comparison of the simulation and experimental wall temperature of (**a**) Evaporator and (**b**) Condenser at the heating input power of 40 W.

5. Results and Discussion

The numerical simulation was performed to observe the evaporation and condensation phenomenon in the thermosyphon. Furthermore, the wall temperature profile, the thermal resistance, and the effective thermal conductivity were obtained during the start-up of the thermosyphon under heating inputs of 40 W, 60 W, and 80 W.

5.1. The Evaporation and Condensation Process of the Numerical Simulation of the Thermosyphon

Pool boiling and vapor-liquid distribution in the evaporator section and the liquid film in the lower region of the condenser section are visualized in Figures 4 and 5, respectively. The blue color represents that the liquid volume fraction is 1, indicating that only liquid exists. As evidenced by the liquid volume fraction (value: 0) represented by the red color, only vapor exists. The evaporator section is represented by "We" in Figure 4, 60% of which is occupied by the liquid pool at 0.1 s.

The heating power transferred the heat into the liquid pool through the walls of the evaporator section. Evaporation occurred at the nucleation site, where liquid reached the saturation temperature at 0.3 s. Vapor bubbles formed and departed from the inner walls of the evaporator section at 0.5 s. As the heating input increased, more vapor bubbles formed and then started to grow upwards and to coalesce, at 1 s and 1.5 s respectively. Finally, the vapor bubbles rose up to the interface between the liquid phase and vapor phase, where they broke up and passed the vapor phase. Moreover, the quantities, size, and shape of the vapor bubbles changed along with the heating input.

Conversely, saturated vapor condensed when it rose up to the condenser section through the adiabatic section. The adiabatic and condenser sections are represented by "Wa" and "Wc" in Figure 5. Condensed liquid film formed on the inner walls of the condenser section, where heat was transferred to the external environment. At the beginning, the film was discontinuous at 5 s, 6 s, 7 s, and 8 s, and the heating input was 40 W (Figure 5a). Based on continuous transportation of vapor from the evaporator section, more liquid was condensed on the inner walls of the condenser section. Eventually, at the heating input of 40 W, a continuous liquid film formed (visualized in Figure 5a) at 9 s and 10 s, and it returned to the liquid pool of the evaporator section through the adiabatic section under gravity.

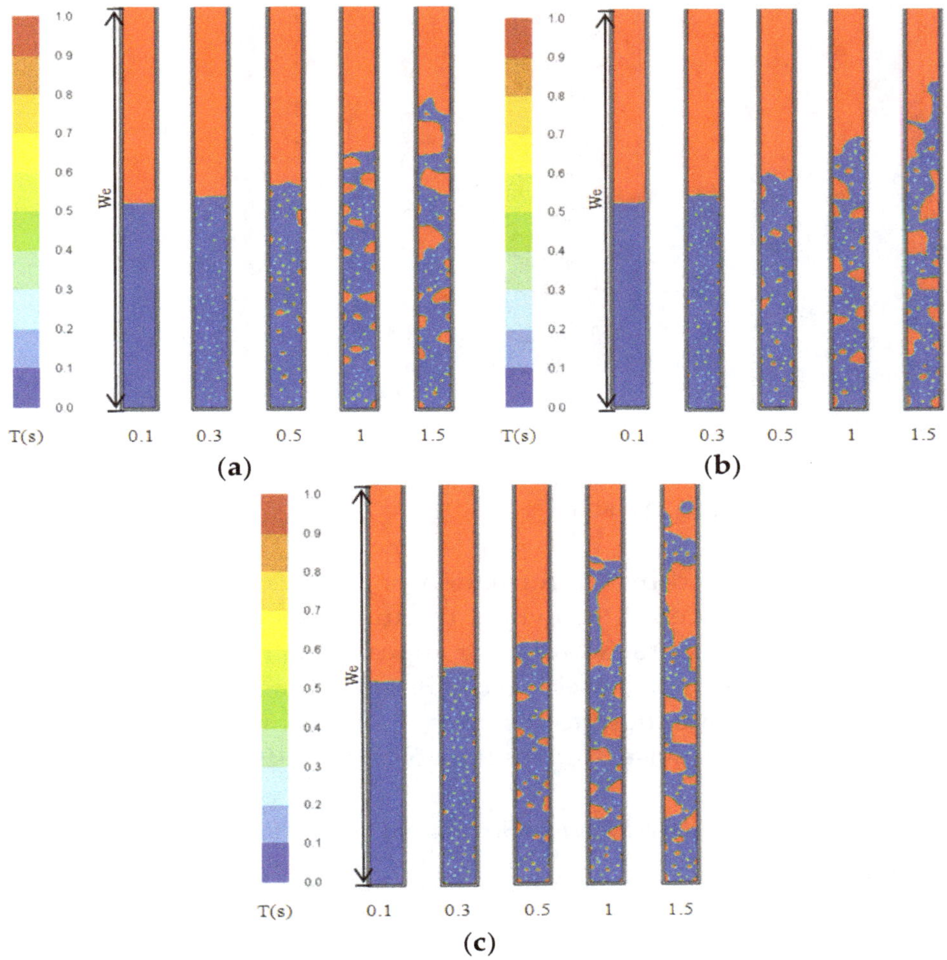

Figure 4. Vapor–liquid distribution of the evaporator section at different heating input powers of (**a**) Q = 40 W, (**b**) Q = 60 W, and (**c**) Q = 80 W.

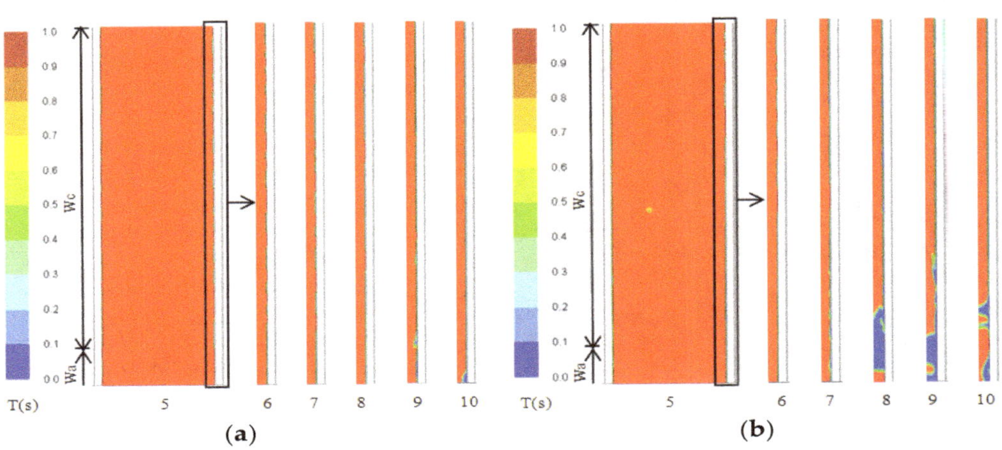

Figure 5. *Cont.*

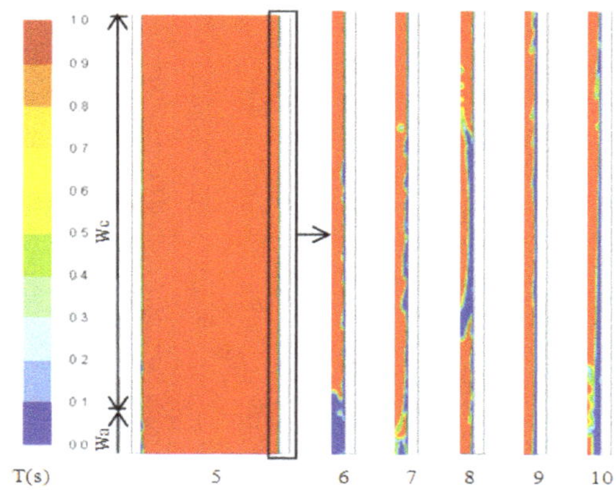

(c)

Figure 5. Liquid film in the lower region of condenser section at different heating input powers of (**a**) Q = 40 W, (**b**) Q = 60 W, and (**c**) Q = 80 W.

Apparently, the time required by the formation of a continuous liquid film is shortened by increasing the heating input power. Meanwhile, the thickness of the liquid film is increased by raising the heating input power at 10 s. More vapor appears and reaches the condenser section with rising heating input power, which increases the condensed liquid and accelerates the formation of a continuous liquid film. The simulation results prove that this numerical model can reproduce the difference in pool-boiling and film-wise condensation between different heating inputs.

5.2. The Wall Temperature Profile of the Numerical Simulation of the Thermosyphon

The average wall temperatures, versus time, for evaporator, adiabatic, and condenser sections are specified in Figure 6. With a rising power input, the average wall temperature of the evaporator section increased between 0 s and 19 s (Figure 6a). By the heat dissipation of the condenser section, the wall temperature of the evaporator section was relatively constant after 19 s, indicating that the thermosyphon needs 19 s to reach stable operation. The wall temperature of the evaporator also increased by elevating the heating input. Meanwhile, at the heating input of 80 W, its wall temperature volatility was also increased. Probably, geyser boiling, which was caused by the larger bubbles in Figure 4c, reduced the stability of the thermosyphon. Compared to the evaporator section, the average wall temperature of the adiabatic section barely varies (Figure 6b). Therefore, the center temperature of the adiabatic section can be specified as the saturation temperature of the thermosyphon.

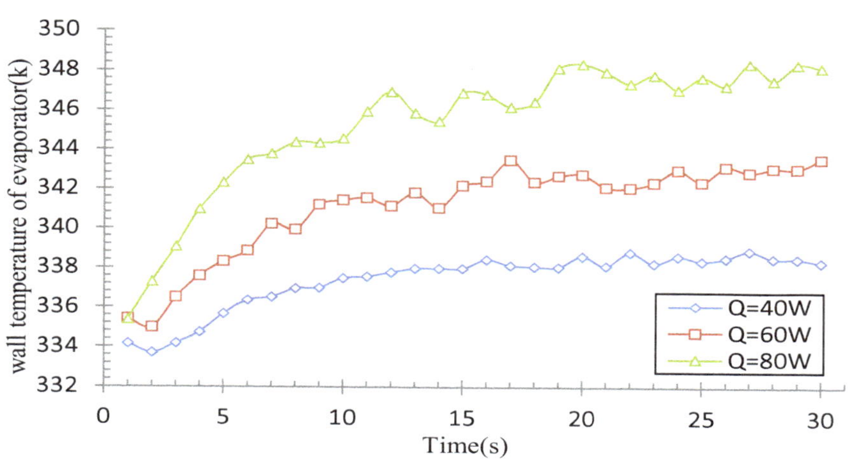

(a)

Figure 6. *Cont.*

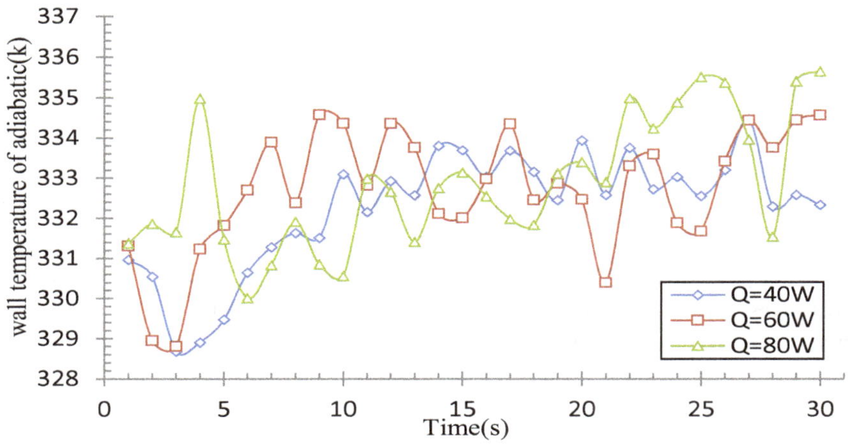

(**b**)

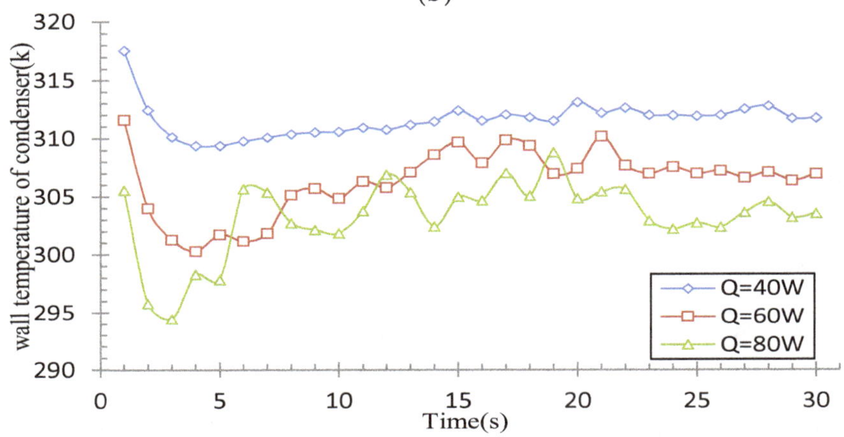

(**c**)

Figure 6. Average wall temperature profiles for (**a**) Evaporator, (**b**) Adiabatic, and (**c**) Condenser sections at different heating input powers.

As shown in Figure 6c, the average wall temperature of the condenser section temporarily decreased from 0 s to 4 s at the heating input of 40 W and 60 W, because a constant heat flux boundary condition was applied on the walls of the condenser section at the beginning of the numerical simulation. The heat of the vapor phase region was transferred to the external environment while that generated by boiling had not yet been transferred to the condenser section. The average wall temperature of the condenser section began to increase when the heat generated by boiling started after 4 s at the heating inputs of 40 W and 60 W. Therefore, this thermosyphon started to work at 4 s at the heating inputs of 40 W and 60 W. In addition, the condenser section started to increase at 3 s at the heating input of 80 W, suggesting that the thermosyphon can work in advance by increasing the heating input.

5.3. The Thermal Resistance of the Numerical Simulation of the Thermosyphon

The thermal resistance of the whole thermosyphon can be defined as follows:

$$R_{ave} = \frac{(T_e - T_c)}{Q_e} \quad (19)$$

where T_e and T_c are the average wall temperatures of the evaporator and the condenser, respectively, and Q_e represents the power input of the evaporator. Based on Figure 7, the thermal resistance increases with extended time and then gradually levels off. In addition, the minimum of thermal resistance is 0.552 K/W, and its decrease is slowed down with increasing heating input (Table 1). r_1 and r_2 have the following forms:

$$r_1 = \frac{R_{40} - R_{60}}{R_{40}} \times 100\% \quad (20)$$

$$r_2 = \frac{R_{60} - R_{80}}{R_{60}} \times 100\% \quad (21)$$

where R_{40}, R_{60}, and R_{80} represent the thermal resistances at the heating inputs of 40 W, 60 W, and 80 W, respectively. The thermosyphon thermal resistances at various power inputs have been compared in this paper; the thermosyphon thermal resistance gradually decreased by increasing the power inputs.

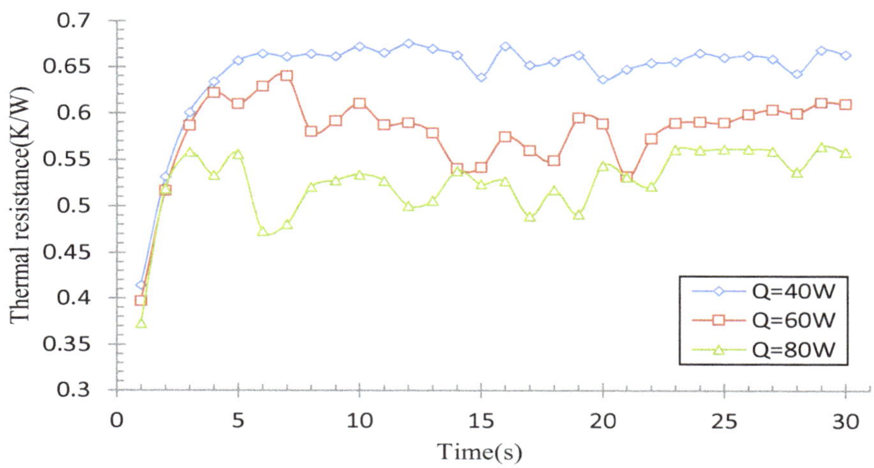

Figure 7. Thermal resistance variations of the thermosyphon at different heating input powers.

Table 1. Thermal resistances of heat pipes at different input powers.

	40 W	60 W	80 W	r_1 (%)	r_2 (%)
R (K/W)	0.656	0.582	0.552	11.3	5.15

5.4. The Effective Thermal Conductivity of the Numerical Simulation of the Thermosyphon

To evaluate the thermal performance of the thermosyphon filled with working fluid, the heat transfer capability is treated as a solid cylinder with equal size. The thermal conductivity of this solid cylinder is specified as the effective one of the thermosyphon with the following correlation:

$$\lambda_{eff} = \frac{4L^2}{d_o^2}[1/(\frac{\ln(d_o/d_i)}{2(L_e/L)} + \frac{\lambda_w}{h_e d_o(L_e/L)} + \frac{\lambda_w}{h_c d_o(L_c/L)} + \frac{\ln(d_o/d_i)}{2(L_c/L)})]\lambda_w \quad (22)$$

where L, L_e, and L_c are the lengths of the total thermosyphon, evaporator, and condenser respectively; d_o and d_i are the outer diameter and inner diameter respectively; and λ_w is the thermal conductivity of the wall. h_e and h_c represent the heat transfer coefficients of the evaporator and the condenser with the following forms, respectively:

$$h_e = \frac{Q}{\pi d_o L_e(T_e - T_s)} \quad (23)$$

$$h_c = \frac{Q}{\pi d_o L_c(T_s - T_c)} \quad (24)$$

where T_e and T_c are the average wall temperatures of the evaporator and the condenser, respectively, and Q represents the power input.

The effective thermal conductivity, versus time, is displayed in Figure 8. Apparently, it is highest initially, and it then decreases with prolonged time. Eventually, the effective thermal conductivity hardly fluctuates. As shown in Table 2, with r_3 and r_4 calculated by (17-1) and (17-2), the effective thermal conductivity increases to a maximum ($2.07 \times 106\ W/m{\cdot}K^{-1}$) with the rising of the heat flux, but the increment amplitude changes little.

$$r_3 = \frac{\lambda_{eff60} - \lambda_{eff40}}{\lambda_{eff40}} \times 100\% \tag{25}$$

$$r_4 = \frac{\lambda_{eff80} - \lambda_{eff60}}{\lambda_{eff60}} \times 100\% \tag{26}$$

where λ_{eff40}, λ_{eff60}, and λ_{eff80} represent the thermal resistances at the heating inputs of 40 W, 60 W, and 80 W, respectively.

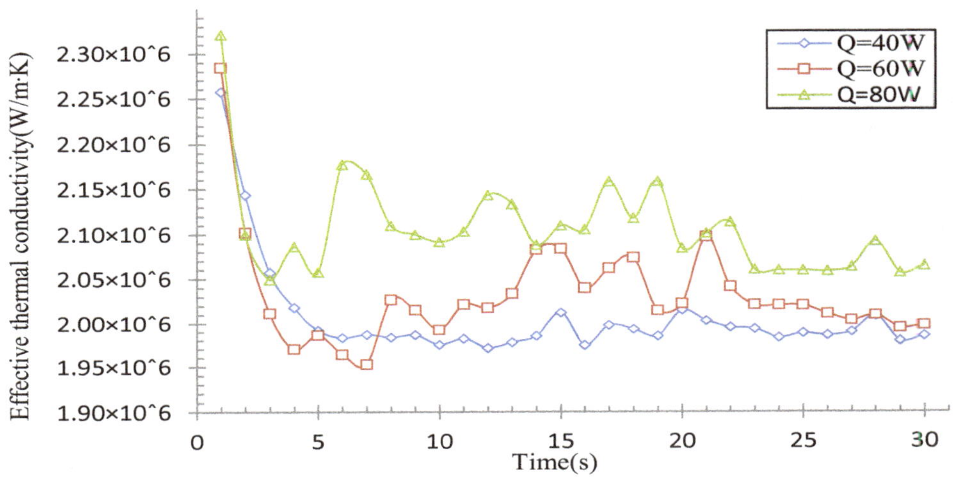

Figure 8. Effective thermal conductivity variations of the thermosyphon at different heating input power.

Table 2. Effective thermal conductivities of heat pipes at different input powers.

	40 W	60 W	80 W	r_3 (%)	r_4 (%)
λ_{eff} (W/m · k)	1.993×10^6	2.032×10^6	2.072×10^6	1.96	1.97

6. Conclusions

A transient two-dimensional model was built with the VOF technique to simulate the complex phenomenon in a thermosyphon. In order to calculate the heat and mass transfer, UDFs were associated with the governing equations of FLUENT by source terms. The CFD results confirmed that this model managed to reproduce pool boiling in the evaporator section and the formation of liquid film in the condenser section during the start-up of thermosyphon. Moreover, the simulation results showed that the quantities, size, shape, and position of vapor bubbles produced by boiling in the evaporator changed along with the heating power at the same time. With rising heating power, the time required by the formation of continuous liquid film was shortened, and the thickness also increased.

The average wall temperatures for condenser, adiabatic, and evaporator sections were investigated at different heating inputs during the operation of the thermosyphon. The prediction results showed that the thermosyphon reached a steady state after 19 s. Variations of the condenser wall temperature suggested that the heat pipes filled with pure water started to work in advance at appropriate heating power, which was around 3 s in the fastest case. We can increase heating power properly to reduce the start-up time and increase the effective thermal conductivity of the thermosyphon.

The thermal performance of the thermosyphon was evaluated by studying the thermal resistance and effective thermal conductivity. With increasing heating power, the thermal resistance decreased to a minimum value (0.552 K/W), whereas the effective thermal conductivity rose to a maximum (2.07 × 106 W/m·K).

Author Contributions: All authors contributed to the paper. Z.Z. designed the research, conceived the idea of the study, and wrote this paper; Y.Z. (Yong Zhang) and Y.Z. (Yanrui Zhang) conducted the numerical simulation and data analyses; Y.Z. (Yimeng Zhou) and H.H. discussed experimental results; all authors performed the experiment.

Acknowledgments: The authors gratefully acknowledge the support of the Jiangsu Marine and Fishery Science and Technology Innovation and Extension project (HY2017-8) and Zhenjiang funds for the Key Research and Development project (GY2016002-1).

Nomenclature

C	Constant of evaporation/condensation
d	Diameter (m)
E	Energy source term (W/m^3)
e	Internal energy (J/kg)
F_{CSF}	Volumetric surface tension force (N)
g	Acceleration of gravity (m/s^2)
I	Unit tensor
k	Surface curvature (1/m)
LH	Latent heat (J/kg)
L	Length (m)
M	Mass transfer rate (kg/s)
p	Pressure (Pa)
T	Temperature (K)
Greek symbols	
α	Volume fraction
λ	Thermal conductivity (W/m K)
μ	Dynamic viscosity (Pa s)
ρ	Density (kg/m^3)
σ_{vl}	Surface tension coefficient (N/m)
Subscript	
CSF	Continuum surface force
c	Condenser
e	Evaporator
eff	Effective
i	Inner
l	Liquid phase
mix	Mixture
o	Out
sat	Saturation
v	Vapor phase
W	Wall

References

1. Faghri, A. Heat Pipe Science and Technology. *Fuel Energy Abstr.* **1995**, *36*, 285. [CrossRef]
2. Weng, Y.C.; Cho, H.P.; Chang, C.C.; Chen, S.L. Heat pipe with PCM for electronic cooling. *Appl. Energy* **2011**, *88*, 1825–1833. [CrossRef]

3. Vasiliev, L.; Lossouarn, D.; Romestant, C.; Alexandre, A.; Bertin, Y.; Piatsiushyk, Y.; Romanenkov, V. Loop heat pipe for cooling of high-power electronic components. *Int. J. Heat Mass Transf.* **2008**, *52*, 301–308. [CrossRef]
4. Zhang, M.; Liu, Z.; Ma, G.; Cheng, S. Numerical simulation and experimental verification of a flat two-phase thermosyphon. *Energy Convers. Manag.* **2009**, *50*, 1095–1100. [CrossRef]
5. Farsi, H.; Joly, J.L.; Miscevic, M.; Platel, V.; Mazet, N. An experimental and theoretical investigation of the transient behavior of a two-phase closed thermosyphon. *Appl. Therm. Eng.* **2003**, *23*, 1895–1912. [CrossRef]
6. Zhang, P.L.; Wang, B.L.; Shi, W.X.; Li, X.T. Experimental investigation on two-phase thermosyphon loop with partially liquid-filled downcomer. *Appl. Energy* **2015**, *160*, 10–17. [CrossRef]
7. Tong, Z.; Liu, X.H.; Jiang, Y. Three typical operating states of an R744 two-phase thermosyphon loop. *Appl. Energy* **2017**, *206*, 181–192. [CrossRef]
8. Tong, Z.; Liu, X.H.; Jiang, Y. Experimental study of the self-regulating performance of an R744 two-phase thermosyphon loop. *Appl. Energy* **2017**, *186*, 1–12. [CrossRef]
9. Naresh, Y.; Balaji, C. Experimental investigations of heat transfer from an internally finned two phase closed thermosyphon. *Appl. Therm. Eng.* **2017**, *112*, 1658–1666. [CrossRef]
10. Jafari, D.; Di Marco, P.; Filippeschi, S.; Franco, A. An experimental investigation on the evaporation and condensation heat transfer of two-phase closed thermosyphons. *Exp. Therm. Fluid.* **2017**, *88*, 111–123. [CrossRef]
11. Chen, L.; Deng, B.L.; Zhang, X.R. Experimental investigation of CO_2 thermosyphon flow and heat transfer in the supercritical region. *Int. J. Heat Mass Transf.* **2013**, *64*, 202–211. [CrossRef]
12. Annamalai, A.S.; Ramalingam, V. Experimental investigation and computational fluid dynamics analysis of air cooled condenser heat pipe. *Therm. Sci.* **2011**, *15*, 759–772. [CrossRef]
13. Legierski, J.; Wiecek, B.; de Mey, G. Measurements and simulations of transient characteristics of heat pipes. *Microelectron. Reliab.* **2006**, *46*, 109–115. [CrossRef]
14. De Schepper, S.C.K.; Heynderickx, G.J.; Marin, G.B. Modeling the evaporation of a hydrocarbon feedstock in the convection section of a steam cracker. *Comput. Chem. Eng.* **2009**, *33*, 122–132. [CrossRef]
15. Lin, Z.; Wang, S.; Shirakashi, R.; Winston Zhang, L. Simulation of a miniature oscillating heat pipe in bottom heating mode using CFD with unsteady modeling. *Int. J. Heat Mass Transf.* **2013**, *57*, 642–656. [CrossRef]
16. Alizadehdakhel, A.; Rahimi, M.; Alsairafi, A. CFD modeling of flow and heat transfer in a thermosyphon. *Int. Commun. Heat Mass* **2010**, *37*, 312–318. [CrossRef]
17. Fadhl, B.; Wrobel, L.; Jouhara, H. Numerical modeling of the temperature distribution in a two-phase closed thermosyphon. *Appl. Therm. Eng.* **2013**, *60*, 122–131. [CrossRef]
18. Jouhara, H.; Fadhl, B.; Wrobel, L. CFD modeling of a two-phase closed thermosyphon charged with R134a and R404a. *Appl. Therm. Eng.* **2015**, *78*, 482–490. [CrossRef]
19. Brackbill, J.; Kothe, D.; Ca, Z. A continuum method for modeling surface tension. *J. Comput. Phys.* **1995**, *100*, 335–354. [CrossRef]
20. Li, J.L.; Li, C.X.; Hu, R.X. *Master Fluent 6.3 Flow Field Analysis*, 1st ed.; Chemical Industry Press: Beijing, China, 2009; pp. 16–55.

8

Numerical Models for Viscoelastic Liquid Atomization Spray

Lijuan Qian [1,*], Jianzhong Lin [1,2,*] and Fubing Bao [1]

[1] College of Mechanical and Electrical Engineering, China Jiliang University, Hangzhou 310018, China; dingobao@cjlu.edu.cn

[2] State Key Laboratory of Fluid Power and Mechatronic Systems, Zhejiang University, Hangzhou 310027, China

* Correspondence: qianlj@cjlu.edu.cn (L.Q.); mecjzlin@zju.edu.cn (J.L.).

Academic Editor: Bjørn H. Hjertager

Abstract: Atomization spray of non-Newtonian liquid plays a pivotal role in various engineering applications, especially for the energy utilization. To operate spray systems efficiently and well understand the effects of liquid rheological properties on the whole spray process, a comprehensive model using Euler-Lagrangian approaches was established to simulate the evolution of the atomization spray for viscoelastic liquid. Based on the Oldroyd model, the viscoelastic linear dispersion relation was introduced into the primary atomization; an extended viscoelastic version of Taylor analogy breakup (TAB) model was proposed; and the coalescence criteria was modified by rheological parameters, such as the relaxation time, the retardation time and the zero shear viscosity. The predicted results are validated with experimental data varying air-liquid mass flow ratio (*ALR*). Then, numerical calculations are conducted to investigate the characteristics of viscoelastic liquid atomization process. Results showed that the evolutionary trend of droplet mean diameter, Weber number and Ohnesorge number of viscoelastic liquids along with axial direction were qualitatively similar to that of Newtonian liquid. However, the mean size of polymer solution increased more gently than that of water at the downstream of the spray, which was beneficial to stable control of the desirable size in the applications. As concerned the effects of liquid physical properties, the surface tension played an important role in the primary atomization, which indicated the benefit of selecting the solvents with lower surface tension for finer atomization effects, while, for the evolution of atomization spray, larger relaxation time and zero shear viscosity increased droplet Sauter mean diameter (*SMD*) significantly. The zero shear viscosity was effective throughout the jet region, while the effect of relaxation time became weaken at the downstream of the spray field.

Keywords: viscoelastic fluid; atomization spray; numerical modeling

---

1. Introduction

Atomization of non-Newtonian liquid has become increasingly prevalent in various engineering applications, such as injection of gelled fuels for propulsion system [1], spray coating for painting [2,3] manufacture of pharmaceutical tablets [4], and materials processing for suspension plasma spray [5].The non-Newtonian droplets can provide attractive and variegated properties to meet the specific requirements. The constitution relation for non-Newtonian liquids are complicated since the relationship between shear stress and rate of strain exhibits non-linear behavior. For the widespread shear-thinning liquids, there are several rheological models to describe the variation of viscosity along with strain rate [6,7], such as power-law model, Carreau model and Oldroyd model [8]. Among them, Oldroyd model with three constants can effectively describe the rheological characteristics for viscoelastic liquids.

For the atomization of non-Newtonian liquid, the external two-phase flow outside of the nozzle orifice can be generally divided into two parts, the near-field region and the far-field domain, where specific physical phenomena govern the flow.

In the near-field region, the primary atomization in which the liquid stream disintegrates into ligaments and drops by interacting with the gas takes place. The studies on primary atomization can be classified into three approaches: experimental measurements, theoretical analysis and modeling calculations.

Based on the techniques of high speed camera and phase Doppler anemometry, experiments provided visualized approaches to reveal the morphology and detailed information of the primary atomization such as breakup length, geometry, spray angle and so on [9]. Dumouchel [10] has reviewed the experimental investigate dedicated to the primary atomization step for Newtonian liquids. Aliseda et al. [11] extracted the images of liquid jet break-up which is closed to the nozzle orifice from high-speed visualizations for several Weber and Reynolds numbers for viscous and non-Newtonian liquids. They observed that at lower Weber numbers the Kelvin–Helmholtz instability grew slowly and several intact wavelengths were observed prior to break-up, In contrast at larger weber number, the Rayleigh-Taylor (R-T) instability was arresting. Acceleration of the interface resulted in the R-T instability creating ligaments of fluid which eventually break-up into droplets. This experimental finding could inspire us to use the R-T instability to describe the breakup of cylindrical ligaments. Harrison et al. [12] experimentally examined the spray cone angle with three types of polymer solutions and studied the cone angle as a function of type of polymer and reduced concentration. Chao et al. [13] evaluated the effect of polymeric additives on spray structure from photographs of the sprays. Their studies show the correlations between additive concentrations and the formation of droplets. These experimental investigations are very helpful for understanding the features of viscoelastic liquid breakup and give some guidance to build up models.

The theoretical analysis of primary atomization mainly focused on the derivation of dispersion relations between the surface wave growth rate and the wave number, which is based on two instability analysis: Rayleigh-Taylor (R-T) and the Kelvin-Helmholtz (K-H) instability. Various breakup modes resulted in different dispersion relations. Sirignano and Mehring [14] have reviewed the basic mechanism of distortion and disintegration of liquid streams. For the non-Newtonian liquid, especially viscoelastic, the linear instability dispersion relationship is more complex than the Newtonian liquid. Middleman [15] firstly performed a linearized stability analysis on viscoelastic jet to predict the stream disintegration. Goren and Gorttlieb [16] expanded the Weber's Newtonian liquid linear stability analysis by involving an equivalent liquid viscosity and including an unrelaxed axial tension into the dispersion equation which can be used to explain the stabilization of the viscoelastic jet. Joseph et al. [17] took the high relative velocity between liquid and the acceleration of droplets into consideration and obtained the dispersion relation based on R-T instability. Yang et al. [18] used linear stability analysis to investigate the instability of a three-dimensional viscoelastic liquid jet surrounded by a swirling air stream.

For the modeling of primary atomization, the direct numerical simulation (DNS) of multiphase flows which is based on the one-fluid formalism coupled with interface tracking algorithms [19–22] is a promising approach to thoroughly investigate the whole process. As reviewed by Villermaux [20] and Gorokhovski et al. [21], the DNS of turbulent primary atomization process is still in its infancy owing to the high computational cost and big numerical challenges. In addition, most of the numerical studies are focused on the Newtonian liquids, few investigations have devoted to viscoelastic liquids since the complicated rheological relations further enhanced the difficulties [22]. Instead, based on some simplifications and assumptions of the gas-liquid interface, and involving the instability analysis, we can find a compromising way to establish a viable physical model and predict the size of ligaments/droplets after primary atomization directly and effectively.

At the far-field domain, the droplets produced by primary atomization are unstable in the turbulent spray, and will undergo a series of events such as secondary breakup, collision and coalescence. For non-Newtonian spray, the effects of viscoelasticity on the droplets evolution and distribution in the downstream field have attracted numerous industrial applications [23,24]. For different applications, the requirements of droplet size and distribution are different, for instance, finer droplets are desirable for combustions and painting spray [25]. On the contrary, in pesticide sprays [26] and anti-misting jet, too small drops are undesirable. The evolution of droplets at the downstream of the spray is critical to the final outcomes. However in the downstream of the spray, there is a scarcity in literature to investigate the evolution and distribution of atomized drops, especially for the atomization of viscoelastic liquids.

In the experimental aspect, most studies devoted to determine the features of non-Newtonian droplet secondary breakup. Rivera [27] used the high-speed photography to find that Non-Newtonian drop secondary breakup mode is qualitatively similar to those observed for Newtonian drops. The only major difference is that the bag stretches more before it ruptures and the resulting breakup fragments persist longer than their Newtonian counterparts. Dechelette et al. [28] experimentally investigated the combined effect of polymers and soluble surfactants on dynamics of jet breakup and formation of satellite drop. Ma et al. [29] found the particle size distribution was fit to the Rosin-Rammler function through 3-D phase Doppler methods. Hartranft and Settles [25] indicated that extensional viscosity had the dominant influence on sheet breakup features compared to Weber number, which means the traditional Weber number may not qualify to describe the breakup of non-Newtonian fluids. These experimental investigations can support us to extend the breakup mode of Newtonian drop to the viscoelastic version.

In the modeling aspects, for the secondary atomization, some investigations have devoted to build up physical models to illustrate droplet breakup. Xue et al. [30] used a two-dimensional computational approach to predict the effect of coupled surfactant and non-Newtonian mechanisms on the formation of satellite drops. Rivera [27] extended the Taylor analogy breakup (TAB) model to describe the inelastic non-Newtonian liquid with power-law model and validated the model with experimental data. It is noted that most of previous models are unable to account for the whole spray pattern and its evolution without considering the poly-dispersion of droplets as well as collision and coalescence [31]. There is very few investigations devoted to establish the numerical models accounting for the evolution of viscoelastic liquid spray [32], due to the complexity of rheological behavior and statistical difficulties.

As such, the motivation of this paper is to logically expand the above mentioned research efforts and aim specially to shed light on the simulations of the downstream of the spray for viscoelastic liquid. The numerical model will consider the primary breakup, secondary breakup, droplet tracking and collision. For primary breakup, Goren and Joseph's dispersion relationship will be used to describe the linear instability of viscoelastic liquid. For secondary atomization, the TAB is extended to the viscoelastic version based on the rheological analysis. For droplet collision, the criterion of coalescence is modified by the rheological properties. The numerical calculations will try to give some detailed information about the evolutions of droplet distribution, size distribution and critical dimensionless numbers (such as Weber number and Ohnesorge number). Results will involve three parts. Firstly, the primary atomization model is validated and compared by experimental data. Then, the evolution of spray field is investigated in detailed by comparison of polymer solution and water. Finally, the effects of rheological properties on spray performance are discussed.

2. Mathematical Models

The mathematical model considers two parts. As shown in Figure 1, the first one is the primary breakup sub-model, which predicts gas velocity at the nozzle exit and the droplet mean size. The second sub-model calculates the evolutions of the gas flow and the droplets spray accounting for secondary breakup, collision and coalescence, momentum transfer. The initial droplets characteristics in the second sub-model are provided by the first sub-model calculation. In droplet-gas two-phase flow, the two-way coupling of gas and droplet phase is considered, while the effect of primary breakup jet on the environment gas is neglected.

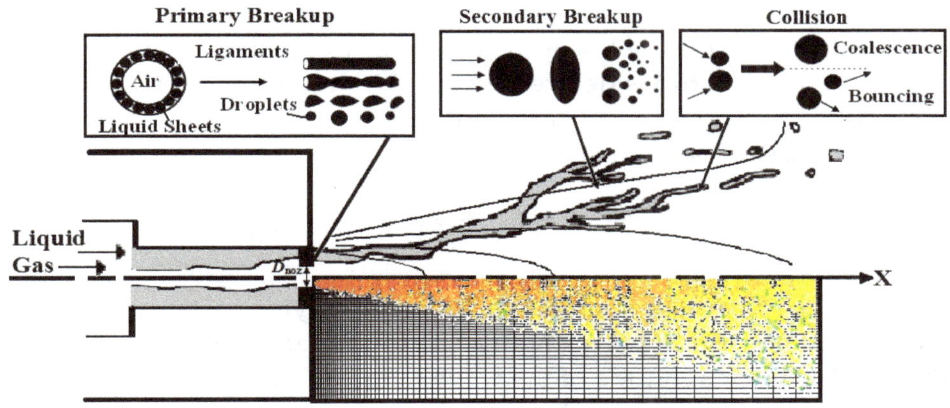

Figure 1. Schematic of the effervescent atomization spray.

2.1. Sub-Model of Primary Atomization

The primary atomization is a process that the liquid stream disintegrates into ligaments and drops by interacting with the gas. It is a complex gas-liquid flow. In order to simplify the process, we extracted the main features of the primary atomization from experimental observations.

At present study, effervescent atomizer is chosen since it has so far been found effective for high viscosity liquids [33], in which the gas is introduced into the liquid at low pressure to form a bubbly two-phase mixture upstream of the orifice. As indicated by experimental observation, the flow pattern closed to the nozzle exit is similar to annular. Then, as illustrated in Figure 1, the primary breakup sub-model for effervescent atomizer can be established through three steps.

The first step is estimation of the annular sheets as well as the cylindrical ligaments. The momentum equation of the annular flow combined with the state equation can be written as:

$$RT\ln\left(\rho_g RT\right)+\frac{1}{2}\left(\frac{\dot{m}_l ALR}{\rho_g \pi r_g^2}\right)^2=\text{const} \tag{1}$$

where ρ_g is the gas density, r_g is the radius of gas flow, $\dot{m}_l$ is the mass flow rate of liquid, and ALR is the air-liquid ratio by mass. The radius of gas flow can be written in terms of orifice radius r_{orific} and void fraction ϑ, $r_g=\sqrt{\vartheta}r_{\text{orifice}}$. The interface velocity slip ratio v_{slip} under different flow rate is expressed as [34]:

$$v_{\text{slip}}=\sqrt{\frac{\rho_l}{\rho_g}\frac{\sqrt{\vartheta}}{1+C(1-\vartheta)}} \tag{2}$$

where C is the experimental coefficient of the mass flow rate scaling. The interface velocity slip ratio and void fraction are related by:

$$1+\frac{\rho_g sr}{\rho_l ALR}=\frac{1}{\vartheta} \tag{3}$$

By solving Equations (1)–(3), ρ_g, ϑ and v_{slip} can be calculated for different operating conditions. The thickness of annular liquid sheet is then calculated from $\delta_{lig}=4(r_o-r_g)/\sqrt{\pi}$, which is also the diameter of the typical cylindrical ligament. Besides, the gas velocity inside the nozzle orifice can be expressed as $V_{g1}=\dot{m}_g/(A_{noz}\times\vartheta\times\rho_g)$, where A_{noz} is the area of the nozzle orifice and $\dot{m}_g$ is the gas flow rate. According to the energy conservation equation, the gas velocity at the nozzle exit can be derived from $V_{g2}=\sqrt{{V_{g1}}^2+2RT\log(P_{in}/P_{out})}$, where R is the gas constant, T is the temperature, P_{in} is the pressure inside the nozzle, and P_{out} is the pressure outside the nozzle.

Secondly, the formed ligaments are approximated by cylindrical jets. The ligaments will breakup into fragments at the wavelength of the most rapidly growing wave. The dispersion relations based on instability analysis are used to determine the critical wave number. Here, Joseph's dispersion relationship [17] is used to describe the linear instability of viscoelastic liquid.

Joseph's dispersion relationship is derived by considering the acceleration of a drop exposed to a high speed air stream. The equations of ligament motion coupled with boundary conditions are solved to deduce the expression of the disturbed interface:

$$-\left[1+\frac{1}{\alpha^2}(-\dot{a}k+\frac{\sigma k^3}{\rho_l})\right]-4\frac{k^2}{\alpha}\frac{\mu_{ve}}{\rho_l}+4\frac{k^3}{\alpha^2}(\frac{\mu_{ve}}{\rho_l})^2(\sqrt{k^2+\alpha\rho_l/\mu_{ve}}-k)=0\left(\rho_l>>\rho_g\right) \quad (4)$$

Equation (4) is an approximate analysis of Rayleigh-Taylor instability based on viscoelastic potential flow which gives the critical wavelength and growth rate to within less than 10% of the exact theory. In Equation (4), k is the magnitude of the wave number, α is the amplification rate, and σ is the liquid surface tension. μ_{ve} is the effective shear viscosity of the liquid, $\mu_{ve}=\mu_0(1+\alpha\lambda_2)/(1+\alpha\lambda_1)$ (Section 2.2.2), for viscoelastic liquid, μ_{ve} is large and $\alpha\rho_l/(k^2\mu_{ve})<<1$, then $(\sqrt{k^2+\alpha\rho_l/\mu_{ve}}-k)$ can be assumed to be $\alpha\rho_l/2k\mu_{ve}$. $\dot{a}$ is the acceleration of the liquid ligament:

$$\dot{a}=\frac{F_{c_d}}{m_{lig}}=\frac{c_d\rho_g(u_g-u_l)^2}{\rho_l\delta_{lig}} \quad (5)$$

where c_d is the drag coefficient. As the ligament Reynolds number $Re_{lig}=\rho_g\delta_{lig}|u_g-u_l|/\mu_g$ is between 3 and 1000, here $c_d=24(1+{Re_{lig}}^{2/3}/6)/Re_{lig}$. Then, Equation (4) reduces to be a three-degree polynomial of α as follows and can be solved analytically:

$$\lambda_1\rho_l\alpha^3+(\rho_l+2\mu_0\lambda_2k^2)\alpha^2+(2\mu_0k^2-\dot{a}k\rho_l\lambda_1+\sigma k^3\lambda_1)\alpha-\dot{a}k\rho_l+\sigma k^3=0 \quad (6)$$

Based on the above dispersion relation, the critical breakup wave number that corresponds to the maximum growth rate can be used to calculate the wavelength of a typical ligament fragment. Finally, assuming that each fragment is stabilized to one spherical droplet under the influence of surface tension, the Sauter mean drop diameter SMD (defined as the ratio of volume–surface mean diameter $\sum_i n_pd_p^3/\sum_i n_pd_p^2$) can then be calculated from the conservation of mass:

$$SMD=\left[3\pi\frac{{\delta_{lig}}^2}{k}\right]^{1/3} \quad (7)$$

2.2. Sub-Model of Droplet-Gas Two-Phase Flow

The second sub-model is based on hybrid Euler-Lagrangian coordinate system to describe the evolution of spray. The flow field is axisymmetric, and the parcels are tracked in three-dimensional coordinate.

2.2.1. Gas Phase

For the turbulent gas spray, the conservation equations of mass, momentum and energy are written as follows:

$$\begin{cases}\frac{\partial\rho_g}{\partial t}+\nabla\cdot\left(\rho_g\mathbf{u}_g\right)=0\\ \frac{\partial\left(\rho_g\mathbf{u}_g\right)}{\partial t}+\nabla\cdot\left(\rho_g\mathbf{u}_g\mathbf{u}_g\right)=-\nabla\left(p+\frac{2}{3}\rho_g\kappa\right)+\nabla\cdot[\sigma_s]+\mathbf{F}\\ \frac{\partial\left(\rho_ge\right)}{\partial t}+\nabla\cdot\left(\rho_ge\mathbf{u}_g\right)=-p\nabla\cdot\mathbf{u}_g-\nabla\cdot\mathbf{q}+\rho_g\varepsilon+Q\end{cases} \quad (8)$$

where $\mathbf{u}_g$ is the gas velocity vector, $\mathbf{F}$ is the momentum transfer between the gas and the discrete liquid phase, $\mathbf{K}$ is the turbulent kinetic energy per unit mass, σ_s is the viscous stress tensor, $\mathbf{q}$ is the heat flux vector, ε is the viscous dissipation rate, and Q is the heat transfer between the gas and the liquid phase. Turbulence effects were modeled by the standard $k-\varepsilon$ model. F and Q are defined as: $\mathbf{F} = -(\alpha_g V_{cell})^{-1} \sum_P N_P \mathbf{F}_\mathrm{p}$, $Q = (\alpha_g V_{cell})^{-1} \sum_p N_p [Q_p + \mathbf{F}_\mathrm{p}(\mathbf{u}_g - \mathbf{u}_l)]$, where α_g is the fraction of gas in the computational cell, V_{cell} is the cell volume, N_p is the number of droplets in this parcel, and $\boldsymbol{F}_p$ is the force acting on each droplet, Q_p is the heat flux through surface of each droplet, and $\mathbf{u}_l$ is the droplet velocity.

For the discrete droplets, the momentum equation can be given as: $\mathbf{F}_\mathrm{p} = \pi r_l^2 \rho_l C_D |\mathbf{u}_g - \mathbf{u}_l| (\mathbf{u}_g - \mathbf{u}_l)/2$, where $\mathbf{F}_\mathbf{p}$ is the force exerted on droplet, m_l is the mass of droplet, r_l is the radius of droplet, g is the gravity force, and C_D is the coefficient of the drag force and can be expressed as: $C_D = 24/Re_l + 6(1 + Re_l^{-1/2}) + 0.4$. Here the droplets are assumed to be spherical, and the effect of droplet deformation on drag coefficient is not taken into consideration.

2.2.2. Liquid Phase

Secondary Atomization

In this section, we first analyze the constitutive relations of the Newtonian and viscoelastic liquids, introducing more physical properties to describe the viscoelastic contributions. Then, the TAB model which is based on Taylor's analogy between an oscillating droplet and a spring-mass system was extend to a non-Newtonian version to describe the deformation and distortion of viscoelastic liquids.

For the Newtonian fluid, the shear stress τ_{ij} can be expressed as:

$$\tau_{ij} = -\mu d_{ij} \tag{9}$$

where μ is the Newtonian viscosity, and d_{ij} is the rate of deformation tensor. In cylindrical coordinates, d_{ij} with (i,j) component can be taken as: $d_{ij} = \partial u_r/\partial z + \partial u_z/\partial r$.

For the viscoelastic fluid, according to Jeffreys' study [35], the rheological model can be expressed with three constants $(\mu, \lambda_1, \lambda_2)$ as follows:

$$\tau_{ij} + \lambda_1 \frac{\partial \tau_{ij}}{\partial t} = -\mu(d_{ij} + \lambda_2 \frac{\partial d_{ij}}{\partial t}) \tag{10}$$

Using the kinetic theory of transient mechanical response in small amplitude motions, the variables (u_r, u_z) in Equations (9) and (10) can be written as $f(r,z,t) = F(r)e^{ikz+\alpha t}$, where k is the wave number, and α is the growth rate of a perturbation. As deduced by Middleman [15], Equations (9) and (10), respectively become the following:

$$T_{rz}(r) = -\mu(ikU_r(r) + \frac{dU_z(r)}{dr}) \tag{11}$$

$$T_{rz}(r) = -\mu \frac{1+\alpha\lambda_2}{1+\alpha\lambda_1}(ikU_r(r) + \frac{dU_z(r)}{dr}) \tag{12}$$

By comparing the above two dynamical equations, it concludes that they are identical, the Newtonian viscosity μ can be replaced by $\mu_{ve} = \mu(1+\alpha\lambda_2)/(1+\alpha\lambda_1)$ for the viscoelastic model. Based on the above analysis, the original TAB model proposed by Amsden et al. [36] can be used as a foundation for the viscoelastic version. The derivation of viscoelastic version of TAB model include three aspects as below.

First is the modification of deformation equation. The TAB model is based on the Raleigh–Taylor instability analysis, describing the droplet distortion by a forced, damped, harmonic oscillator in which the forcing term is given by the aerodynamic interaction between the droplet and gas; the damping is due to the liquid viscosity, and the restoring force is supplied by the surface tension. Droplet distortion is denoted by the deformation parameter, $\zeta = 2\tau/r$, where τ denotes the change of radial cross-section from its equilibrium position and r is the initial drop radius. The deformation equation of ζ in terms of the normalized distortion parameter is:

$$\ddot{\zeta} + \frac{5\mu_l}{\rho_l r^2}\dot{\zeta} + \frac{8\sigma}{\rho_l r^3}\zeta = \frac{2\rho_g |U_{rel}|^2}{3\rho_l r^2} \tag{13}$$

where ρ_l is the density, μ_l is the viscosity, σ is the surface tension, U_{rel} is the relative drop–gas velocity, and the subscripts g and l denote the gas and liquid, respectively. It is assumed that the external aerodynamic force is normal to the drop and the turbulence fluctuations of the surrounding gas are ignored. In reality, there are various modes when droplets breakup as they are subjected to the stresses of the surrounding turbulence gas. However, it is impossible to describe all secondary breakup modes in a single model, therefore only the fundamental mode of oscillation is considered in TAB model.

For a viscoelastic version of TAB model, the constant Newtonian shear viscosity μ_l was replaced by a viscoelastic viscosity μ_{ve}. In the expression of μ_{ve}, the growth rate of a perturbation α can be taken by the rate of strain $\dot{\gamma}$:

$$\mu_{ve} = \mu_0 \frac{1 + \dot{\gamma}\lambda_2}{1 + \dot{\gamma}\lambda_1} \tag{14}$$

where λ_1 is the fluid relaxation time, λ_2 is the retardation time, and μ_0 is the zero shear viscosity. The rate of strain $\dot{\gamma}$ is approximated as $\dot{\gamma} = U_{rel}/2r_0$. As indicated in Equation (14), on the one hand, μ_{ve} will approach zero shear viscosity μ_0 when the rate of strain $\dot{\gamma}$ is close to 0. On the other hand, μ_{ve} will approach $\mu_0\lambda_2/\lambda_1$ when $\dot{\gamma}$ becomes larger. Therefore, when $\lambda_2 < \lambda_1$ and the rate of strain is increasing, $\mu_{ve} < \mu_0$, which means that the liquid is shear thinning. For dilute polymer solutions, the relationship of retardation time and relaxation time is $\lambda_2 = \lambda_1 \eta_{solvent}/\eta_{solution}$ [37], where $\eta_{solvent}$ and $\eta_{solution}$ are the viscosity of solvent and solution, respectively.

Second, the breakup criterion is changed. In TAB model, drop breakup occurs when the normalized drop distortion $\zeta(t)$ exceeds the critical value of 1. For Newtonian droplet, the stability limit has been determined experimentally to be $We_g = \rho_g U_{rel}^2 r/\sigma = 6$. For high viscous liquid, the effect of viscosity on the stability should be related to the breakup criterion. As proposed by Brodkey [38], the critical Weber number can be modified by introducing the Ohnesorge number:

$$We_{g.crit} = 6(1 + 1.077Oh^{1.6}),\ Oh = \mu_l/\sqrt{2\rho_l \sigma r} \tag{15}$$

when $Oh \leq 0.1$, the value of this expression approximates to 6, which means the case of low viscosity for Newtonian liquid is recovered. When Oh is higher, the stability limit is increased and scales as $\mu_l^{1.6}$ in the asymptotic limit. Based on the Brodkey's empirical relation, an effective Weber number is introduced to instead of the regular one [38]:

$$We_g^{eff} = \frac{We_g}{1 + 1.077Oh^{1.6}} \tag{16}$$

for the viscoelastic liquid, μ_{ve} is used instead of μ_l in Oh number.

Third, the creation of the product droplets is also revised. The population dynamics is used to define the rate of droplet creation. For each breakup event, with the mass conservation principle, it is assumed that the number of product droplets is proportional to the number of critical parent drops, where the proportionality constant depends on the drop breakup regime:

$$\frac{d}{dt}\overline{m}(t) = -3K_{bu}\overline{m}(t) \tag{17}$$

where $\overline{m}(t)$ denotes the mean mass of the product drop distribution, and the breakup frequency K_{bu} depends on the drop breakup regimes. Breakup frequency $K_{bu} = 0.05\omega$ as suggested by Amsden et al. [36]

is used. The drop oscillation frequency ω is given by $\omega^2 = 8\sigma/(\rho_l r^3) - 25\mu_l^2/(4\rho_l^2 r^4)$, where μ_l is also replaced by μ_{ve} for viscoelastic version.

Droplets Collision

Droplet collision is calculated by a statistical method, rather than a deterministic method since the real collisions process is highly stochastic. All the droplets in one computational parcel behave in the same manner, they either do or do not collide, which is depended on the probability of the collision whether larger than a random number. For collision result, an impact parameter, $\psi = \chi/(r_1 + r_2)$ is involved to judge the outcomes, where χ is the distance between the center of one drop and the relative velocity vector U_{rel}, r_1 and r_2 are the radii of small and large droplets, respectively. The criteria, χ_{cr} determines the transition boundary: drops coalescence when $\chi \leq \chi_{cr}$, and bounce when $\chi > \chi_{cr}$. χ_{cr} can be expressed as:

$$\chi_{cr}{}^2 = (r_1 + r_2)^2 \min[1.0, 2.4f(\gamma)/We_l] \tag{18}$$

where the collision Weber number for Newtonian liquid defined as: $We_l = \rho_l|\boldsymbol{U}_{rel}|r_1/\sigma_l$. For viscoelastic liquid, $We_l^{eff} = We_l/(1 + 1.077Oh^{1.6})$ is used instead of We_l. $f(\gamma)$ is a function of the radius ratio γ: $f(\gamma) = \gamma^3 - 2.4\gamma^2 + 2.7\gamma$ ($\gamma = r_2/r_1 \geq 1$).

According to the conservation of momentum, the drop velocity after a bouncing collision is [36]: $u_{1new} = \{u_1 m_1 + u_2 m_2 + m_2(u_1 - u_2)[(\chi - \chi_{cr})/(r_1 + r_2 - \chi_{cr})]\}/(m_1 + m_2)$. The droplet velocity after coalescence is calculated as $u_{new} = (u_1 m_1 + u_2 m_2)/(m_1 + m_2)$. m_1 and m_2 are the mass of the small and large droplets, respectively; and u_1 and u_2 are the velocity of the small and large droplets, respectively. The new radius after droplet coalescence is $r_{new} = (r_1^3 + r_2^3)^{1/3}$.

3. Numerical Setup

Figure 2 shows the computational mesh and boundary conditions of numerical model. The radius and axial lengths of the computational regime are 6 cm and 15 cm respectively, and 2π in the circular direction. The computational domain is in a cylindrical coordinate system with size of 56 (radial) $\times$ 65 (axial) $\times$ 32 (azimuthal). Finer meshes are used for the core region of the spray. The grid dependency has been validated in our previous studies [39,40] and the computational results are convergence under present mesh size.

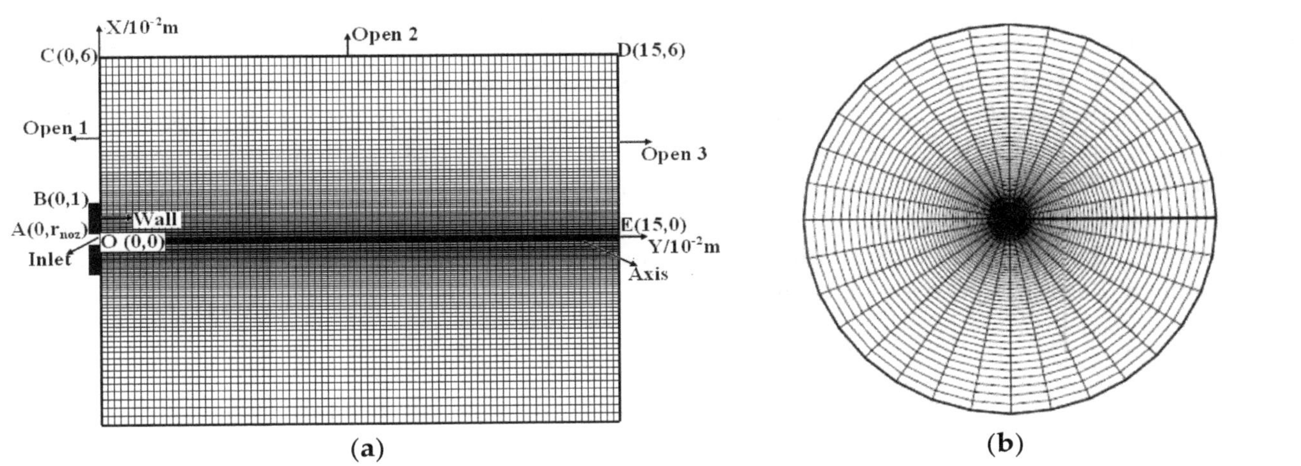

Figure 2. Geometry and computational mesh: a middle section plane. (**a**) Middle section plane; and (**b**) Cross section plane.

As presented in Table 1, there are five types of boundary conditions involved for the gas field. The parcels are introduced in the system at the nozzle exit, which has the original size calculated from primary breakup sub-model with Rosin-Rammler distribution: $F_{R-R}(D) = 1 - \exp\{-(D/SMD_0)^{q_{co}}\}$, where q_{co} indicates the value of the width of the distribution. The Rosin-Rammler distribution is a prevalent size distribution used in the atomization field [41], and the experimental data also supported the Rosin-Rammler function for non-Newtonian fluid atomization spray [31]. The number of parcels should be large enough to reduce statistical fluctuation. Usually, a total number of 10,000 computational parcels are injected into the flow for each case.

Table 1. Five types of boundary conditions involved in the model.

Type		Location	Conditions
Open	1	Line BC	$v(x,0,\theta) > 0$: P, T, u are assigned the same as that in ambient $v(x,0,\theta) \leq 0$: $\partial u/\partial y = 0$, $\partial\phi/\partial y = 0$ ($\phi = P, T$)
	2	Line CD	$u(6,y,\theta) < 0$: P, T, v are assigned the same as that in ambient $u(6,y,\theta) \geq 0$: $\partial\phi/\partial x = 0$, $\partial v/\partial y = 0$
	3	Line DE	$v(x,15,\theta) < 0$: P, T, u are assigned the same as that in ambient $v(x,15,\theta) \geq 0$: $\partial u/\partial y = 0$, $\partial\phi/\partial y = 0$
Wall		Line AB	$u, v, w = 0$; $T = 300$ K; turbulent wall function
Inlet		Line OA	Data from primary breakup sub-model
Axis		$x = 0$	$\xi\|_{x=0} = (\sum_{i=1}^{N} \xi_i)/N$ ($\xi = u, v, T$, ξ_i: ξ at Δx from the centerline); $w = 0$
Periodic		$\theta = 2\pi$	Circular direction: periodic boundary conditions

4. Results and Discussion

4.1. Validation of Primary Atomization

Figure 3 shows a comparison of experimental data [42] and model predictions for polymer solution atomization at $P_{inj} = 0.35$ MPa, $D_{noz} = 1$ mm and ALR ranging from 0.02 to 0.11. The physical and rheological properties are listed in Table 2. The polymer solution were formulated by adding poly(ethylene oxide) to solvent having a common composition of 60% glycerine–40% water solution by mass. Polyethylene oxide (PEO) molecular weights of 100,000 were used at mass concentrations of 0.001%, 0.01% and 0.15% [42]. The dispersion relations proposed by Joseph are solved analytically. Results show that, with an increase in air–liquid mass flow ratio (ALR), droplet SMD will decline significantly, especially at lower ALR. The agreement between the predictions and measurements is qualitative achieved within 5%–20%. Joseph's model is effective to capture the influence of rheological properties on the breakup process of polymer solution and to quantitatively predict the resulting droplet sizes. However the shortcoming is that the divergence may occur when the liquid ligament is not similar to the flattened drop. Therefore, at lower ALR, the divergence is obviously since the liquid ligaments is slender.

Table 2. Physical properties of the fluids in Figure 3 [42].

Code	Polymer Concentration	Zero Shear Viscosity η_0	Relaxation Time λ_1	Surface Tension σ
		mPa·s	s	10^{-3} J/m^2
1	0.001%	11	2.8×10^{-7}	65.1
2	0.01%	11.1	1.4×10^{-5}	62.6
3	0.15%	15.4	2.06×10^{-4}	62.6

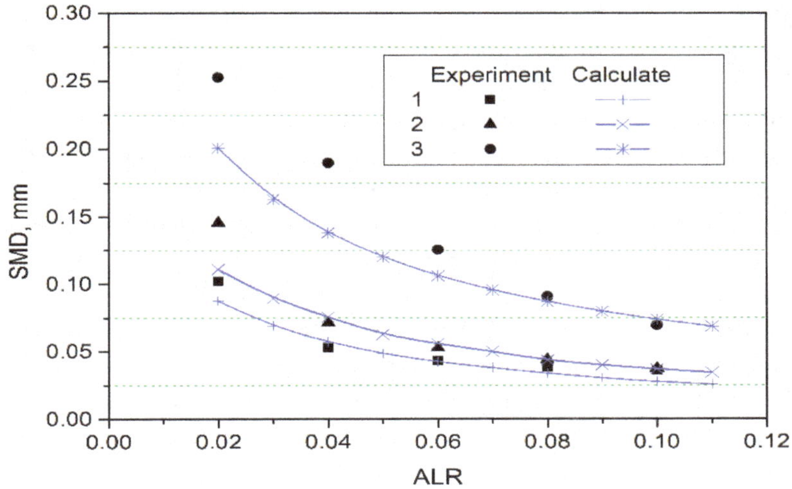

Figure 3. Comparison of experimental data and model predictions for the Sauter mean diameter (*SMD*) versus air-liquid mass flow ratio (*ALR*) at $P_{inj} = 0.35$ MPa, $D_{noz} = 1$ mm.

4.2. Evolution of Spray Field

Among the investigations of the far-field region, the particle size evolution along with the axial distance is the primary interest. Figure 4 firstly validates the numerical results of water's *SMD* with experimental data [43] and then compares the downstream variation of *SMD* for water and polymer solution. Both experiment and simulations were conducted based on the same operating conditions. The operating parameters and liquid physical properties are listed in Table 3.

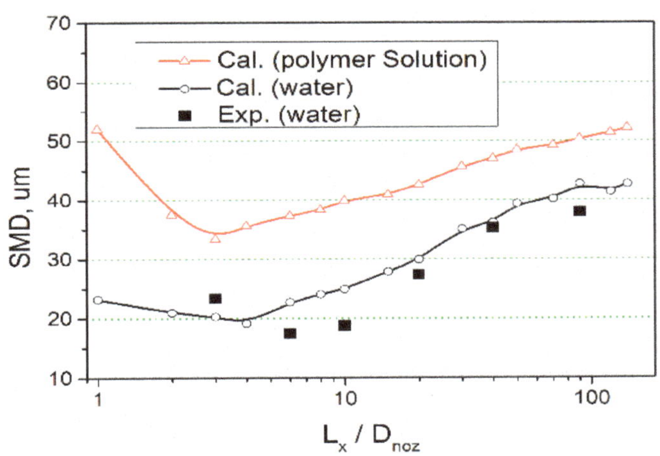

Figure 4. Droplet Sauter mean diameter along with the dimensionless axial distance for water and polymer solution.

Table 3. Operating parameters and liquid physical properties of baseline case.

Parameters	Value	Parameters	Value
D_{noz} (mm)	1	η_0 (10^{-3}Pa·s)	15
σ (10^{-3}J/m^2)	65	$\eta_{solvent}$ (10^{-3}Pa·s)	10
ALR	0.06	λ_1 (s)	2.0×10^{-6}
$\dot{m}_l$ (g/s)	10	P_{inj} (10^6 Pa)	0.35

As shown in Figure 4, an agreement between the simulation results and experimental data has been obtained for the case of water, and the divergence of the experiment and simulation is between 2.5% and 25%. In Figure 4, the viscoelastic liquid added the difficulties in the breakup of ligaments and droplets, impairing the atomization effect by increasing the drop size in the whole process, which is consistent with the results of other researchers [12,44]. Further, in both the cases of water and polymer solution, the droplet diameter first decreases and then increases with the axial distance. This phenomenon can be explained by the competition between breakup and coalescence effects. Figure 5 shows the occurrence number of droplets breakup and collision. Close to the nozzle exit, with high kinetic energy and large velocity difference between the droplet and atomization gas, turbulence breakup dominates, causing the drop size to rapidly decrease, while, further downstream, as the kinetic energy decay and small velocity difference, turbulence breakup ends and droplet coalescence plays an important role, causing the drop size to gradually increase.

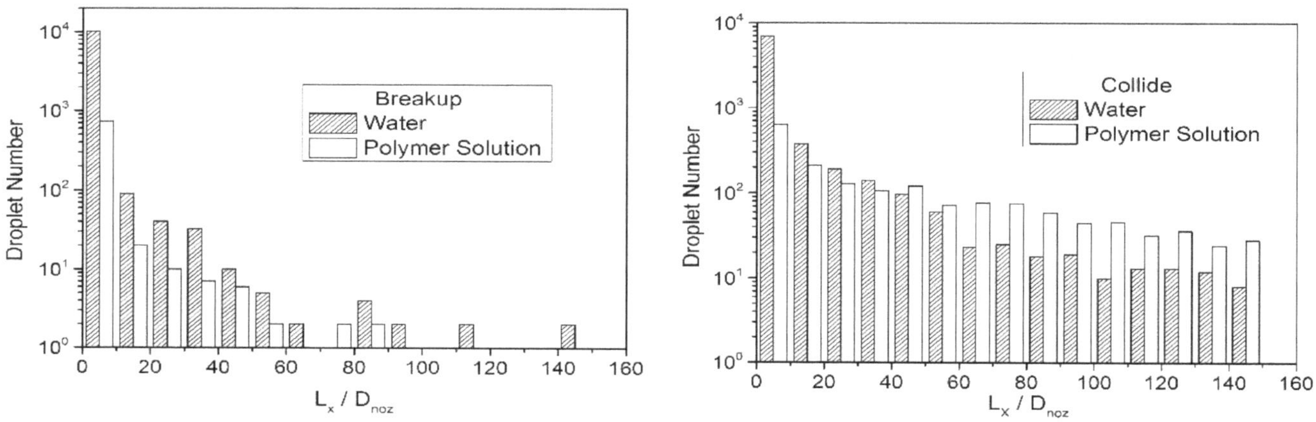

Figure 5. Occurrence number of droplets breakup and collision.

Besides the common and well-known points, there are two noteworthy differences between these two types of liquids. First, in Figure 4, the gap between the mean size of water and polymer solution is shortened along with the development of the jet. Before the nadir of the curve, the case of polymer solution decays more steeply than that of water. In contrast, after the nadir of the curve, the case of water rises more obviously than that of polymer solution. On the rising part of the curves, the increasing degree of water is about 123%, while the case of polymer solution is less than 57% (this value is calculated from $(SMD_{L_x/D_{noz}=140} - SMD_{L_x/D_{noz}=4})/SMD_{L_x/D_{noz}=4}$). This phenomenon indicates that the mean size of polymer solution increases more gently than that of water at the axial distance $(5\text{–}100)L_x/D_{noz}$, which is benefit to control the desirable size in industrial applications. Secondly, in Figure 5, the occurrence numbers for water are higher than polymer solutions for both breakup and collision process in the upstream of the spray. This is because droplets breakup time and frequency depend on the competition between the aerodynamic force and restoring force. An increase in liquid viscosity will delay and weaken the breakup process, resulting in higher droplet size. The decrease of the frequency of droplet secondary breakup would also lead to the number of product droplets decreasing, therefore the collision occurrence number of polymer solutions is obviously less than that of water in the upstream region. However, when the axial distance beyond $40L_x/D_{noz}$, the collision occurrence number for polymer solution exceeds that of water. Since the droplets collision process were influenced by the droplet size and concentration rate. Concentrated drop distribution causes the collision happening frequently. The outcome of collision was also affected by the liquid viscosity. For the polymer solution droplets, impact parameter increased due to higher viscosity, leading to an increase in possibility of droplet coalescence after collision. However, the change of collision occurrence number did not influence the trend of *SMD* in Figure 4 significantly, due to the occurrence number is too small compared to the total particle number.

The most important factors influencing the droplet breakup can be grouped into two dimensionless numbers, the Weber number (We) and the Ohnesorge number (Oh). The average Weber number is defined as $We = \rho_g |\overline{u}_{rel}|^2 SMD/\sigma_l$. Here the modified weber number (Equation (16)) is used for the polymer solution. The average Ohnesorge number is expressed as $Oh = \mu_l/\sqrt{\rho_l \sigma_l SMD}$. For viscoelastic liquid, the constant viscosity μ_l was replaced by a viscoelastic viscosity μ_{ve}.

The evolutions of Weber and Ohnesorge number along with the axial direction are illustrated in Figure 6. Weber number indicates the ratio of the inertial force to the surface tension force, in which the trend is basically determined by the gas-liquid relative velocity. At the nozzle exit, We is higher due to the large difference between the velocity of droplet and gas, causing the secondary breakup of droplets. In this period, We of polymer solution is larger than that of water due to the larger SMD at the beginning. Then at downstream of the spray, We dramatically declines as the kinetic energy dissipated and We of polymer solution is less than that of water since We for viscoelastic liquid is modified by the effective viscosity. The effective Weber number decreased as liquid viscosity increasing. The Ohnesorge number is the ratio of an internal viscosity force to an interfacial surface tension force, which significantly increases with the increase of liquid viscosity. For both water and polymer solution, the Oh curves have a peak at about 4 mm from the nozzle exit, which is corresponding to the nadir of the curve of droplet size (as shown in Figure 4). In short, the case of polymer solution has a greater magnitude of change for both We and Oh due to introducing the changes in viscosity.

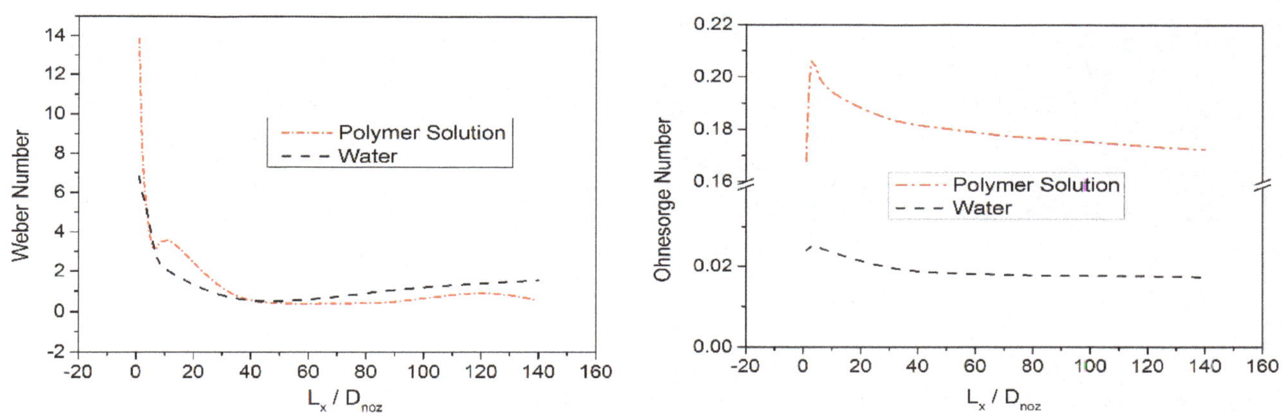

Figure 6. Evolutions of Weber and Ohnesorge number along with axial direction for water and polymer solution.

In addition to drop size and critical numbers, liquid viscosity can also affect the spray pattern. Figure 7 shows the spatial distributions of droplet size and number for water and polymer solution, respectively. The cross-sectional plane is located at the axial distance of 50 mm. It can be seen from Figure 7 that the spatial distribution of droplets becomes wider for water. The distribution range of water case is about 40% larger than the polymer solution. It can be explained as the ligaments and fragments are more likely to adhere due to viscoelastic properties. Furthermore, finer droplets are easier for the gas to accelerate and carry them outward of the centerline, resulting in the spray angle will expand. In turn, the distribution will also affect the droplet size. Concentrated drop distribution causes the collision happening frequently and the droplet size increasing, which is in accordance with that indicated in Figure 5.

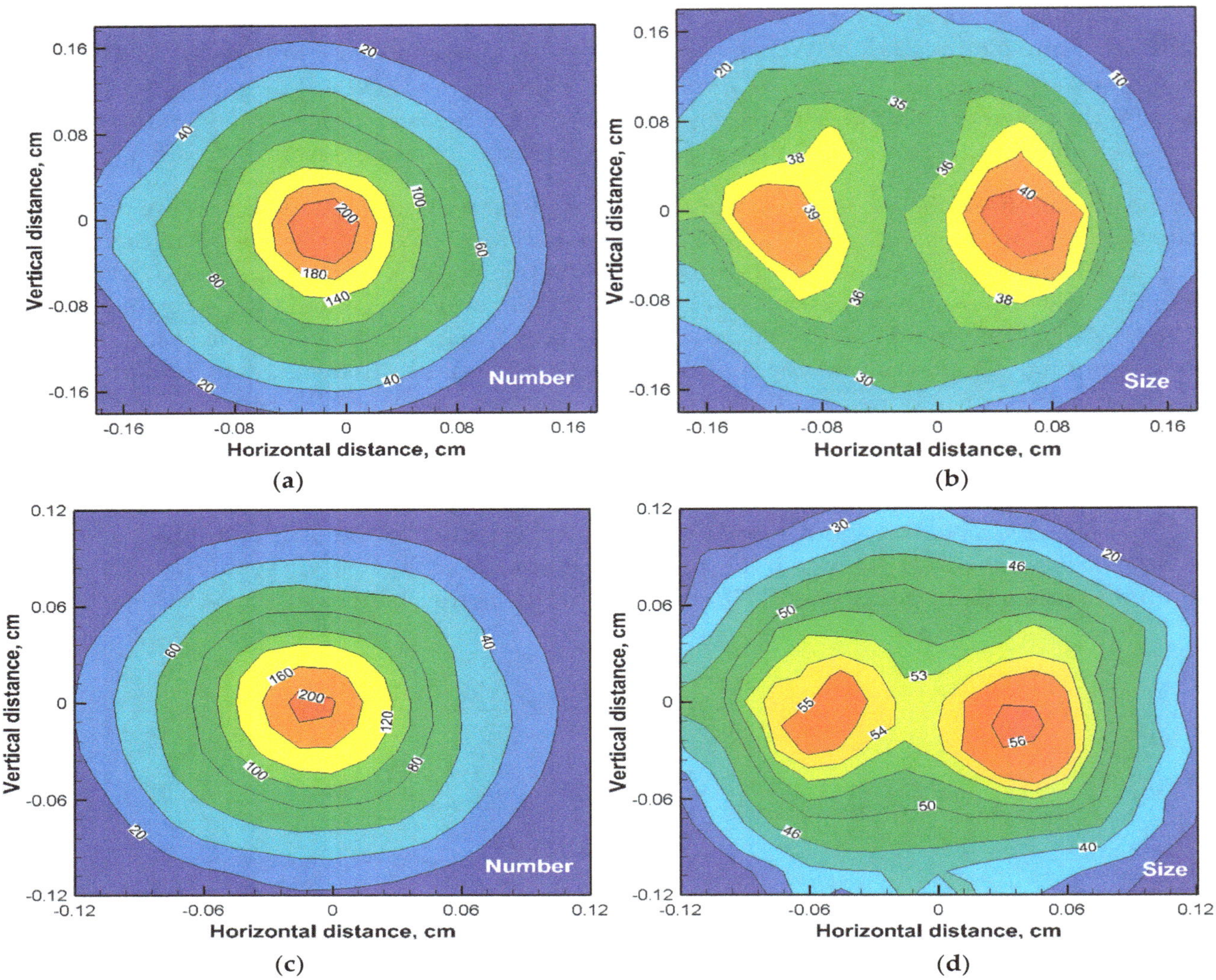

Figure 7. Spatial distribution of the drop number and size at the axial distance of 50 mm. (**a**) Number distribution of water; (**b**) Size distribution of water ; (**c**) Number distribution of polymer solution; (**d**) Size distribution of polymer solution.

4.3. Effects of Viscoelastic Parameters

Figure 8 shows the analysis of the effects of rheological properties on primary atomization. Figure 9 explores the evolutions of droplet size along with axial direction for various liquid properties. For polymer solutions, the rheology properties can be adjusted through the addition of polymer molecules. The liquid physical properties of base case are as the same as that of code 2 in Table 2. All cases in Figures 8 and 9 share the same operating conditions except the specified parameters in the legend of the figure. In order to compare the behavior of Newtonian and viscoelastic liquid, the calculated results of water spray have been added into Figures 8 and 9.

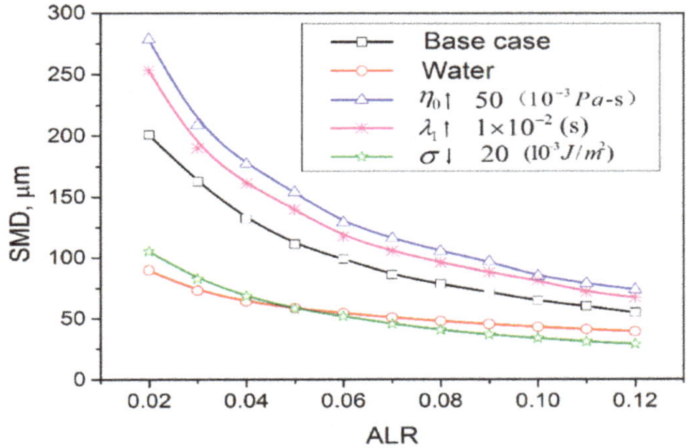

Figure 8. Droplet *SMD* versus *ALR* for various liquid properties.

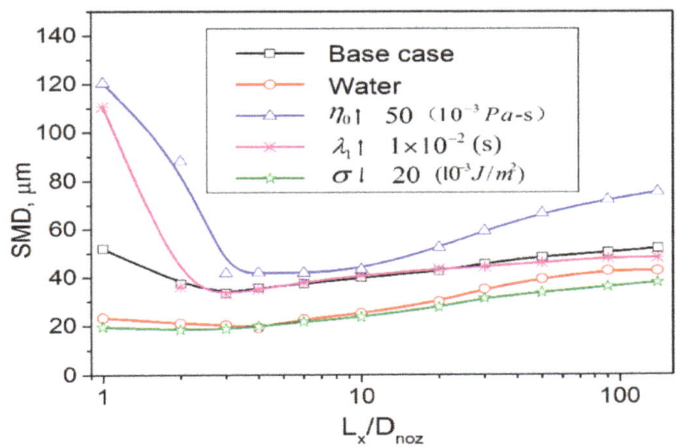

Figure 9. Droplet *SMD* along the axial distance for various liquid properties.

Obviously, from Figures 8 and 9, the increases in relaxation time and zero-shear viscosity cause the atomization quality deteriorates since the growth rate of the disturbance wave reduces and the liquid becomes stringier to resist the disruptive force, while small surface tension brings the droplet size to decrease significantly. The case with surface tension of 20×10^{-3} J/m^2 is almost identical to the case of water. In both instability analysis and TAB model, the surface tension can resist the occurrence and development of instability and smooth the disturbance. That is, a decrease in surface tension brings the ligaments and droplet less stability. This finding shows the importance of selecting solvents with lower surface tension for the polymer solution when finer atomization effects are desirable. Moreover, the effects of relaxation time become weaken at the downstream of the spray field. The case with shorter relaxation time approaches to the base case since the influence of relaxation time depends on the gas-liquid relative velocity. When the relative velocity declined rapidly, the effective viscosity of viscoelastic liquid is mainly up to zero shear viscosity.

5. Conclusions

A comprehensive mathematical model has been proposed to simulate the outflow of the effervescent atomization for viscoelastic liquid, accounting for primary atomization droplet tracking, secondary breakup and droplet collision. The computational models have included two parts. The first one is the primary breakup sub-model. With the simplification and assumptions of the atomization of liquid stream, the primary breakup sub-model has been established based on the viscoelastic linear dispersion relation. The second sub-model can calculate the evolutions of the droplets spray, in which the gas-liquid two-phase flow has been simulated by the Euler-Lagrangian approach. Both the droplet secondary atomization model and collision model have been extended to the viscoelastic version by introducing rheological parameters. The main findings of the present study can be summarized in four points:

(1) Based on the comparison with experimental data and the predicted results of other dispersion relation, the primary model using Joseph's relation effectively captured the influence of rheological properties on the breakup process of polymer solution and to quantitatively predict the resulting droplet sizes.

(2) The evolutionary trend of droplet mean diameter, Weber number and Ohnesorge number of viscoelastic liquids along with axial direction were qualitatively similar to that of Newtonian liquids. The *SMD* for the case of polymer solutions were obviously larger than that of water throughout the spray field. While, the gap of *SMD* between water and polymer solution became shorten along with the development of the jet. This phenomenon indicated that the mean size of polymer solution increased more gently than that of water at the axial distance $(5\text{–}100) L_x / D_{noz}$, which benefited the stable control of the desirable size in the applications.

(3) The changes of Weber and Ohnesorge number for polymer solution were more dramatic than that of water, since the viscosity of the viscoelastic liquid was influenced by the relative velocity of gas and droplets. Besides, the distribution range of water was about 40% larger than that of polymer solution in present study, which indicated the spray angle became narrow for the viscoelastic liquids.

(4) In primary atomization, the surface tension was the dominating physical property for viscoelastic liquid rather than relaxation time and zero shear viscosity. When the finer atomization effects were desirable, it was important to select solvents with lower surface tension for the polymer solutions, while, concerning the evolution of atomization spray, larger relaxation time and zero shear viscosity increased droplet *SMD* significantly. The zero shear viscosity was effective throughout the jet region, while the effect of relaxation time became weaken at the downstream of the spray field.

Acknowledgments: This work was supported by the Major Program of the National Natural Science Foundation of China with Grant No. 11632016, and the National Natural Science Foundation of China with Grant No. 11372006 and 11402259. Lijuan Qian would also like to thank Zhejiang Provincial Natural Science Foundation (LY17A020008) and Young Talent Cultivation Project of Zhejiang Association for Science and Technology (2016YCGC013).

Author Contributions: Lijuan Qian performed the codes and wrote the paper; Fubing Bao analyzed the data; Jianzhong Lin contributed ideas and suggestions.

References

1. Mallory, J.A. Jet Impingement and Primary Atomization of Non-Newtonian Liquids. Ph.D. Thesis, Purdue University, West Lafayette, IN, USA, 2012.
2. Yoo, H.; Kim, C. Generation of inkjet droplet of non-Newtonian fluid. *Rheol. Acta* **2013**, *52*, 313–325. [CrossRef]
3. Hoath, S.D.; Vadillo, D.C.; Harlen, O.G.; Mcllroy, C.; Morrison, N.F.; Hsiao, W.; Tuladhar, T.R.; Jung, S.; Martin, G.D.; Hutchings, I.M. Inkjet printing of weakly elastic polymer solution. *J. Non-Newton. Fluid Mech.* **2014**, *205*, 1–10. [CrossRef]
4. Petersen, F.J.; Wørts, O.; Schæfer, T.; Sojka, P.E. Effervescent atomization of aqueous polymer solutions and dispersions. *Pharm. Dev. Technol.* **2001**, *6*, 201–210. [CrossRef] [PubMed]
5. Pawlowski, L. Suspension and solution thermal spray coatings. *Surf. Coat. Technol.* **2009**, *203*, 2807–2829. [CrossRef]
6. Bartolo, D.; Boudaoud, A.; Narcy, G.; Bonn, D. Dynamics of non-newtonian droplets. *Phys. Rev. Lett.* **2007**, *99*, 174502. [CrossRef] [PubMed]

7. Schowalter, W.R. *Mechanics of Non-Newtonian Fluids*; Pergamon Press: Oxford, UK, 1978.
8. Oldroyd, O.G. Non-Newtonian effects in steady motion of some idealized elastic viscous fluids. *Proc. R. Soc.* **1958**, *279*, A245.
9. Li, L.K.B.; Dressler, D.M.; Green, S.I.; Davy, M.H.; Eadie, D.T. Experiments on air-blast atomization of viscoelastic liquids, Part 1: Quiescent conditions. *At. Sprays* **2009**, *19*, 157–190. [CrossRef]
10. Dumouchel, C. On the experimental investigation on primary atomization of liquid streams. *Exp. Fluids* **2008**, *45*, 371–422. [CrossRef]
11. Aliseda, A.; Hopfinger, E.J.; Lasheras, J.C.; Kremer, D.M.; Berchielli, A.; Connolly, E.K. Atomization of viscous and non-newtonian liquids by a coaxial, high-speed gas jet: Experiments and droplet size modeling. *Int. J. Multiph. Flow* **2008**, *34*, 161–175. [CrossRef]
12. Harrison, G.M.; Mun, R.P.; Cooper, G.; Boger, D.V. A note on the effect of polymer rigidity and concentration on spray atomization. *J. Non-Newton. Fluid Mech.* **1999**, *85*, 93–104. [CrossRef]
13. Chao, K.K.; Child, C.A.; Grens, E.A., II; Williams, M.C. Anti-misting action of polymeric additives in jet fuels. *AIChE J.* **1984**, *30*, 111–120. [CrossRef]
14. Sirignano, W.A.; Mehring, C. Review of theory of distortion and disintegration of liquid streams. *Prog. Energy Combust. Sci.* **2000**, *26*, 609–655. [CrossRef]
15. Middleman, S. Stability of a viscoelastic jet. *Chem. Eng. Sci.* **1965**, *20*, 1037–1040. [CrossRef]
16. Goren, S.L.; Gorttlieb, M. Surface-tension driven breakup of viscoelastic liquid threads. *J. Fluid Mech.* **1982**, *120*, 245–266. [CrossRef]
17. Joseph, G.E.; Funda, B.T. Rayleigh-Taylor instability of viscoelastic drops at high Weber number. *J. Fluid Mech.* **2002**, *453*, 109–132. [CrossRef]
18. Yang, L.J.; Tong, M.X.; Fu, Q.F. Linear stability analysis of a three-dimensional viscoelastic liquid jet surrounded by a swirling air stream. *J. Non-Newton. Fluid Mech.* **2013**, *191*, 1–13. [CrossRef]
19. Fuster, D.; Bagué, A.; Boeck, T.; Le Moyner, L.; Leboissetier, A.; Popinet, S.; Scardovelli, R.; Zaleski, S. Simulation of primary atomization with an octree adaptive mesh refinement and VOF method. *Int. J. Multiph. Flow* **2009**, *35*, 550–565. [CrossRef]
20. Villermaux, E. Fragmentation. *Annu. Rev. Fluid Mech.* **2007**, *39*, 419–446. [CrossRef]
21. Gorokhovski, M.; Herrmann, M. Modeling primary atomization. *Annu. Rev. Fluid Mech.* **2008**, *40*, 343–366. [CrossRef]
22. Zhu, C.; Ertl, M.; Weigand, B. Numerical investigation on the primary breakup of an inelastic non-Newtonian liquid jet with inflow turbulence. *Phys. Fluids* **2013**, *25*, 083102. [CrossRef]
23. Stelter, M.; Brenn, G.; Durst, F. The influence of viscoelastic fluid properties on spray formation from flat-fan and pressure-swirl atomizers. *At. Sprays* **2002**, *12*, 299–327. [CrossRef]
24. Dexter, R.W. Measurement of extensional viscosity of polymer solutions and its effects on atomization from a spray nozzle. *At. Sprays* **1996**, *6*, 167–191. [CrossRef]
25. Hartranft, T.J.; Settles, G.S. Sheet atomization of non-Newtonian liquids. *At. Sprays* **2003**, *13*, 191–221. [CrossRef]
26. Mun, R.P.; Young, B.W.; Boger, D.V. Atomization of dilute polymer solutions in agricultural spray nozzles. *J. Non-Newton. Fluid Mech.* **1999**, *83*, 163–178. [CrossRef]
27. Rivera, C.L. Secondary Breakup of Inelastic Non-Newtonian Liquid Drops. Ph.D. Thesis, Purdue University, West Lafayette, IN, USA, 2010.
28. Dechelette, A.; Campanellab, O.; Corvalanc, C.; Sojka, P.E. An experimental investigation on the breakup of surfactant-laden non-Newtonian jets. *Chem. Eng. Sci.* **2011**, *66*, 6367–6374. [CrossRef]
29. Ma, Y.; Bai, F.; Chang, Q.; Yi, J.; Jiao, K.; Du, Q. An experimental study on the atomization characteristics of impinging jets of power law fluid. *J. Non-Newton. Fluid Mech.* **2015**, *217*, 49–57. [CrossRef]
30. Xue, Z.; Carlos, M.C.; Dravid, V.; Sojka, P.E. Breakup of shear-thinning liquid jets with surfactants. *Chem. Eng. Sci.* **2008**, *63*, 1842–1849. [CrossRef]
31. Mallory, J.; Sojka, P.E. On the primary atomization of non-Newtonian imping jets: Volume II linear stability theory. *At. Sprays* **2014**, *24*, 525–554. [CrossRef]
32. Ashgriz, N. *Handbook of Atomization and Sprays: Theory and Applications*; Springer: New York, NY, USA, 2011.
33. Broniarz-Press, L.; Ochowiak, M.; Woziwodzki, S. Atomization of PEO aqueous solutions in effervescent atomizers. *Int. J. Heat Fluid Flow* **2010**, *31*, 651–658. [CrossRef]

34. Lund, M.T.; Jian, C.Q. The influence of atomizing gas molecular weight on low mass flow-rate effervescent atomization performance. *J. Fluids Eng.* **1998**, *120*, 750–754. [CrossRef]
35. Bird, R.B. Useful non-Newtonian models. *Annu. Rev. Fluid Mech.* **1976**, *8*, 13–34. [CrossRef]
36. Amsden, A.A.; O'Rourke, P.J.; Butler, T.D. *KIVA-II: A Computer Program for Chemically Reactive Flows with Sprays*; Technical Report LA-11560-MS; Los Alamos National Laboratory: Los Alamos, NM, USA, 1980.
37. Denn, M.M. Issues in viscoelastic fluid mechanics. *Annu. Rev. Fluid Mech.* **1999**, *22*, 13–32. [CrossRef]
38. Brodkey, R.S. *The Phenomena of Fluid Motions, Addison-Wesley Series in Chemical Engineering*; Addison-Wesley: New York, NY, USA, 1967.
39. Xiong, H.B.; Lin, J.Z.; Zhu, Z.F. Three-dimensional simulation of effervescent atomization spray. *At. Sprays* **2009**, *19*, 1–16. [CrossRef]
40. Lin, J.Z.; Qian, L.J.; Xiong, H.B.; Chan, T.L. Effects of operating conditions on droplet deposition onto surface of atomization impinging spray. *Surf. Coat. Technol.* **2009**, *203*, 1733–1740. [CrossRef]
41. Lefebvre, A.H. *Atomization and Sprays*; Hemisphere Publishing Corporation: Carlbad, CA, USA, 1989.
42. Geckler, S.C.; Sojka, P.E. Effervescent atomization of viscoelastic liquids: Experiments and modeling. *J. Fluids Eng.* **2008**, *130*, 061303. [CrossRef]
43. Liu, L.S. Experimental and Theoretical Investigation on the Characteristics and Two-Phase Spray Flow Field of Effervescent Atomizers. Ph.D. Thesis, Tianjing University, Tianjin, China, 2001.
44. Mansour, A.; Chigier, N. Air-Blast atomization of Non-Newtonian liquids. *J. Non-Newton. Fluid Mech.* **1995**, *58*, 161–194. [CrossRef]

9

Designing Incidence-Angle-Targeted Anti-Cavitation Foil Profiles using a Combination Optimization Strategy

Di Zhu [1,2], Ruofu Xiao [1,2,*], Ran Tao [2,3] and Fujun Wang [1,2]

1 College of Water Resources and Civil Engineering, China Agricultural University, Beijing 100083, China; zhu_di@cau.edu.cn (D.Z.); wangfj@cau.edu.cn (F.W.)
2 Beijing Engineering Research Center of Safety and Energy Saving Technology for Water Supply Network System, China Agricultural University, Beijing 100083, China; randytao@mail.tsinghua.edu.cn
3 Department of Energy and Power Engineering, Tsinghua University, Beijing 100083, China
* Correspondence: xrf@cau.edu.cn

Abstract: In hydraulic machinery, the surface of the blade can get damaged by the cavitation of the leading-edge. In order to improve the cavitation performance, the anti-cavitation optimization design of blade leading-edge is conducted. A heuristic-parallel locally-terminated improved hill-climbing algorithm, which is named as the global dynamic-criterion (GDC) algorithm was proposed in this study. The leading-edge shape of NACA 0009-mod foil profile was optimized by combining the GDC algorithm, CFD prediction, Diffusion-angle Integral (DI) design method and orthogonal test. Three different optimal foil geometries were obtained for specific incidence angles that 0, 3, and 6 degrees. According to the flow field analyses, it was found that the geometric variation of the optimized foil fits the incoming flow better at the respective optimal incidence angles due to a slighter leading-edge flow separation. The pressure drops become gentler so that the cavitation performance get improved. Results show that the GDC algorithm quickly and successfully fits the target condition by parallel running with the ability against falling into local-best tarps. The $-C_{pmin}$ of the optimal foils was improved especially by +11.4% and +14.5% at 3 and 6 degrees comparing with the original foil. This study provided a reference for the anti-cavitation design of hydraulic machinery blades.

Keywords: leading edge; global optimization; cavitation inception; orthogonal test; CFD simulation

1. Introduction

Leading-edge (LE) cavitation is the most common type of cavitation in hydraulic machinery [1,2]. The occurrence of LE cavitation is strongly related to the pressure changes generated by fluid flow around the geometry [3,4]. The direct effect of geometric shape on LE cavitation will be considered in the design of blades or other flow-around bodies. The cavity or traveling bubbles of LE cavitation often covers the surface [5] which may cause direct material damage due to the collapse of cavitation bubbles [6]. Therefore, it is very important to investigate and improve the LE cavitation (including delaying its occurrence, reducing its size, etc.) for the design of hydraulic machinery [7–9]. Compared with the design of the flow passage, the blade will exert a direct force on the fluid medium while working and generate a pressure difference between the two sides of blade surface [10–12]. The blade cavitation, especially the LE cavitation characteristics, has become a key factor restricting the blade design under specific conditions. In the past, many researchers have studied the occurrence and characteristics of LE cavitation [13–15]. Generally, cavitation occurs when pressure drops below the saturation pressure. Because of the flow separation and local pressure drop at the blade LE, cavitation often occurs at the LE [16]. Experimental studies found that the development of the cavity had a negative effect on the flow separation due to inter-phase surface tension [17]. The interaction effect causes a special relationship between the separation position and cavitation inception position in the laminar flow case around a smooth body [18]. This relationship will be different in turbulent flow case, rough surface case and other different boundary layer cases [19]. Generally, the flow separation and pressure drop caused by the geometric characteristics of blade LE will regularly affect the occurrence and development of cavitation at a certain incidence angle [20].

Foil was commonly used as the simplified model of hydraulic machinery blades in engineering [21]. Meanline-symmetrical foils can further reduce the complexity of geometric parameters and improve the pertinence of the research [22]. The smooth symmetrical NACA0009 foil profile can be used to study the laminar flow under a certain Reynolds number. NACA0009 and its modified profiles are popular in the former studies of hydrodynamics and hydraulic machinery including the LE cavitation, tip-clearance cavitation, wake and hydraulic damping, excitation and response [23,24]. Also, numerous researches have been carried out on the LE cavitation of hydraulic machineries [25–30].

Based on the study of LE cavitation on foils and turbo blades, there have been abundant researches on the mechanism, occurrence, development and variation. However, there is still a problem that has not been solved perfectly that how to obtain a hydraulic geometry for improving the LE cavitation. Optimization is an effective way to solve this problem by properly defining the optimized target and choosing the reasonable algorithm. In this study, the foil shape with better LE cavitation characteristics was set as the optimization target. Feasible algorithms including genetic algorithm, ant colony algorithm, hill climbing algorithm, etc. can be used. Each of them has its own advantages and disadvantages and requires improvements or adjustments in different cases [31–35]. Huang et al. [36] used the genetic algorithm to drive the computational fluid dynamics simulations, guiding the configuration of a suction jet and a blowing jet on the airfoil's upper surface. Liu et al. [37] selected an adaptive simulated annealing algorithm to solve the energy performance calculation model. The weighted average efficiency of the impeller after the three-condition optimization has increased by 1.46% than that of original design. Liu et al. [38] used a multi-objective optimization design system to develop an ultrahigh-head runner with good overall performance. Compared to the initial runner, the preferred runner's efficiency under turbine mode is increased by about 0.7% and the pump efficiency by about 0.6%, while the runner's cavitation is greatly promoted. Liu et al. [39] proposed the hydraulic design method of controllable blade angle for rotodynamic multiphase pump with impeller and diffuser. The orthogonal optimization method was employed to optimize the geometry parameters. The distributions of gas volume fraction and the pressure became more uniform after optimization, and improved the transporting performance of the multiphase pump. In this study, the global dynamic-criterion (GDC) algorithm which can run in parallel, stop from dropping into local-best trap was chosen. Finally, reducing the number of effective tests and reducing time cost is the top priority of engineering optimization. Therefore, the strategy combining the GDC algorithm, CFD prediction, Diffusion-angle Integral (DI) method [40] for foil thickness geometry design and orthogonal test was used for optimization. The new optimization algorithm proposed in this paper can complete the optimization process quickly and simply with high optimization efficiency. Based on the strategy above, the optimal hydraulic shape and improved cavitation performance of NACA0009 foil can be obtained at different specific incidence angles. This research shows an example for designing the anti-cavitation turbo blade of hydraulic machinery including axial-flow pump, mixed-flow pump, centrifugal pump and other bladed pumps.

2. The Studied Hydrofoil Object

The NACA0009-mod profile was used as the objective hydrofoil [24]. As shown in Figure 1, it is a symmetrical foil, m is the meanline direction and t is the thickness. The total meanline length is $m_{total} = L$. The thickness t distribution along m direction can be expressed as:

$$\begin{cases} \frac{t}{L} = a_0\left(\frac{m}{L}\right)^{\frac{1}{2}} + a_1\left(\frac{m}{L}\right) + a_2\left(\frac{m}{L}\right)^2 + a_3\left(\frac{m}{L}\right)^3 & 0 \le \frac{m}{L} \le 0.5 \\ \frac{t}{L} = b_0 + b_1\left(1-\frac{m}{L}\right) + b_2\left(1-\frac{m}{L}\right)^2 + b_3\left(1-\frac{m}{L}\right)^3 & 0.5 \le \frac{m}{L} \le 1 \end{cases} \tag{1}$$

where

$$\begin{cases} a_0 = +0.1737 \\ a_1 = -0.2422 \\ a_2 = +0.3046 \\ a_3 = -0.2657 \end{cases} \quad \begin{cases} b_0 = +0.0004 \\ b_1 = +0.1737 \\ b_2 = -0.1898 \\ b_3 = +0.0387 \end{cases}. \tag{2}$$

The hydrofoil was built based on Equations (1) and (2), and another trailing edge modification. The TE modification was cut on the trailing edge to $l = 0.9091L$. The TE was processed by circular arc, and the radius is $r = 0.0168L$. The final hydrofoil model is shown in Figure 1. The incidence angle was defined as α. The 2D flow domain ($1.5l \times 7.5l$) for computational fluid dynamics (CFD) simulations was built as shown in Figure 2. The commercial software ANSYS CFX was used for numerical simulation.

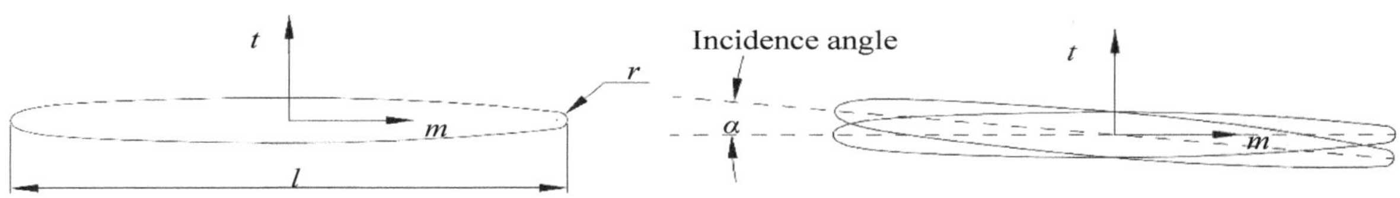

Figure 1. Hydrofoil model and its parameters.

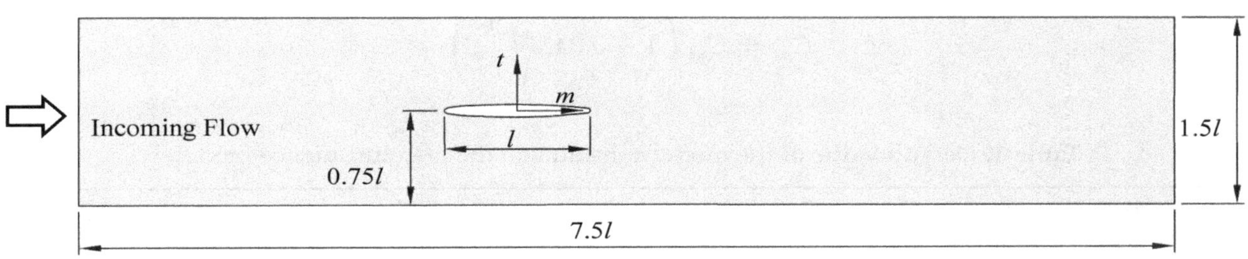

Figure 2. Flow domain for computational fluid dynamics (CFD) simulation.

3. Methods

3.1. Numerical Model of Turbulent Flow

In this study, the Reynolds-averaged Navier–Stokes (RANS) equations were used to solve the turbulent flow. Considering the turbulence isotropic assumption and rotating/curvature insensitivity of eddy viscosity models, the v^2–f model, which can be more suitable especially for the viscous flow separation from a curved hydrofoil surface, was used. The v^2–f model was based on the turbulence kinetic energy k equation, dissipation rate ε equation, velocity variance scale $\overline{v^2}$ equation, and the elliptic relaxation function f. It can be expressed as [41–43]:

$$\frac{\partial(\rho k)}{\partial t} + \frac{\partial(\rho u_i k)}{\partial x_i} = P - \rho\varepsilon + \frac{\partial}{\partial x_j}\left[\left(\mu + \frac{\mu_t}{\sigma_k}\right)\frac{\partial k}{\partial x_j}\right] + S_k \tag{3}$$

$$\frac{\partial(\rho k)}{\partial t} + \frac{\partial(\rho u_i k)}{\partial x_i} = \frac{C'_{\varepsilon 1}P - C_{\varepsilon 2}\rho\varepsilon}{T} + \frac{\partial}{\partial x_j}\left[\left(\mu + \frac{\mu_t}{\sigma_\varepsilon}\right)\frac{\partial \varepsilon}{\partial x_j}\right] + S_\varepsilon \tag{4}$$

$$\frac{\partial\left(\rho\overline{v^2}\right)}{\partial t}+\frac{\partial\left(\rho\overline{v^2}u_i\right)}{\partial x_i}=\rho kf-6\rho\overline{v^2}\frac{\varepsilon}{k}+\frac{\partial}{\partial x_j}\left[\left(\mu+\frac{\mu_t}{\sigma_k}\right)\frac{\partial\overline{v^2}}{\partial x_j}\right]+S_{\overline{v^2}} \tag{5}$$

$$f-L^2\frac{\partial^2 f}{\partial x_j^2}=(C_1-1)\frac{2k-3\overline{v^2}}{3kT}+C_2\frac{P}{\rho k}+\frac{5\overline{v^2}}{kT}+S_f \tag{6}$$

where C_1, C_2, $C_{\varepsilon1}$, $C'_{\varepsilon1}$, $C_{\varepsilon2}$, C_η, C_μ, and C_L are model constants; σ_k and σ_ε are the turbulent Prandtl numbers; and S_k, S_ε, $S_{\overline{v^2}}$, and S_f are source terms. Term T is the turbulent time scale and L is the turbulent length scale:

$$T'=\max\left[\frac{k}{\varepsilon},6\sqrt{\frac{\nu}{\varepsilon}}\right] \tag{7}$$

$$T=\min\left[T',\frac{\alpha}{\sqrt{3}}\frac{k}{\overline{v^2}C_\mu\sqrt{2S^2}}\right] \tag{8}$$

$$L'=\min\left[\frac{k^{3/2}}{\varepsilon},\frac{1}{\sqrt{3}}\frac{k^{3/2}}{\overline{v^2}C_\mu\sqrt{2S^2}}\right] \tag{9}$$

$$L=C_L\max\left[L',C_\eta\left(\frac{v^3}{\varepsilon}\right)^{1/4}\right] \tag{10}$$

where α is a model constant, ν is the kinematic viscosity, and S is the strain rate tensor. Thus, the turbulent viscosity μ_t can be expressed as

$$\mu_t=\rho C_\mu\overline{v^2}T. \tag{11}$$

All the default values of the model constants are listed in Table 1. Based on $C_{\varepsilon1}$, the value of $C'_{\varepsilon1}$ can be calculated by

$$C'_{\varepsilon1}=C_{\varepsilon1}\left(1+0.045\sqrt{\frac{k}{\overline{v^2}}}\right). \tag{12}$$

Table 1. Default value of the model constants in the v^2–f turbulence model.

Constant	α	C_1	C_2	$C_{\varepsilon1}$	$C_{\varepsilon2}$	C_η	C_μ	C_L	σ_k	σ_ε
Value	0.6	1.4	0.3	1.4	1.9	70	0.22	0.23	1.0	1.3

3.2. CFD Setup

Based on the setting of the turbulent flow, the flow domain was discretized using unstructured mesh elements as shown in Figure 3. The mesh scheme was determined by an independence check and had 64,868 nodes in total. Considering the usage of the v^2–f turbulence model, the near-wall mesh in the boundary layer was checked and refined with three prism layers (first layer height 1×10^{-4} mm, growth rate 1.2). The y^+ value was in the range of 0.08–6.86, which fits the requirement of a direct near-wall solution. In the simulation, the Reynolds number Re was set to 5×10^5 with the incidence angle between 0 and 6 degrees. The boundary conditions were set as follows: Firstly, the velocity inlet boundary was set at the inflow, the velocity was perpendicular to the inlet boundary, and the pressure followed the Neumann condition. Secondly, the pressure boundary was set at the outflow with average 0 Pa, and the velocity followed the Neumann condition. Thirdly, the no-slip condition was applied on the wall boundaries including the upper domain boundary, lower domain boundary, and the foil surface. Moreover, to simplify the 3D case to 2D, symmetry boundaries were given perpendicular to the 2D domain plane. The fluid medium was set as water at 20 °C with density 1×10^3 kg/m^3 and dynamic viscosity 1.01×10^{-3} Pa·s. The steady-state simulation was conducted with a maximum iteration number of 1000, and the convergence criterion was set to 1×10^{-5}. The transient-state simulation was conducted based on the steady-state simulation. For a better convergence performance and timely flow regime resolution, the total time and time step were determined based on the Reynolds number and Courant number. Finally, the total time was set as 2 s with a constant time step of 2×10^{-5} s which can be suitable in this case.

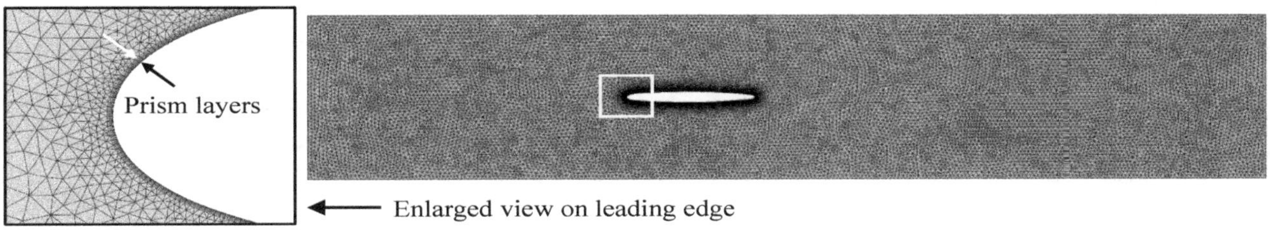

Figure 3. Schematic map of mesh elements with an enlarged view on LE.

3.3. Brief Introduction to the Diffusion Angle Integral Method

The Diffusion-angle Integral (DI) method [40] can improve the cavitation performance by changing the geometry of the foil LE. First, the DI method requires the geometric deconstruction of the foil, as shown in Figure 4. The shape of the arc and diffusion section of the foil changes rapidly. So, the DI method was mainly used to design the geometry of the foil LE. In this study, the length of the design section was $0.15l$ along the length of the meanline, which is divided into circular and integral sections. Through the DI method, the number of design parameters can be simplified to three, which greatly improves the design efficiency. The DI method is mainly divided into five steps as follows:

1. Providing the long/short axis ratio $R_{ab} = a_{LE}/b_{LE}$;
2. Based on R_{ab}, scaling the ellipse arc into an arc;
3. Providing the diffusion angle γ_s and calculating the scaled LE arc r_{LE};
4. Providing the thickness integral coefficient B (the change rate of Part 2 in Figure 4) and integrating the thickness diffusion part;
5. Based on R_{ab}, scaling the designed arc back to an ellipse arc.

Based on the steps above, the LE ellipse arc can be calculated under coordinate scaling:

$$a_{LE} = R_{ab} \frac{t_{O'} - 2m_{O'} \tan \gamma_s}{2(\sin \gamma_s \tan \gamma_s - \tan \gamma_s + \cos \gamma_s)} \tag{13}$$

$$b_{LE} = \frac{a_{LE}}{R_{ab}} \tag{14}$$

where $t_{O'}$ is the thickness at O' and $m_{O'}$ is the m position at O'. The increase in t at the thickness diffusion section can be calculated by

$$\Delta t = C_s \int_{m_A}^{m_{O'}} \tan \gamma(m) dm \tag{15}$$

where m_A is the m position at point A, C_s is the scale factor, and $\gamma(m)$ is the thickness integral expression which is defined as

$$\gamma(m) = \left(\frac{m_{O'} - m}{m_{O'} - m_A} \right)^B. \tag{16}$$

The variation law at thickness diffusion section can be controlled by the coefficient B. Applying DI method can be simple by following the steps above but the detailed mathematical deduction process of the DI method is complex. Hence, only the brief introduction is put here and the detail of DI method can be found in Ref. [40].

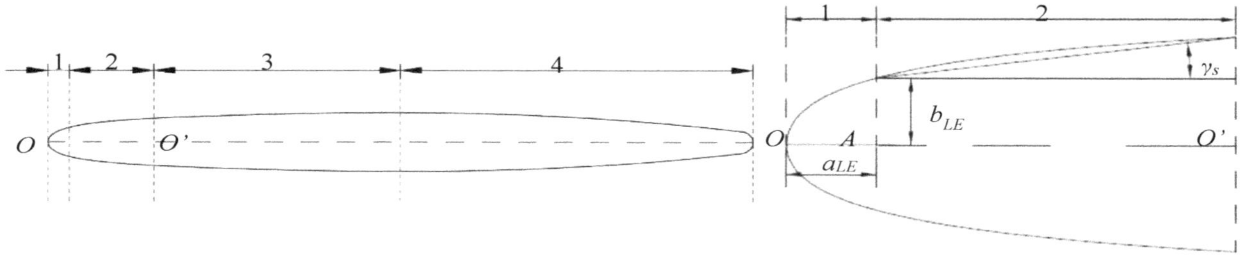

Figure 4. Geometry deconstruction and parameters of the Diffusion angle Integral method: 1—Elliptical arc section; 2—Thickness diffusion section; 3—Thickness transition section; 4—Thickness reduction section.

3.4. Orthogonal Testing

The orthogonal test method can be used instead of large scale comprehensive test with fewer test times. In this paper, 25 tests (3 factors and 5 levels) were used in the design of orthogonal test. The optimum values of 3 parameters in the DI method under different incidence angles were analyzed. The factor levels of R_{ab}, γ_s, B are shown in Table 2.

Table 2. Factor levels.

Level	Factor		
	R_{ab}	γ_s [degree]	B
1	1	1	1
2	1.5	2	2
3	2	3	3
4	2.5	4	4
5	3	5	5

In the process of flow, the dimensionless pressure coefficient Cp is used to characterize the surface pressure of the foil:

$$C_p = \frac{p - p_\infty}{\frac{1}{2}\rho V_\infty^2} \tag{17}$$

where p is the foil surface pressure, p_∞ is the reference position average pressure (calculation domain inlet), and V_∞ is the reference position average speed (calculation domain inlet). Typically, the number of cavitation number C_σ is defined as:

$$C_\sigma = \frac{p_\infty - p_v}{\frac{1}{2}\rho V_\infty^2}, \tag{18}$$

when cavitation begins, we have $p_{min} = p_v$. Therefore, the minimum pressure coefficient C_{pmin} on the foil surface is negatively correlated with the inception cavitation number $C_{\sigma i}$:

$$-C_{pmin} = C_{\sigma i}. \tag{19}$$

The minimum pressure coefficient of the foil surface C_{pmin} was chosen as the evaluation index. As shown in Table 3, the minimum pressure coefficients $C_{p\alpha}$ of foils with different incidence angles of 0, 3, and 6 degrees were obtained by numerical simulation. Then, the range analyses were conducted based on the results of orthogonal test as shown in Table 4. K_1~K_5 and k_1~k_5 are the polar difference values and averaged polar difference values, respectively. R is the range value based on orthogonal test.

Table 3. Orthogonal test table with factors, levels and test values.

Order Number	R_{ab}	γ_s [deg]	B	C_{p0}	C_{p3}	C_{p6}
1	1	1	1	−1.4987	−2.889	−4.4240
2	1	2	2	−1.4316	−2.8391	−4.4486
3	1	3	3	−1.3561	−2.8080	−4.5451
4	1	4	4	−1.2756	−2.7677	−4.5856
5	1	5	5	−1.1910	−2.7169	−4.5187
6	1.5	1	2	−1.0299	−2.1063	−3.5227
7	1.5	2	3	−0.9875	−2.0877	−3.5436
8	1.5	3	4	−0.9407	−2.0723	−3.5969
9	1.5	4	5	−0.8888	−2.0526	−3.5870
10	1.5	5	1	−0.8322	−2.0367	−3.6896
11	2	1	3	−0.7890	−1.7245	−3.1430
12	2	2	4	−0.7610	−1.7185	−3.1881
13	2	3	5	−0.7299	−1.7103	−3.2384
14	2	4	1	−0.6940	−1.7091	−3.2879
15	2	5	2	−0.6526	−1.7020	−3.3610
16	2.5	1	4	−0.6439	−1.5140	−3.0710
17	2.5	2	5	−0.6262	−1.5180	−3.1082
18	2.5	3	1	−0.6048	−1.5189	−3.1723
19	2.5	4	2	−0.5783	−1.5251	−3.2313
20	2.5	5	3	−0.5467	−1.5277	−3.3146
21	3	1	5	−0.5469	−1.4048	−3.1355
22	3	2	1	−0.5357	−1.4083	−3.1153
23	3	3	2	−0.5216	−1.4109	−3.2224
24	3	4	3	−0.5027	−1.4163	−3.2881
25	3	5	4	−0.4768	−1.4230	−3.3829

Table 4. Range analysis results.

ParamETERS	C_{p0}			C_{p3}			C_{p6}		
	R_{ab}	γ_s [deg]	B	R_{ab}	γ_s [deg]	B	R_{ab}	γ_s [deg]	B
K_1	−9.0567	−6.3002	−6.0576	−14.144	−9.8917	−9.8353	−20.095	−15.252	−15.524
K_2	−6.5729	−6.1712	−6.1082	−10.546	−9.8318	−9.8667	−15.855	−15.314	−15.609
K_3	−5.3795	−6.0521	−6.0855	−8.8359	−9.8091	−9.8531	−14.194	−15.587	−15.642
K_4	−4.7500	−5.8966	−6.0104	−7.9727	−9.7685	−9.7886	−13.778	−15.725	−15.619
K_5	−4.3877	−5.7268	−5.8853	−7.5234	−9.7216	−9.6791	−13.880	−15.924	−15.409
k_1	−1.8114	−1.2601	−1.2115	−2.8289	−1.9783	−1.9671	−4.0190	−3.0503	−3.1048
k_2	−1.3146	−1.2342	−1.2216	−2.1093	−1.9664	−1.9733	−3.1711	−3.0629	−3.1217
k_3	−1.0759	−1.2104	−1.2171	−1.7672	−1.9618	−1.9706	−2.8388	−3.1175	−3.1284
k_4	−0.9500	−1.1793	−1.2021	−1.5945	−1.9537	−1.9577	−2.7556	−3.1450	−3.1238
k_5	−0.8775	−1.1454	−1.1771	−1.5047	−1.9443	−1.9358	−2.7760	−3.1848	−3.0817
R	0.9338	0.1147	0.0446	1.3242	0.0340	0.0375	1.2634	0.1345	0.0467

It can be seen from Tables 2 and 3 that R_{ab} has the greatest influence on the minimum pressure coefficient C_p of the foil surface. When the incidence angle is 0 degree, the optimal parameter combination is $R_{ab} = 3$, $\gamma_s = 5$, and $B = 4$. When the incidence angle is 3 degrees, the optimal parameter combination is $R_{ab} = 3$, $\gamma_s = 2$, and $B = 1$. When the incidence angle is 6 degrees, the optimal parameter combination is $R_{ab} = 2.5$, $\gamma_s = 1$, and $B = 4$. The optimal parameter combination given by the orthogonal testing is the basis for the next step in finding the optimal design.

4. Optimization

4.1. Global Dynamic Criterion Algorithm

To further optimize the design, an improved hill climbing algorithm with heuristic parallel characteristics was proposed in this paper. The algorithm was used to search for the distribution of multiple local optimal values in a certain area. It can optimize the design parameters of foil LE.

The optimal value of the design parameters of foil LE is searched quickly and efficiently. The minimum pressure coefficient of foil surface can be improved. The algorithm includes the following steps:

(1) The algorithm is based on hill climbing algorithm. Firstly, 10 parameter combinations denoted as F_n ($n = 1, 2, 3 \ldots, 10$) are randomly generated near the optimal values obtained by the orthogonal experiments of 3 parameters R_{ab}, γ_s, B in the DI method. The corresponding foil is numerically simulated for each set of parameters F_n. Three different cases of incidence angle 0, 3 and 6 degrees are calculated respectively. The minimum pressure coefficient $C_{p\alpha}$ of the foil LE shape under each incidence angle is obtained. Therefore, the minimum pressure coefficient obtained from the LE parameters of each group is defined as C_{pmin}:

$$C_{pmin} = I \times C_{p0} + j \times C_{p3} + k \times C_{p6} \qquad (i + j + k = 1). \tag{20}$$

When focus on cavitation performance at 0 degree of incidence angle, here are $i = 0.8$, $j = 0.1$, $k = 0.1$. For 3 degrees, here are $i = 0.1$, $j = 0.8$, $k = 0.1$. For 6 degrees, here are $i = 0.1$, $j = 0.1$, $k = 0.8$.

(2) We then set the initial decision condition C_{pt}. We decide whether the C_{pmin} obtained from each set of parameters F_n satisfies the criteria. If $C_{pmin} \geq C_{pt}$, we keep this set of parameters and search in the neighborhood of each parameter. The search range should not exceed 5% of the total parameter range. If $C_{pmin} < C_{pt}$, we discard this set of parameters, and re-generate a group for the next round of the search. The optimal value of each search is T_n ($n = 1, 2, 3, \ldots, 10$).

(3) When searching in a small neighborhood, the new set of parameters F_n' will be used for the foil LE design. A new two-dimensional foil is thus obtained. Three different cases of incidence angle—0, 3, and 6 degrees—are calculated. The minimum pressure coefficient $C_{p\alpha}$ of the foil LE shape under each incidence angle is obtained. Taking the case of 0 degree as an example, C'_{pmin} is obtained using Equation (16). For each group of F_n', the optimal T_1 value obtained from the first search is the criterion for the second search. If $C'_{pmin} \geq T_1$, we keep this set of parameters and search in the neighborhood of the new set of parameters F_n'. The search range should not exceed 5% of the total parameter range. If $C'_{pmin} < T_1$, we discard this set of parameters, and re-generate a group for the next round of the search.

(4) For the third search round, it is necessary to continue to change the criteria T_3. It is changed to the weighted value of the optimal value T_1 of the first search round and the optimal value T_2 of the second search round. T_3 is defined as:

$$T_3 = p \times T_1 + q \times T_2 \qquad (p + q = 1) \tag{21}$$

The purpose of changing the criteria is to continuously improve the goal of optimization. Here, $p = 0.4$ and $q = 0.6$.

(5) In order to avoid falling into the local optimal solution, we set a small probability to terminate the current search. The termination probability Y and iteration number x have a certain functional relationship. The function has the following characteristics. With increasing x, the termination probability Y increases. When iteration number $x \to +\infty$, termination probability $Y \to 1$. According to the above characteristics, the function can be written as

$$\begin{cases} Y(x) = 0, x < 3 \\ Y(x) = \left[-e^{\left(\frac{b-x}{a}\right)} + c\right]\%, x \geq 3 \end{cases} \tag{22}$$

where a, b, and c are constants. The function can adjust the specific values of a, b, and c according to the maximum iteration number, so as to adjust the distribution of the termination probability. Here, the values in this study are as follows: $a = 2.8$, $b = 10$, and $c = 1$.

(6) The maximum iteration number in this article is set to 50 to have a balance between searching time and improvement effect. Repeat the above steps until reach the maximum iteration number, and finally get the global optimal solution. The optimum design parameters of foil LE at 0 degree, 3 degrees and 6 degrees of incidence angles are obtained.

4.2. Comparison of Foil Geometry

After optimization, the optimal geometries were get respectively on 0 degree, 3 degrees and 6 degrees. As shown in Figure 5, differences can be found in the range of 0–0.15 m/l. The optimal geometry for 0 degree has the smallest thickness values within 0–0.15 m/l. The optimal geometry for 6 degrees has the largest thickness values within 0–0.15 m/l. The differences of geometry showed the adaptability of incoming flow striking and local separation caused by incidence angle. The mechanism was analyzed in detail in the next section.

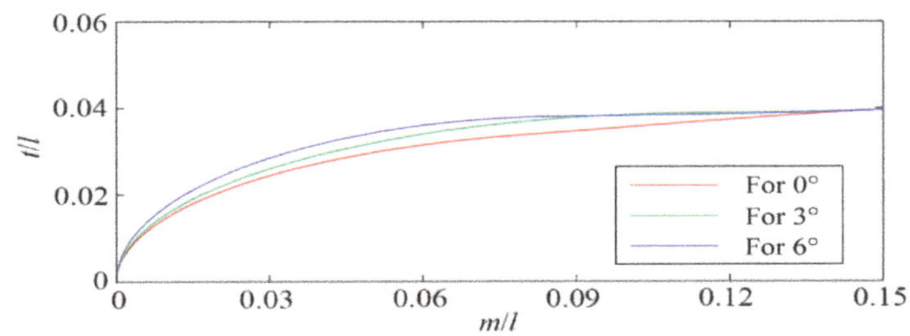

Figure 5. Schematic diagram of the optimal foil LE shape for 0, 3, and 6 degrees.

5. Results and Discussion

5.1. Comparison of $-C_{pmin}$ and Lift/Drag Ratio

The $-C_{pmin}$ of the original foil and the optimal three foils were compared, as shown in Figure 6. It can be seen from the Figure that the $-C_{pmin}$ of the optimal foils and the original foil at the 0 degree is similar with only difference of 1.2%. However, with the increase of the incidence angle, the optimal foils are lower than the original foil at 3 degrees and 6 degrees, which decreases by 11.4% and 14.5% respectively. It shows that the cavitation performance of the optimal foil at 3°~6° is improved. Comparing the three optimal foils, it can be seen that the $-C_{pmin}$ of the optimal for 0 degree' foil are lower in the range of 0~3 degrees, but higher at 6 degrees' incidence angle. However, the $-C_{pmin}$ of the optimal for 6 degrees' foil are lower at 6 degrees' incidence angle, but higher in the range of 0~3 degrees. Generally, the application of the DI method and GDC algorithm on optimizing the foil geometry can effectively improve the cavitation performance at large incidence angle. The foil with better cavitation performance can be obtained by focusing on a certain incidence angle, but the performance of both large and small incidence angles is difficult to be considered at the same time. Figure 7 compared the lift/drag ratio F_L/F_D of the original foil and the optimal foils. From the comparison of F_L/F_D, it can be seen that the F_L/F_D of the optimal foils are slightly lower than that of the original foil, but the overall difference is very small. The minimum difference is 0.1% and the maximum difference is 5%. The results showed that the performance of foil is not affected after optimization.

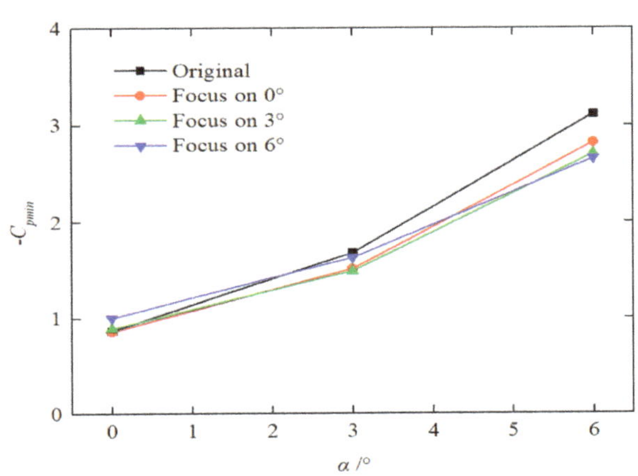

Figure 6. Comparison of $-C_{pmin}$ values.

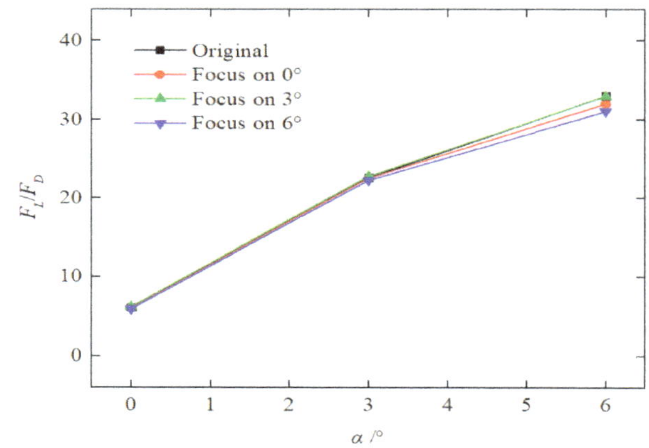

Figure 7. Comparison of lift/drag ratio F_L/F_D.

5.2. Pressure Distribution on the Foil Surface

Figure 8 shows a comparison of the surface pressure distribution curves of the three optimal foils under different incidence angles. It can be seen from the figure that the position of the lowest pressure point obtained by different LE shapes at the same incidence angle is relatively small. The lowest pressure coefficient has a certain difference. In Figure 8a, where $\alpha = 0$ degree, the optimal foil for 0 degree has the gentlest pressure drop on LE, while the optimal foil for 6 degrees has the most sudden pressure drop on LE. This shows that the optimal foil for 0 degree fits the incidence angle of 0 degree the best. The opposite relationship can be found in Figure 8c. However, the situation becomes complex in Figure 8b. At $\alpha = 3$ degrees, the optimal foil for 3 degrees fitted the incidence angle of 3 degrees the best. The optimal foil for 0 degree also performed well at the incidence angle of 3 degrees. The optimal foil for 6 degrees was the worst and had the minimum $-C_{pmin}$. Generally, the optimal design for a specific incidence angle has a gentler pressure drop on the LE. The cavitation scale would consequently be smaller after the design optimization.

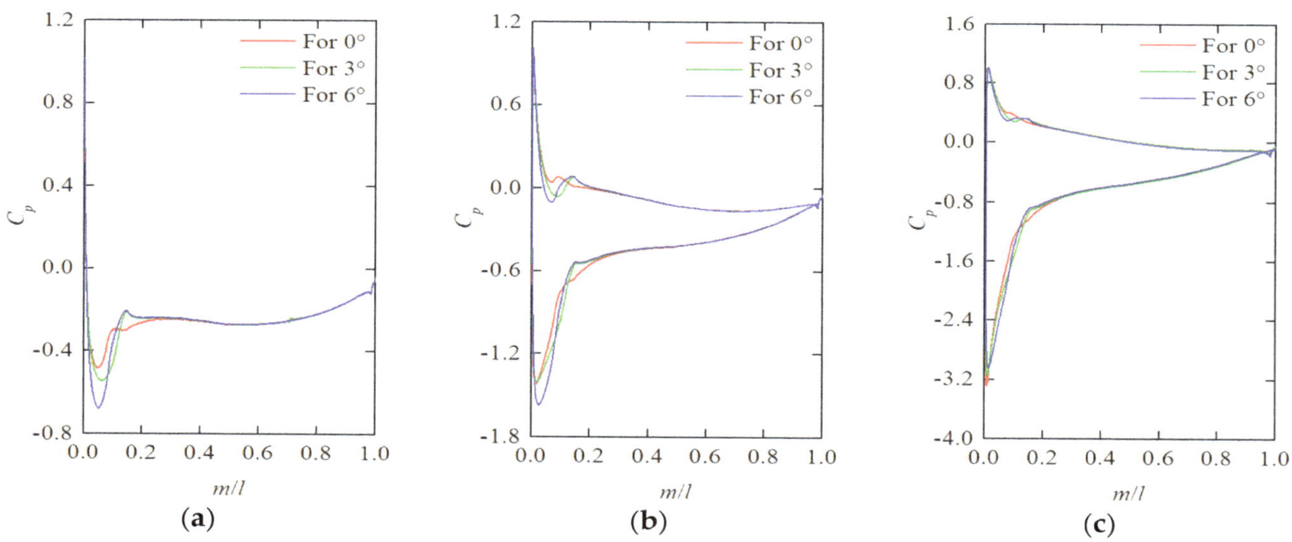

Figure 8. Surface pressure distribution curves of the three optimal foils under different incidence angles: (**a**) $\alpha = 0°$; (**b**) $\alpha = 3°$; (**c**) $\alpha = 6°$.

5.3. Flow Field around the Foil

The pressure and velocity vectors at the LE of the three optimal foils were compared and analyzed, as shown in Figure 9 and Figure 10. It can be seen from Figure 9 that when the incidence angle is 0 degree, the low-pressure area of the LE of the foil optimal for 0 degree is smaller. With the change of the LE geometry, the low-pressure zones of the foils optimal for 3 degrees and for 6 degrees gradually increase. After increasing the incidence angle, the pressure distributions of the three optimal foils' LE are similar. By analyzing the velocity vector, it is found that the three foils each have the smallest flow separation at their respective optimal incidence angle. This is because the gradient of geometric change is slower under the corresponding incidence angle. The local separation of the optimal foil LE is improved obviously. The pressure drop near the LE slows down. The shape of the foil LE can be adapted to the direction of incoming flow, making the flow more suitable for the foil. It is thus shown that the cavitation performances of the optimal foils at various incidence angles can be significantly improved by using the DI method and the GDC algorithm.

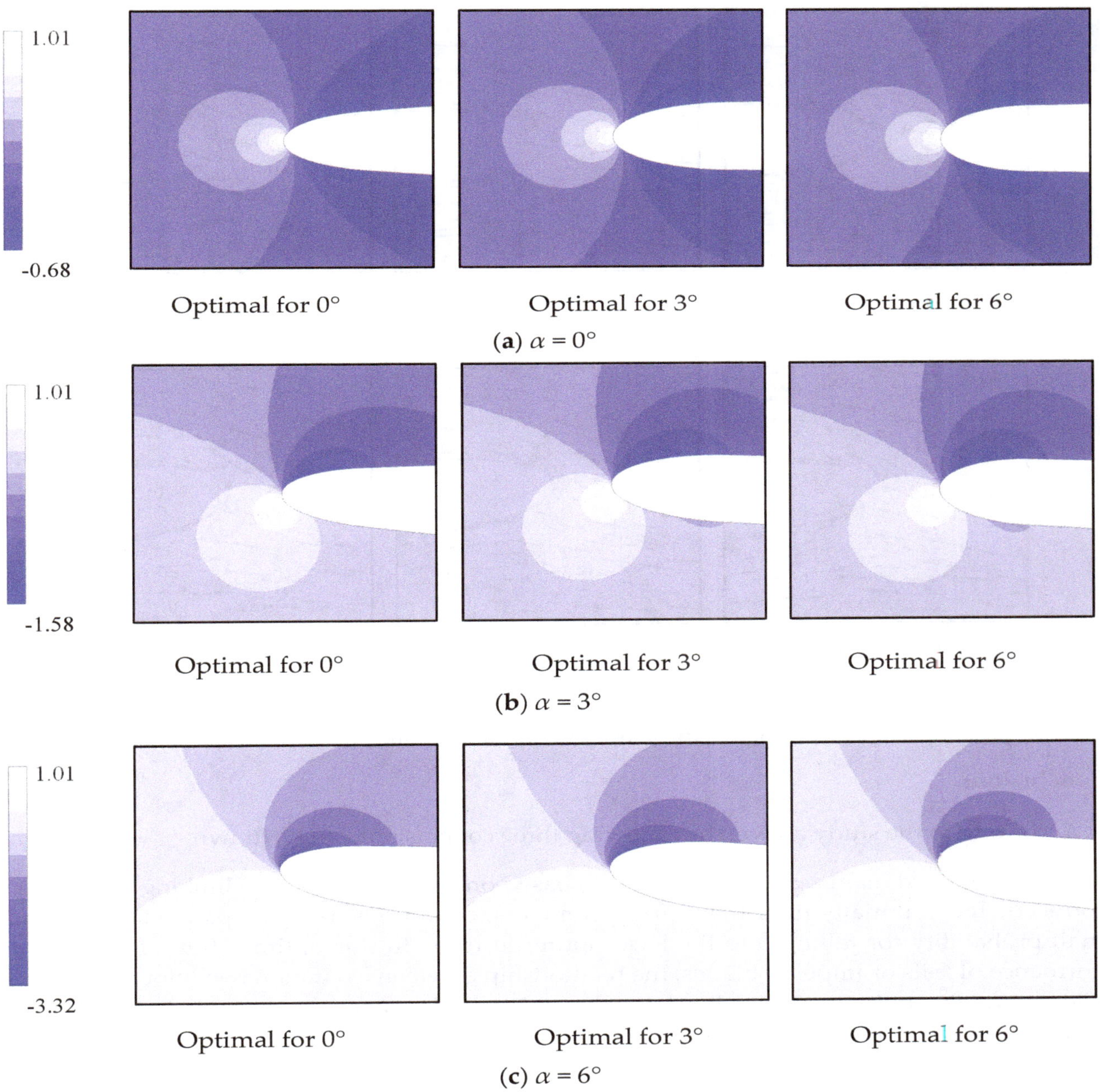

Figure 9. Contour of the dimensionless pressure coefficient C_p.

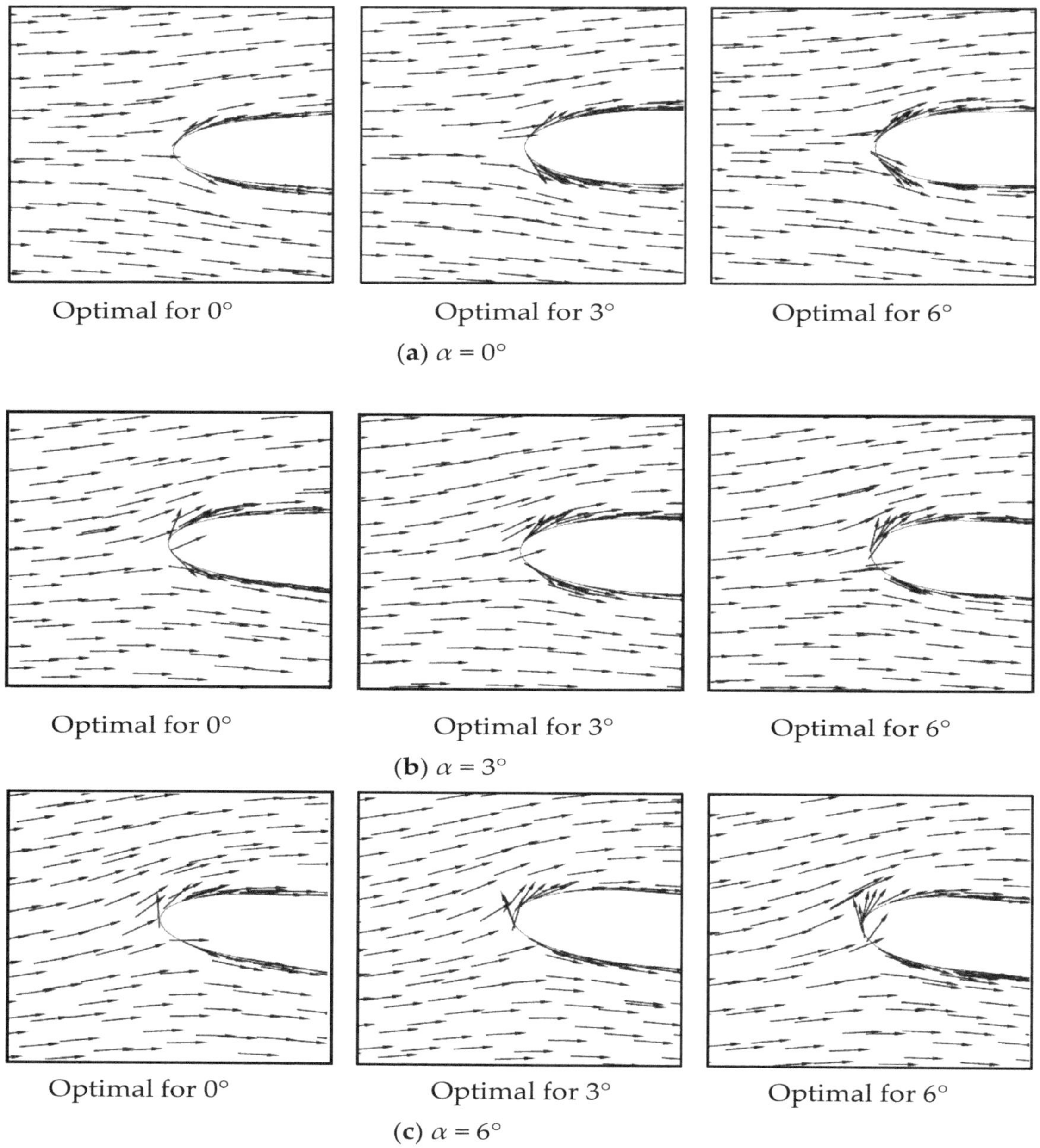

Figure 10. Velocity vectors around the foils.

6. Conclusions

According to the study above, the following three conclusions can be drawn:

(a) The global dynamic-criterion algorithm based on the improved hill-climbing algorithm was introduced. It can initially filter the input parameters/conditions, run in a parallel mode, and set a small probability for falling into the local optimum trap. In this optimization of the cavitation performance of foils or impeller blades, the relationship between cavitation behavior and foil/blade profile is strongly nonlinear. Thus, the global dynamic criterion algorithm provided a reasonable and convenient solution for this cavitation optimization problem.

(b) Several typical methods and some new methods were combined for the optimization. The typical methods include the CFD simulation for turbulent flow and the orthogonal test. The new methods include the Diffusion-angle Integral method and the global dynamic-criterion algorithm. They worked together to search the optimal geometry for improving the cavitation performance for 0, 3 and 6 degrees' incidence angles.

(c) The front 15% geometry was re-designed with slight changing on geometry and impressive enhancement on cavitation performance. The $-C_{pmin}$ of the optimal foils was improved by 1.2%, 11.4% and 14.5% at 0 degree, 3 degrees and 6 degrees comparing with the original foil. In the design range, different parameter values of Diffusion-angle Integral method caused different geometries. The geometries fitted the incoming flow better with smaller scale LE separation and gentler pressure drop. The cavitation performance was enhanced at different incidence angles.

Above all, the geometry re-design around leading-edge can efficiently improve the cavitation performance of hydrofoil. This study provided a successful work for applying anti-cavitation design to the impellers of hydraulic turbomachinery. This study will be helpful for readers who need the improvement of cavitation performance for foils and impellers.

Author Contributions: Methodology, D.Z. and R.T.; Investigation, D.Z.; Writing-Original Draft Preparation, D.Z.; Writing-Review & Editing, R.X.; Supervision, F.W.; Funding Acquisition, R.T., R.X. and F.W.

References

1. Hammitt, F.G. *Cavitation and Multiphases Flow Phenomena*, 1st ed.; McGraw-Hill: New York, NY, USA, 1980; ISBN 9780070259072.
2. Luo, X.; Ji, B.; Tsujimoto, Y. A review of cavitation in hydraulic machinery. *J. Hydrodyn.* **2016**, *28*, 335–358. [CrossRef]
3. Pan, Z.Y.; Yuan, S.Q. *Fundamentals of Cavitation in Pumps*, 1st ed.; Jiangsu University Press: Zhenjiang, China, 2013.
4. Bishop, R.J.; Totten, G.E. Effect of pump inlet conditions on hydraulic pump cavitation: A review. *ASTM Spec. Tech.* **2001**, *339*, 318–332.
5. Lauterborn, W.; Bolle, H. Experimental investigation of cavitation-bubble collapse in the neighbourhood of a solid boundary. *J. Fluid Mech.* **1975**, *72*, 391–399. [CrossRef]
6. Wang, G.; Liu, S.; Shintani, M. Study on Cavitation Damage Characteristics around a Hollow-Jet Valve. *JSME Int. J.* **1999**, *42*, 649–657. [CrossRef]
7. Visser, F.C.; Backx, J.J.M.; Geerts, J. Pump impeller lifetime improvement through visual study of leading-edge cavitation. In Proceedings of the International Pump Users Symposium, Houston, TX, USA, 3–5 March 1998.
8. Tao, R.; Xiao, R.; Zhu, D. Predicting the inception cavitation of a reversible pump-turbine in pump mode. In Proceedings of the 9th International Symposium on Cavitation, Lausanne, Switzerland, 6–10 December 2015.
9. Ruan, H.; Luo, X.; Liao, W. Effects of low pressure edge thickness on cavitation performance and strength for pump-turbine. *Trans. Chin. Soc. Agric. Eng.* **2015**, *31*, 32–39.
10. Adhikari, R.; Vaz, J.; Wood, D. Cavitation Inception in Crossflow Hydro Turbines. *Energies* **2016**, *9*, 237. [CrossRef]
11. Anaka, T.; Tsukamoto, H. Transient behavior of a cavitating centrifugal pump at rapid change in operating conditions-Part 2: Transient phenomena at pump startup/shutdown. *J. Fluids Eng.* **1999**, *121*, 850–856.
12. Balasubramanian, R.; Sabini, E.; Bradshaw, S. Influence of impeller leading edge profiles on cavitation and suction performance. In Proceedings of the 27th International Pump Users Symposium, Houston, TX, USA, 12–15 September 2011.
13. Obeid, S.; Jha, R.; Ahmadi, G. RANS Simulations of Aerodynamic Performance of NACA 0015 Flapped Airfoil. *Fluids* **2017**, *2*, 2. [CrossRef]

14. Wang, G.; Ostoja-Starzewski, M. Large eddy simulation of a sheet/cloud cavitation on a NACA0015 hydrofoil. *Appl. Math. Model.* **2007**, *31*, 417–447. [CrossRef]
15. Neill, G.D.; Reuben, R.L.; Sandford, P.M. Detection of incipient cavitation in pumps using acoustic emission. *J. Process. Mech. Eng.* **1997**, *211*, 267–277. [CrossRef]
16. Fukaya, M.; Okamura, T.; Tamura, Y. Prediction of cavitation performance of axial flow pump by using numerical cavitating flow simulation with bubble flow model. In Proceedings of the 5th International Symposium on Cavitation, Osaka, Japan, 1–4 November 2003.
17. Barre, S.; Rolland, J.; Boitel, G. Experiments and modeling of cavitating flows in venturi: Attached sheet cavitation. *Eur. J. Mech. B/Fluids* **2009**, *28*, 444–464. [CrossRef]
18. Acosta, A.J.; Tsujimoto, Y.; Yoshida, Y. Effects of leading edge sweep on the cavitating characteristics of inducer pumps. *Int. J. Rotating Mach.* **2007**, *7*, 397–404. [CrossRef]
19. Numachi, F.; Oba, R.; Chida, I. Effect of surface roughness on cavitation performance of hydrofoils-report 1. *J. Basic Eng.* **1965**, *87*, 495–502. [CrossRef]
20. Yao, Z.; Xiao, R.; Wang, F. Numerical investigation of cavitation improvement for a francis turbine. In Proceedings of the 9th International Symposium on Cavitation, Lausanne, Switzerland, 1–5 December 2015.
21. Kinnas, S.A. Non-Linear Corrections to the Linear Theory for the Prediction of the Cavitating Flow around Hydrofoils. Ph.D. Thesis, Massachusetts Institute of Technology, Boston, MA, USA, 1985.
22. Tao, R.; Xiao, R.; Farhat, M. Effect of leading edge roughness on cavitation inception and development on a thin hydrofoil. *J. Drain. Irrig. Mach. Eng.* **2017**, *35*, 921–926.
23. Bouziad, Y.A. Physical modelling of leading edge cavitation: Computational methodologies and application to hydraulic machinery. *EPFL* **2005**, *3353*. [CrossRef]
24. Ausoni, P. Turbulent vortex shedding from a blunt trailing edge hydrofoil. *EPFL* **2009**. [CrossRef]
25. Coutier-Delgosha, O.; Reboud, J.L.; Delannoy, Y. Numerical simulation of the unsteady behavior of cavitating flows. *Int. J. Numer. Methods Fluids* **2003**, *42*, 527–548.
26. Kunz, R.F.; Boger, D.A.; Stinebring, D.R. A preconditioned Navier–Stokes method for twophase flows with application to cavitation prediction. *Comput. Fluids* **2000**, *29*, 849–875. [CrossRef]
27. Yang, W.; Xiao, R.; Wang, F. Influence of splitter blades on the cavitation performance of a double suction centrifugal pump. *Adv. Mech. Eng.* **2014**, *6*. [CrossRef]
28. Gopalan, S.; Katz, J. Flow structure and modeling issues in the closure region of attached cavitation. *Phys. Fluids* **2000**, *12*, 895–911. [CrossRef]
29. Luo, X.; Zhang, Y.; Peng, J. Impeller inlet geometry effect on performance improvement for centrifugal pumps. *J. Mech. Sci. Technol.* **2008**, *22*, 1971–1976. [CrossRef]
30. Tan, L.; Cao, S.; Wang, Y. Numerical Simulation of Cavitation in a Centrifugal Pump at Low Flow Rate. *Chin. Phys. Lett.* **2012**, *29*, 014702. [CrossRef]
31. Ting, C.K. On the mean convergence time of multi-parent genetic algorithms without selection. *Adv. Artif. Life* **2005**, 403–412.
32. Akbari, R.; Ziarati, K. A multilevel evolutionary algorithm for optimizing numerical functions. *Int. J. Ind. Eng. Comput.* **2011**, *2*, 419–430. [CrossRef]
33. Whitley, D. A genetic algorithm tutorial. *Stat. Comput.* **1994**, *4*, 65–85. [CrossRef]
34. Colorni, A.; Dorigo, M.; Maniezzo, V. Ant system for job-shop scheduling. *Oper. Res. Stat. Comput. Sci.* **1994**, *34*, 39–53.
35. Cormen, T.H.; Leiserson, C.E.; Rivest, R.L. *Introduction to Algorithms*, 2nd ed.; McGraw-Hill: New York, NY, USA, 2001.
36. Huang, L.; Huang, P.G.; Lebeau, R.P. Optimization of aifoil flow control using a genetic algorithm with diversity control. *J. Aircr.* **2007**, *44*, 1337–1349. [CrossRef]
37. Liu, H.; Wang, K.; Yuan, S. Multicondition Optimization and Experimental Measurements of a Double-Blade Centrifugal Pump Impeller. *J. Fluids Eng.* **2013**, *135*, 111031. [CrossRef] [PubMed]
38. Liu, L.; Zhu, B.; Bai, L. Parametric design of an ultrahigh-head pump-turbine runner based on multiobjective optimization. *Energies* **2017**, *10*, 1169. [CrossRef]
39. Liu, M.; Tan, L.; Cao, S. Design method of controllable blade angle and orthogonal optimization of pressure rise for a multiphase pump. *Energies* **2018**, *11*, 1048. [CrossRef]

40. Tao, R.; Xiao, R.; Wang, F. Improving the cavitation inception performance of a reversible pump-turbine in pump mode by blade profile redesign: Design concept, method and applications. *Renew. Energy* **2019**, *133*, 325–342. [CrossRef]
41. Behnia, M.; Parneix, S.; Shabany, Y. Numerical study of turbulent heat transfer in confined and unconfined impinging jets. *Int. J. Heat Fluid Flow* **1999**, *20*, 1–9. [CrossRef]
42. Durbin, P.A. Separated Flow Computations with the k–ε–v2 Model. *AIAA J.* **1995**, *33*, 659–664. [CrossRef]
43. Parneix, S.; Durbin, P.A.; Behnia, M. Computation of 3-D Turbulent Boundary Layers Using the V2F Model. *Flow Turbul. Combust.* **1998**, *60*, 19–46. [CrossRef]

10

Parametric Investigation using Computational Fluid Dynamics of the HVAC Air Distribution in a Railway Vehicle for Representative Weather and Operating Conditions

Christian Suárez [1], Alfredo Iranzo [2,*], José Antonio Salva [1], Elvira Tapia [2], Gonzalo Barea [3] and José Guerra [2]

1 AICIA, Andalusian Association for Research & Industrial Cooperation, Camino de los Descubrimientos s/n, Edf. Escuela Superior de Ingenieros de Sevilla, 41092 Seville, Spain; chss@us.es (C.S.); jsalva@us.es (J.A.S.)
2 Thermal Engineering Group, Energy Engineering Department, School of Engineering, University of Seville, Camino de los Descubrimientos s/n, 41092 Seville, Spain; etapia@us.es (E.T.); jjguerra@us.es (J.G.)
3 Hispacold, C/Pino Alepo 1, Polígono Industrial El Pino, 41016 Seville, Spain; gbarea@hispacold.es
* Correspondence: airanzo@us.es.

Academic Editor: Bjørn H. Hjertager

Abstract: A computational fluid dynamics (CFD) analysis of air distribution in a representative railway vehicle equipped with a heating, ventilation, air conditioning (HVAC) system is presented in this paper. Air distribution in the passenger's compartment is a very important factor to regulate temperature and air velocity in order to achieve thermal comfort. A complete CFD model, including the car's geometry in detail, the passengers, the luminaires, and other the important features related to the HVAC system (air supply inlets, exhaust outlets, convectors, etc.) are developed to investigate eight different typical scenarios for Northern Europe climate conditions. The results, analyzed and discussed in terms of temperature and velocity fields in different sections of the tram, and also in terms of volumetric parameters representative of the whole tram volume, show an adequate behavior from the passengers' comfort point of view, especially for summer climate conditions.

Keywords: heating, ventilation, air conditioning; computational fluid dynamics; railway vehicle; heat transfer; thermal comfort; tram

---

1. Introduction

Air distribution in passenger's compartment is a very important factor to regulate temperature and air velocity in order to achieve thermal comfort. The study of air distribution plays an important role in the design of new HVAC (heating, ventilation, and air conditioning) equipment and also in the evaluation of existing solutions. Technological solutions such as innovations in air-conditioning and other forms of cooling or ventilation can improve environmental conditions, which is beneficial for human health, comfort, and productivity.

Saving energy and providing thermal comfort are two important goals of HVAC systems. Regarding the first goal, the European Project ECORailS (Energy Efficiency and Environmental Criteria in the Awarding of Regional Rail Transport Vehicles and Services, 2011) [1] for enhanced energy-efficiency of regional rail passenger services is a good example, as its main objective is to reduce the specific energy consumption of European regional passenger rail transport by 15% by 2020. While it is relatively easy to define and estimate the energy consumption of HVAC systems, the evaluation of thermal comfort is much more challenging. The American Society of Heating, Refrigerating and Air-Conditioning Engineers (ASHRAE) defines thermal comfort as "condition of mind that expresses satisfaction with the thermal environment" [2]. As it depends on personal factors (such as metabolic rate or clothing) and environmental factors (such as air velocity, air temperature, air temperature stratification, radiant temperature, radiant temperature asymmetry, relative humidity, or turbulence intensity in the occupied zone), the definition and evaluation of thermal comfort is subjective and complex.

European standard EN-14750-1:2006 [3], relative to air conditioning for urban and suburban rolling stock, establishes a classification of the railway vehicles depending on the number of standing passengers, the average travel time, and the average time between two vehicle stops. According to that definition a railway vehicle is category A for suburban/regional transport and B for the rest. It also defines different comfort parameters and requirements that depend on the vehicle category and are related to the temperature and velocity within the railway vehicle.

Keeping fixed the parameters related to personal factors and the indoor geometry, thermal comfort is directly related to air conditions: velocity, temperature, and temperature stratification. It is evident that different diffusers and different locations of supply inlets and exhaust outlets will affect air distribution and, consequently, also the distribution of the cited thermal comfort parameters. Moreover, for a given HVAC system operating with different conditions (air supply temperatures, air supply flow, climate conditions etc.) these parameters will also change. Hence, it is necessary to understand quantitatively how these different operating conditions will affect local thermal comfort.

In public transport, such as railway vehicles, thermal comfort is even more important than in building applications. In fact, standing passengers inside a cabin for several minutes/hours do not have opportunities of moving. Passenger comfort has to be one of the most important elements in the design of public transport in order to persuade people to choose it instead of other means of transport. This is the reason why, for several years, many studies have focused on the air distribution system in the passenger's cabin.

As discussed by Liu et al., in its literature review paper [4] (focused on the particular case of commercial airliner cabins), two main approaches are available for analyzing air distribution: experimental measurements inside the cabin (with different equipment such as hotwire anemometers and hot-sphere anemometers, particle tracking velocimetry, particle streak velocimetry, particle image velocimetry, and ultrasonic anemometry) and numerical simulations (mainly CFD simulations). R. Lieto [5] studied, both numerically and experimentally, the indoor climate in city busses. He contrasted his numerical results with measurements of the air speed in some selected points in two different situations (a moving vehicle with the door closed and the vehicle at a bus stop with open doors). Zhu et al. [6] also studied, numerically and experimentally, the micro-environmental conditions in public transportation buses, focusing on thermal comfort and also on air quality. In other work, Zhu et al. [7] studied numerically the risk of airborne influenza infection in a bus microenvironment with different ventilation system configurations.

There are more publications regarding air distribution in aircraft cabins that are also focused on thermal comfort and air quality. Bianco et al. [8] studied numerically-transient simulations of the thermal and fluid dynamic fields in the cabin of an executive aircraft, considering 2D and 3D models. In other papers [9–14] the studies have been devoted to air quality and the distribution of pollutants, mainly CO and CO_2.

Although comfort conditions in public transport, such as buses or aircraft cabin, planes have been widely discussed in recent years, to the best of the knowledge of the authors, there are only a few papers specifically related to railway vehicles. Thompson et al. [15] presented a review of cooling systems used in railway vehicles and the advantages of using heat storage in the complex environment of an underground railway. Show [16] discussed a ventilation criterion for ensuring a clean environment, with low carbon-dioxide concentration (lower than 0.1% in air). To obtain such acceptable indoor air-quality inside the train compartment, he mentions two possibilities: to increase the fresh-air supply rate for dilution or to design a better air-distribution system. However, none of these studies in railway vehicles include an analysis of air distribution and its relationship with the passengers' thermal comfort.

To the best of the knowledge of the authors, perhaps due to the complex and time-consuming process of the geometry definition and the high required computational time in the simulations, there are not any previous studies related specifically to air distribution railway vehicles with passengers using CFD techniques. Even though all the modes of transportation share some similarities, such as the air conditioning and the comfort conditions, there are also some differences in the functionality and the passengers' distribution that make these transports different from a thermal analysis point of view. Moreover, as concluded in Yan et al. [17] the computational manikin model approach representing the passengers in the vehicle can affect the temperature fields, and an excessive degree of simplification can be incapable of predicting accurate results.

Experimental measurements, which are often considered reliable to analyze air distribution and thermal comfort, can be technically very difficult and expensive. It is very difficult to obtain results under realistic thermo-fluid conditions or detailed geometry. On the other hand, CFD simulations offer a good alternative. Due to their flexibility and efficiency a lot of scenarios can be studied with a relatively low cost. Furthermore, the results in terms of temperature and velocity fields can be obtained for the whole volume of interest and not only for a specific measured point.

In the present work, a detailed CFD analysis of air distribution and the temperature and velocity profiles in different sections of a representative railway vehicle is performed (including the passengers), in order to discuss the results in terms of thermal comfort.

2. Problem Definition and Methodology

A detailed three-dimensional CFD model of a railway vehicle formed by four cars is developed using the software ANSYS-CFX (ANSYS, Inc., Southpointe, Canonsburg, PA, USA). The tram (39.475 m length and 2.650 m width) is divided in two parts (part one: cars C1 and N, and part two: cars R and C2). In each car there are two doors and eight windows. Although cars C1 and C2 are identical, cars N and R are geometrically different.

In the model, a representative HVAC system and a random distribution of passengers are integrated. In all the studied cases, the tram remains stopped and the driver's cabin is not included. A 3D view of the railway vehicle is shown in Figure 1.

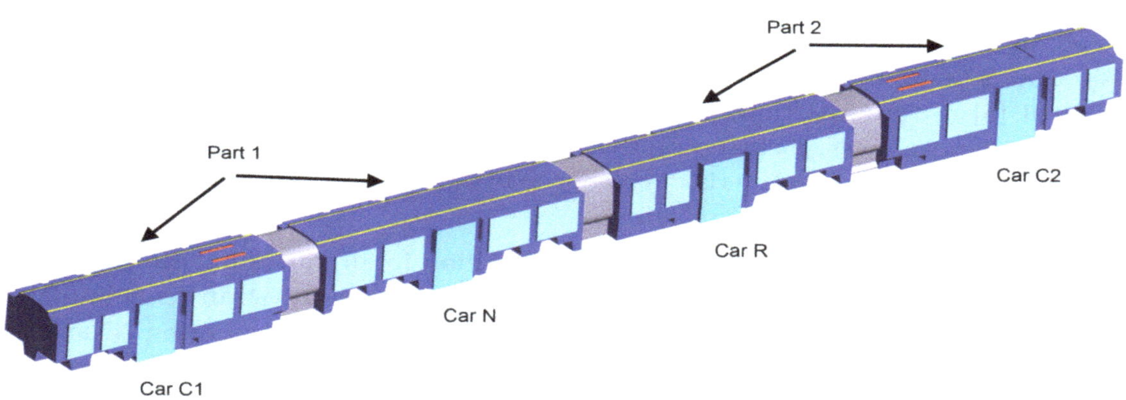

Figure 1. 3D railway vehicle model view.

The boundary conditions related to the HVAC system are shown in Figure 2, for part one of the tram. The conditioned air (3000 m^3/h in each part of the tram) is supplied by two longitudinal forced ventilation inlets located at the ceiling of the tram. In the extreme cars (C1 and C2) there are two forced ventilation exhausts. Air is homogenously supplied along the tram, except near the forced exhaust vents located at the ceiling, where vents are closed in order to avoid the short-circuiting of the supply air.

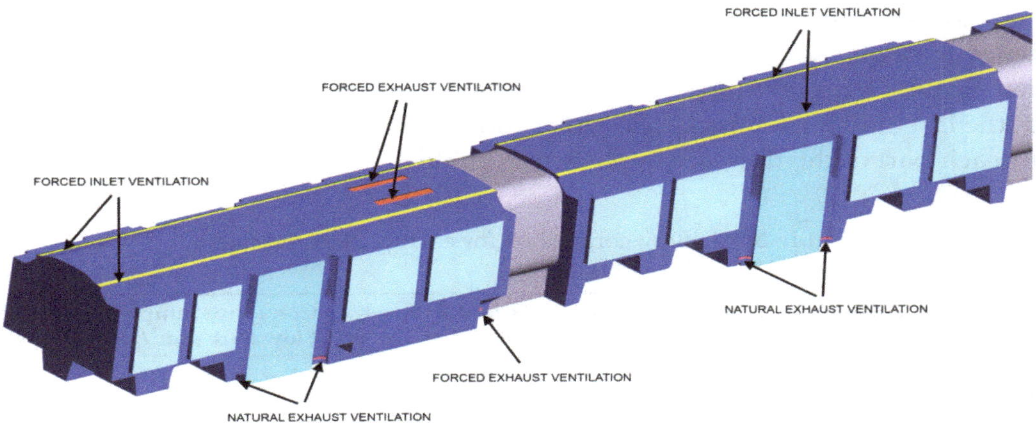

Figure 2. Inlets and exhaust openings, part 1.

In each door there are two natural exhaust openings. In cars C1 and C2 there are, in total, two forced exhaust ventilation systems located at the walls of 150 m^3/h each.

In addition, the HVAC system includes 16 convectors (four in each car) that only work during winter climate conditions. These convectors are of two different models: type 1 (supplied flow 230 m^3/h) and type 2 (supplied flow 200 m^3/h). The number of the different convectors in each car is shown in Table 1.

Table 1. Convector types in each car.

Model	Tram (C1-N-R-C2)	Car C1	Car N	Car R	Car C2
Type 1	10	2	4	2	2
Type 2	6	2	0	2	2
Total	16	4	4	4	4

The cases studied are based on two different passengers' occupation scenarios, one including the passengers (220 in total, 120 standing and 100 seated) and the other without passengers. In the cases including occupation, it is considered a sensible heat load of 78 W/person (17.16 kW in total). There are two lighting rows in the longitudinal direction on each side of the ceiling. The total lighting sensible heat load in the tram is 2.67 kW (homogeneously distributed in the four cars). The overall heat loss coefficient through the tram envelope (including walls, windows, floor and ceiling) is assumed to be 2 $W/(m^2K)$.

Eight different parameters of working conditions have been studied for typical weather conditions in Northern Europe, in an attempt to investigate air distribution, temperature, and velocity fields inside the tram and its effects on thermal comfort. The temperature conditions are summarized in Table 2 for different scenarios.

Table 2. Temperature working conditions.

Case	Season	Passengers (Yes/No)	Supply Temperature (°C)	Exterior Temperature (°C)	Convector Type 1 Supply Temperature (°C)	Convector Type 2 Supply Temperature (°C)
1	SUMMER	No	20.3	40	-	-
2	SUMMER	No	12.4	28	-	-
3	SUMMER	Yes	12.4	28	-	-
4	SUMMER	Yes	12.1	27	-	-
5	WINTER	No	32.6	0	41	32
6	WINTER	No	22.4	−16	43	35
7	WINTER	Yes	22.4	−16	43	35
8	WINTER	Yes	21.6	−6	43	35

Cases 1–4 and 5–8 correspond to summer and winter weather conditions, respectively. Scenarios without passengers (cases 1, 2, 5, and 6) and with passengers (cases 3, 4, 7, and 8) are considered. In all cases, supplied air flow temperature, exterior temperature, and the convectors' supplied temperature are indicated (the last only for winter cases). Solar radiation is not included in the model. The air flow conditions for each part of the tram are described in Table 3.

Table 3. Air flow conditions for each part of the tram.

Case	Supplied Flow (m^3/h)	Fresh Flow (m^3/h)	Forced Exhaust Ceiling v(m^3/h)	Forced Exhaust Walls (m^3/h)	Convector Supply Flow Part 1 (m^3/h)	Convector Supply Flow Part 2 (m^3/h)
1	3000	-	3000	-	-	-
2	3000	1100	1900	150	-	-
3	3000	1100	1900	150	-	-
4	3000	1650	1350	150	-	-
5	3000	-	3000	-	1780	1720
6	3000	1100	1900	150	1780	1720
7	3000	1100	1900	150	1780	1720
8	3000	1650	1350	150	1780	1720

In all the studied scenarios the fresh air flow for each part of the tram is indicated. The difference between the 3000 m^3/h supplied flow in each part of the tram and the fresh flow is the forced flow exhausted by the ceiling extractions (recirculated air flow).

The rest of the air is exhausted by the forced wall extractions located at the walls in cars C1 and C2 and by the natural openings located at the doors. Particularly, in cases 1 and 5 the HVAC system does not supply fresh air, which means that in each part of the tram, the 3000 m^3/h supplied air flow is exhausted by the ceiling extractions. The forced wall extractions located at the walls of cars C1 and C2 are active in all cases except for cases 1 and 5. Finally, the convectors' supplied air flow in the two different parts of the tram (only for winter cases) is also indicated.

For the defined operating conditions, a 3D CFD model is used to predict the temperature and velocity fields of air inside the tram. With this methodology, the conservation equations of mass, momentum and energy (Navier-Stokes equations) are solved using a computer-based tool over the region of interest, with specified conditions on the boundaries of the tram.

The mesh is formed by tetrahedral elements in the fluid volume and prismatic elements near the walls in order to obtain the correct refinement of the viscous kinematic and thermal boundary layer. For the inlet and outlet regions of air ventilation, the mesh has been generated with smaller elements in order to obtain more precision and detail in the characteristics of the air movement in this critical zone for the velocity field.

In the analysis it is considered a steady-state regime, and the sheer stress transport (SST) turbulence model is used. Air is considered to behave like an ideal gas.

The passengers are modeled in two different positions: standing or sitting. For the purpose of this study, the design capacity of the tram is 220 passengers, 120 standing and 100 sitting. In order to reduce the number of nodes and the time consumption during the simulations, a simplified geometry model of the passengers has been defined. A sketch of the passengers' distribution in car C1 is shown in Figure 3.

A similar passenger distribution is defined in the rest of the cars.

Figure 3. Passenger's distribution in car C1.

3. Results

Results are presented in three different sections. Firstly, the air distribution within the tram volume is described briefly, with a focus on the main inlets and outlets of air and pointing out the singularities of the velocity fields in its proximities. Secondly, results are discussed in terms of the temperature fields in different representative planes in the tram (shown in Figure 4).

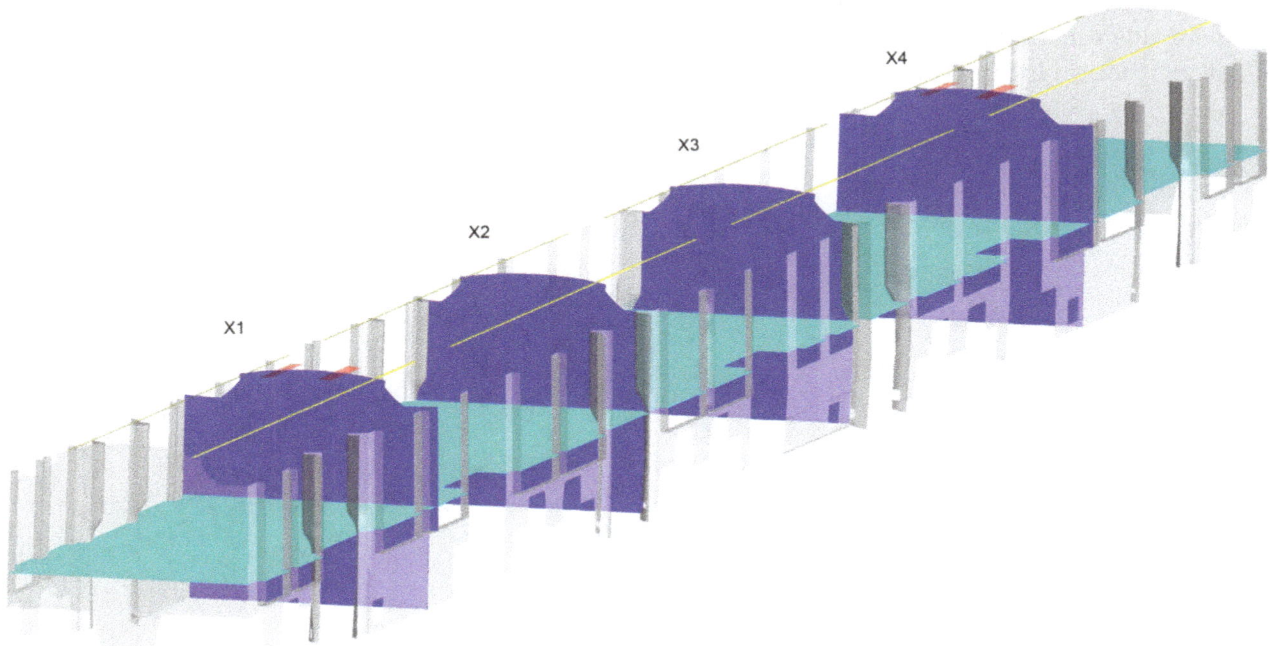

Figure 4. Selected horizontal and vertical planes.

The selected horizontal plane, representative for the occupancy plane, is at a height of +1.1 m. Four different vertical planes (X1, X2, X3, and X4) are selected, one in each car of the tram. While planes X1 and X4 are representative of the forced extraction ventilation located at the ceiling of cars C1 and C2, respectively, planes X2 and X3 are representative of the forced inlet ventilation and the natural exhaust ventilation located at doors of cars N and R, respectively. Finally, the results are also quantitatively analyzed in the tram volume. In order to avoid areas of local discomfort such as the proximities of the tram envelope, the luminaires, or the convectors, it is defined an interior comfort volume that does not consider these zones to evaluate the comfort parameters. In the present work, where the design of the HVAC system is defined in advance, these comfort parameters are useful to compare multiple scenarios with different working conditions, but these parameters could be also used to analyze the impact in air distribution and thermal comfort of different HVAC systems.

3.1. Description of Air Distribution in the Tram

The conditioned air is supplied by the longitudinal inlets located at the cars' ceiling (and in the case of winter season, also by the convectors). Air is homogenously supplied by the longitudinal inlets except near the forced exhaust vents located at the ceiling, where the vents are closed in order to avoid a short circuit in the air distribution. The vents are also closed in the joints between cars. While most of the air is exhausted by the forced ventilation system located at the ceiling in cars C1 and C2, the rest is driven out of the tram by the natural openings located at each door and the forced exhaust ventilation systems located at the walls of cars C1 and C2.

In Figure 5a, the velocity fields for case 1 (without passengers) in the vertical section containing the ceiling extractions in car C1 (plane X1) is shown. It is observed an ascendant air flow and the maximum velocities are reached in the proximities of the ceiling extractions. In Figure 5b, the velocity fields for the same case in the vertical section containing the longitudinal supply inlets and the natural openings located at the doors in car N (plane X2) is shown. It is observed that air is driven from the supply inlets located at the ceiling to the doors' natural openings located near the floor. A circular air flow pattern is reached and a relatively small amount of air is driven out of the tram by the natural exhaust ventilation located at the doors.

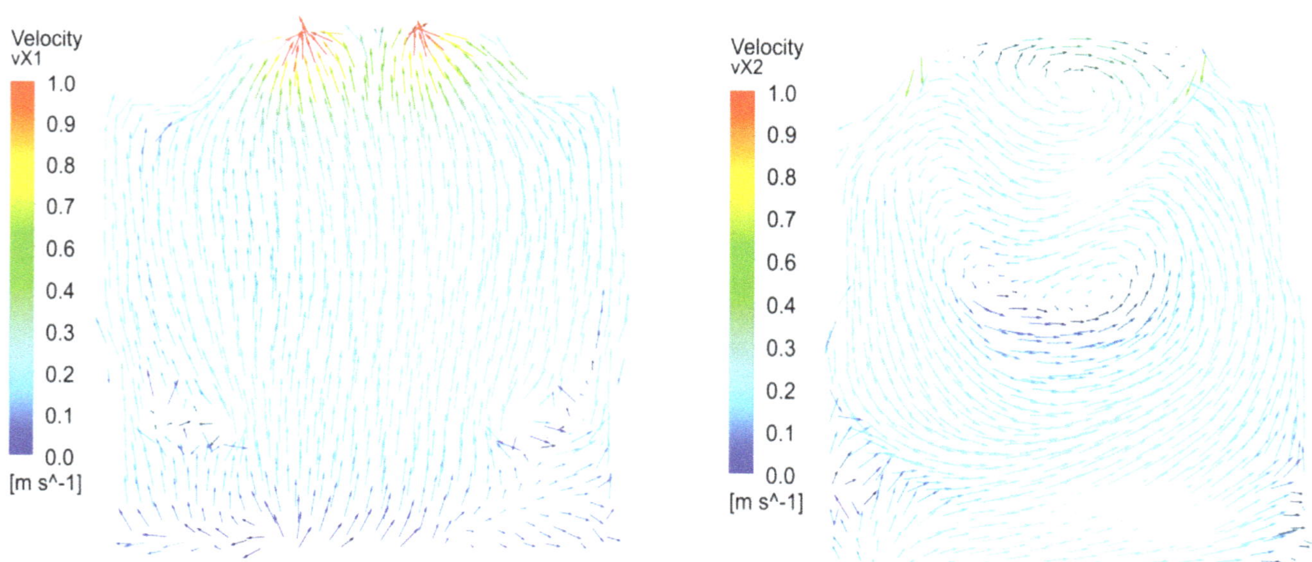

(a) (b)

Figure 5. Velocity fields, case 1, plane X1 (**a**) and plane X2 (**b**).

Similar air flow distribution is observed for cases with occupation, even though the air distribution is more complex. Figure 6 shows the velocity fields for Case 3 (with passengers) in the same sections described in the previous figure.

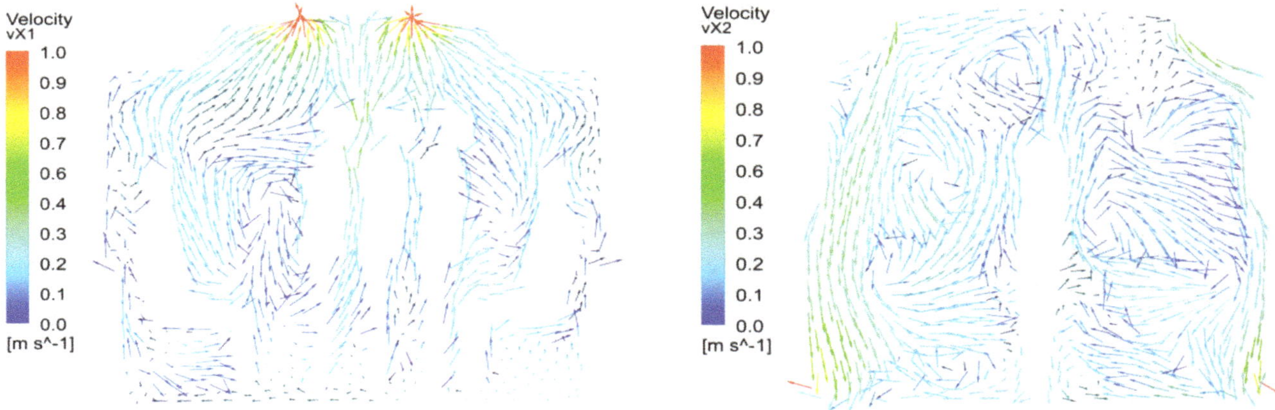

(a) (b)

Figure 6. Velocity fields, Case 3, plane X1 (**a**) and plane X2 (**b**).

In Figure 6a, it is observed again that air is driven to the ceiling forced extractions, reaching maximum velocities values near them. In Figure 6b, air is driven from the supply inlets located at the ceiling to the doors' natural openings located near the floor. Again, the maximum velocities are reached in the proximities of inlets and outlets. Even though the conditioned air is not supplied near the ceiling extractions, as observed in Figures 5a and 6a, air distribution and circulation inside the tram is adequate, with no death zones without air circulation.

3.2. Temperature Distribution

According to the European standard EN-14750-1:2006 [3], the difference of temperature at a horizontal plane of a height of +1.1 m must be lower than 8 °C for B category railway vehicles. The same condition must be satisfied at any vertical plane (X direction). In this work, the requirement is quantified in terms of the percentage area that is in the range $T_{avg} \pm 4$ °C, where T_{avg} is the average temperature at the particular plane. With this definition, a percentage of 100% means that the whole plane satisfies the requirement. The temperature field in the cited horizontal plane is presented in Figures 7 and 8 for the summer cases (Cases 1–4) and the winter cases (Cases 5–8) respectively.

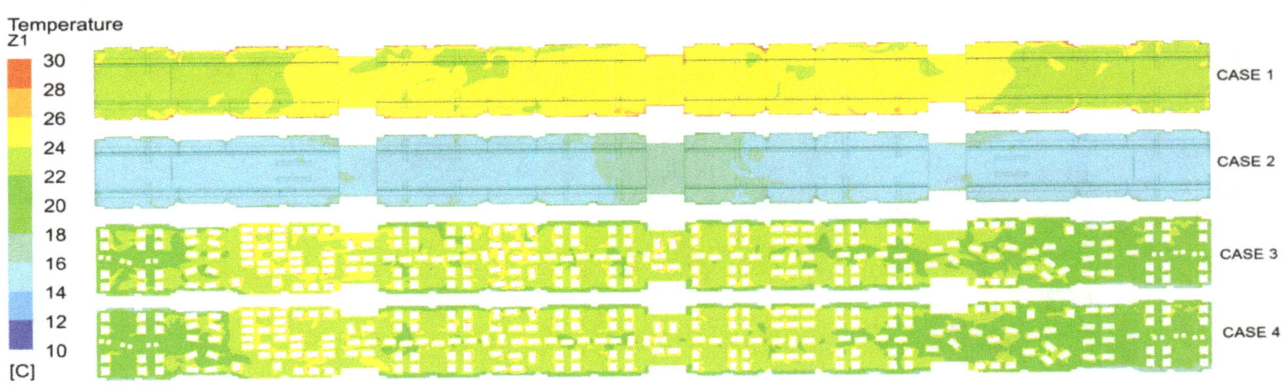

Figure 7. Temperature fields in the horizontal plane at +1.1 m, summer cases (1–4).

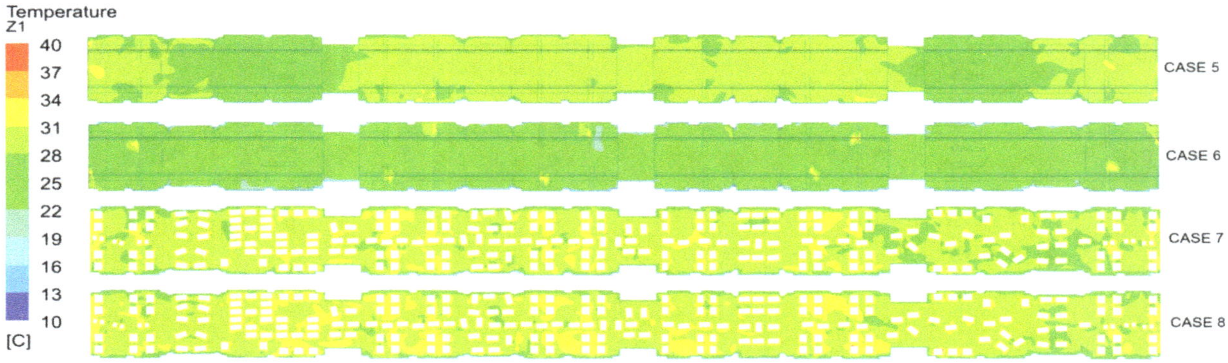

Figure 8. Temperature fields in the horizontal plane at +1.1 m, winter cases (5–8).

The quantitative results at the horizontal plane, summarized in Table 4, show a homogeneously-distributed temperature around the average temperature in all cases, especially for summer cases 1–4 (percentages above 98%) in comparison with winter cases 5–8 (percentages above 96%).

Table 4. Temperature parameters at surfaces.

Comfort Parameters (Plane Z +1.1 m)		
Case	**T_{avg} (°C)**	**T_{avg} ± 4 °C (%)**
1	24.4	100
2	15.7	100
3	22.7	100
4	22.4	98
5	28.6	100
6	24.3	99
7	29.4	97
8	29.9	96

As an example, Figures 9 and 10 show the temperature fields in the vertical planes X1, X2, X3, and X4 for Cases 3 and 7 (representative cases with passengers for summer and winter seasons, respectively).

Figure 9. Temperature fields at vertical planes X1, X2, X3, and X4, summer Case 3.

Figure 10. Temperature fields at vertical planes X1, X2, X3, and X4, winter case 7.

In the representative summer case (Figure 9), it is observed that maximum temperatures are reached near the passengers' proximities and minimum temperatures are reached near the supply inlets. In the representative winter case (Figure 10) the maximum temperatures are also reached near the passengers' proximities and also in the convectors' zone of influence, while the minimum temperatures are reached near the tram envelope (due to the high gradient of temperature with the exterior).

The quantitative results at the vertical planes for the eight cases are summarized in Table 5.

Table 5. Temperature parameters at vertical planes.

	Comfort Parameters (Planes X1, X2, X3, and X4)							
Case	Plane X1 (Car C1)		Plane X2 (Car N)		Plane X3 (Car R)		Plane X4 (Car C2)	
	T_{avg} (°C)	$T_{avg} \pm 4$ °C (%)	T_{avg} (°C)	$T_{avg} \pm 4$ °C (%)	T_{avg} (°C)	$T_{avg} \pm 4$ °C (%)	T_{avg} (°C)	$T_{avg} \pm 4$ °C (%)
1	24.7	100	24.4	100	24.9	100	24.5	100
2	15.8	100	15.7	100	15.6	100	15.6	100
3	24.7	100	24.4	100	24.9	100	24.5	100
4	24.6	99	23.6	98	23.2	99	22.6	100
5	27.3	100	27.4	86	28.2	87	27.4	100
6	23.9	99	23.0	91	24.5	95	23.8	98
7	30.7	95	28.8	83	28.7	90	28.7	96
8	31.7	96	29.2	84	29.7	92	29.8	97

Results show a homogeneously distributed temperature around the average temperature in all cases, especially for summer cases (1–4), with percentages above 98%. In winter cases (5–8), in which the temperature gradients are higher, the percentages in the range are lower (percentages above 83%). In these winter cases, the results are better for sections X1 and X4 (percentages above 95%) in comparison with planes X2 and X3. This is explained by the fact that sections X1 and X4 are located at the extreme cars in which the forced ceiling extractions cause a more homogeneous air distribution.

3.3. *Temperature and Velocity within the Tram Volume*

The previously-analyzed sections were selected in a first approach as representative of the whole tram thermal behavior. These sections included hot areas (such as the proximities of the lighting or the passengers) and other cold areas (such as the proximities of the ceiling air supply inlets or the tram envelope in winter cases) that are not strictly situated in the occupancy zones and can distort the results. Taking into account the whole volume of the tram, the volume of these zones is not negligible. In order to obtain a more realistic approach of the tram thermal behavior in the zone where the passengers are located, the comfort volume is defined as the space between heights +0.5 m and +1.7 m, excluding the convectors' influence zone and also the proximities of the walls of the tram.

Temperature and velocity requirements are quantified in terms of the volume percentage that satisfies the given requirement in the defined comfort volume.

In the case of the temperature requirement, it is calculated the percentage volume in the range $T_{avg} \pm 4$ °C, where T_{avg} is the average temperature at the volume.

In the case of the velocity requirement, it is quantified the percentage volume in which the velocity is lower than the maximum allowed (v_{max}), where v_{max} is defined according to EN-14750-1:2006 [3], and depends on the railway vehicle category and also on the average temperature in the tram.

This maximum allowed velocity v_{max} is shown in Table 6 for different average temperatures for B category trams:

Table 6. EN-14750-1 velocity requirements depending on the average temperature (category B).

T_{avg} (°C)	Maximum Allowed Velocity (m/s)
18	0.30
22	0.35
25	0.70
28	1.4
30	2.0
35	4.0

Taking into account this definition, the results in the defined volume are summarized in Table 7.

Table 7. Velocity requirement depending on the average temperature (category B).

Case	Comfort Parameters (Volume)				
	Temperature		Velocity		
	T_{avg} (°C)	$T_{avg} \pm 4$°C (%)	v_{max} (m/s)	v_{av} (m/s)	$v < v_{max}$ (%)
1	24.5	100	0.60	0.19	100
2	15.8	100	0.30	0.15	99
3	23.1	99	0.42	0.17	97
4	22.9	94	0.42	0.16	96
5	29.0	97	1.60	0.20	100
6	23.7	95	0.54	0.20	98
7	29.0	89	1.60	0.19	100
8	29.6	89	1.80	0.19	100

The quantitative results in the defined comfort volume show a good behavior in terms of the defined comfort parameters for temperature and velocity. In the case of air temperature, results show a homogeneously-distributed temperature around the average temperature in all cases, especially for summer cases 1–4 (percentages above 94%) in comparison with winter cases 5–8 (percentages above 89%). In the case of air velocity, results in the defined comfort volume are excellent. In winter cases 5–8, where the maximum allowed velocities are higher (due to the higher average temperatures in the tram) the percentages are 100% except for case 6 (98%). In summer cases 1–4, the percentages are above 96%.

4. Conclusions

A CFD analysis of air distribution in a railway vehicle equipped with a specific HVAC system has been presented for eight different scenarios representative of typical summer and winter Northern European climate conditions.

Firstly, a description of the CFD model focusing on the HVAC system characteristics is presented and the results are discussed in terms of air distribution. Results show that air distribution and circulation inside the tram is adequate, with no death zones without air circulation.

Temperature fields in different representative horizontal and vertical sections of the tram are also analyzed in terms of the area percentage that is in the range $T_{avg} \pm 4$ °C, where T_{avg} is the average temperature at the particular plane. The quantitative results at the selected sections show a homogeneously distributed temperature around the average temperature in all cases, especially for summer cases 1–4 (percentages above 98%) in comparison with winter cases 5–8 (percentages above 83%). These worse results for winter conditions are explained by the fact that in these cases the difference of temperature between the hot areas (such as the convectors' influence zone or the passengers' proximities) and the cold areas (the proximities of the tram envelope, with very low exterior temperatures) is higher in comparison with the summer cases.

To obtain a more realistic approach representative of the whole tram volume (not depending on the selected sections) the temperature and velocity results are quantified in a comfort volume, defined as the space between heights +0.5 m and +1.7 m, excluding the convectors' influence zone and also the proximities of the walls of the tram. This comfort volume represents the volume occupied by the passengers. The quantitative results in the defined comfort volume corroborate the previous temperature surface results. Again, better results are observed especially for summer cases 1–4 (percentages above 94%) in comparison with winter cases 5–8 (percentages above 89%). In the case of air velocity, the results for the defined comfort volume are excellent for both winter and summer climate conditions.

The main contribution of this work to the state of the art is the detailed CFD analysis of a representative railway vehicle for summer and winter conditions, considering a realistic occupancy level. As indicated in the introduction section, to the best of the knowledge of the authors there are no previous studies specifically related to air distribution in railway vehicles with passengers. The present work introduces this analysis and establishes a methodology and starting point for future studies. It is expected that HVAC systems designers and integrators will increasingly require CFD analysis of the final systems (railway vehicles and other means of public transport) in order to finely adjust the system configuration and provide more satisfactory comfort levels to final users. It is, therefore, necessary to establish an appropriate methodology for such CFD analysis. The design of the air distribution system in the railway coach will also require CFD analysis of the air ducts for ensuring better performance (lower pressure drop and better flow distribution).

Finally, solar radiation is not included in the model in the present work. Even though for winter weather conditions this assumption is conservative in terms of thermal comfort, further analysis must be carried out in future works that analyze the influence of solar radiation in the results, especially for summer weather conditions.

Author Contributions: Christian Suárez ran and post-processed the complete set of simulation cases, analyzed the results, and wrote the manuscript. Alfredo Iranzo critically analyzed the simulation results and revised the manuscript. José Antonio Salva and Elvira Tapia conducted the pre-processing of the simulations cases. Gonzalo Barea established the operating conditions according to the applicable normative and José Guerra critically revised the whole work and coordinated the tasks. All authors read and approved the final manuscript.

References

1. Energy Efficiency and Environmental Criteria in the Awarding of Regional Rail Transport Vehicles and Services. Guidelines for Public Transport Administrations in Europe. ECORailS Project, 2011. Available online: http://www.ecorails.eu/index.php (accessed on 1 September 2016).
2. *ASHRAE Standard 55–1981: Thermal Environmental Conditions for Human Occupancy*; American Society of Heating, Refrigerating, and Air-Conditioning Engineers, Inc.: Atlanta, GA, USA, 1981.
3. *EN 14750-1:2006: Railway Applications—Air Conditioning for Urban and Suburban Rolling Stock—Part 1: Comfort Parameters*; BSI: London, UK, 2006.
4. Liu, W.; Mazumdar, S.; Zhang, Z.; Poussou, S.B.; Liu, J.; Lin, C.-H.; Chen, Q. State-of-the-art methods for studying air distributions in commercial airliner cabins. *Build. Environ.* **2012**, *47*, 5–12.
5. De Lieto Vollaro, R. Indoor Climate Analysis for Urban Mobility Buses: A CFD Model for the Evaluation of thermal Comfort. *Int. J. Environ. Prot. Policy* **2013**, *1*, 1–8. [CrossRef]
6. Zhu, S.; Demokritou, P.; Spengler, J. Experimental and numerical investigation of micro-environmental conditions in public transportation buses. *Build. Environ.* **2010**, *45*, 2077–2088. [CrossRef]
7. Zhu, S.; Srebric, J.; Spengler, J.; Demokritou, P. An advanced numerical model for the assessment of airbone transmission of influenza in bus microenvironments. *Build. Environ.* **2012**, *47*, 67–75. [CrossRef]
8. Bianco, V.; Manca, O.; Nardini, S.; Roma, M. Numerical investigation of transient thermal and fluidynamic fields in a executive aircraft cabin. *Appl. Therm. Eng.* **2009**, *29*, 3418–3425. [CrossRef]
9. Dygert, R.; Dang, T. Mitigation of cross-contamination in an aircraft cabin via localized exhaust. *Build. Environ.* **2010**, *45*, 2015–2016. [CrossRef]
10. Mazumbar, S.; Poussou, S.; Lin, C.; Isukapalli, S.; Plesniak, M.; Chen, Q. Impact of scaling and body movement on contaminat transport in airliner cabins. *Atmos. Environ.* **2011**, *45*, 6019–6028. [CrossRef]
11. Poussou, S.; Mazumdar, S.; Plesniak, M.; Sojka, P.; Chen, Q. Flow and contaminant transport in an airliner cabin induced by a moving body: Model experiments and CFD predictions. *Atmos. Environ.* **2010**, *44*, 2830–2839. [CrossRef]
12. Yan, W.; Zhang, Y.; Sun, Y.; Li, D. Experimental and CFD study of unsteady airbone pollutant transport within an aircraft cabin mock-up. *Build. Environ.* **2009**, *44*, 34–43. [CrossRef]
13. Zhang, T.; Chen, Q. Novel air distribution systems for commercial aircraft cabins. *Build. Environ.* **2007**, *42*, 1675–1684. [CrossRef]
14. Zhang, Z.; Chen, X.; Mazumdar, S.; Zhang, T.; Chen, Q. Experimental and numerical investigation of airflow and contaminant transport in an airliner cabin mockup. *Build. Environ.* **2009**, *44*, 85–94. [CrossRef]
15. Thompson, J.A.; Maidment, G.G.; Missenden, J.F. Modelling low-energy cooling strategies for underground railways. *Appl. Energy* **2006**, *83*, 1152–1162. [CrossRef]
16. Chow, W.K. Ventilation of enclosed train compartments in Hong Kong. *Appl. Energy* **2002**, *71*, 161–170. [CrossRef]
17. Yan, Y.; Li, X.; Yang, L.; Tu, J. Evaluation of manikin simplification methods for CFD simulations in occupied indoor environments. *Energy Build.* **2016**, *127*, 611–626. [CrossRef]

11

Disc Thickness and Spacing Distance Impacts on Flow Characteristics of Multichannel Tesla Turbines

Wenjiao Qi, Qinghua Deng *, Yu Jiang, Qi Yuan and Zhenping Feng

Shaanxi Engineering Laboratory of Turbomachinery and Power Equipment, Institute of Turbomachinery, School of Energy and Power Engineering, Xi'an Jiaotong University, Xi'an 710049, China; qiwenjiao@stu.xjtu.edu.cn (W.Q.); guaiily@stu.xjtu.edu.cn (Y.J.); qyuan@mail.xjtu.edu.cn (Q.Y.); zpfeng@mail.xjtu.edu.cn (Z.F.)

* Correspondence: qhdeng@mail.xjtu.edu.cn

Abstract: Tesla turbines are a kind of unconventional bladeless turbines, which utilize the viscosity of working fluid to rotate the rotor and realize energy conversion. They offer an attractive substitution for small and micro conventional bladed turbines due to two major advantages. In this study, the effects of two influential geometrical parameters, disc thickness and disc spacing distance, on the aerodynamic performance and flow characteristics for two kinds of multichannel Tesla turbines (one-to-one turbine and one-to-many turbine) were investigated and analyzed numerically. The results show that, with increasing disc thickness, the isentropic efficiency of the one-to-one turbine decreases a little and that of the one-to-many turbine reduces significantly. For example, for turbine cases with 0.5 mm disc spacing distance, the former drops less than 7% and the latter decreases by about 45% of their original values as disc thickness increases from 1 mm to 2 mm. With increasing disc spacing distance, the isentropic efficiency of both kinds of turbines increases first and then decreases, and an optimal value and a high efficiency range exist to make the isentropic efficiency reach its maximum and maintain at a high level, respectively. The optimal disc spacing distance for the one-to-one turbine is less than that for the one-to-many turbine (0.5 mm and 1 mm, respectively, for turbine cases with disc thickness of 1 mm). To sum up, for designing a multichannel Tesla turbine, the disc spacing distance should be among its high efficiency range, and the determination of disc thickness should be balanced between its impacts on the aerodynamic performance and mechanical stress.

Keywords: Tesla turbine; fluid dynamics; disc thickness; disc spacing distance; isentropic efficiency

---

1. Introduction

In the last two decades, one of the interests is to develop small-scale turbomachinery due to market demands driving and manufacture techniques making progress [1], which is mainly applied to small power generation unit, such as distributed energy systems using low-grade energy and increasing energy utilization efficiency. In addition, small-scale turbomachinery is also applied as the mobile or small power device, which can be used in unmanned aircraft, miniaturized radio-controlled vehicles [2,3], and so on.

Several research groups designed, manufactured and experimented some small or micro gas turbines to study their feasibility and effectiveness [4–9]. These studies show that two major problems are faced when conventional turbines are scaled down. One is the rapidly increasing rotational speed, and the other is the significantly decreasing flow efficiency. Moreover, Brayton cycle system cannot be realized when the flow efficiency of conventional turbines goes down to a certain amount due to their microminiaturization [10]. However, a Tesla turbine is considered as one of the best alternatives [11].

In 1913, the first Tesla turbine was designed, manufactured and patented by the famous scholar Nikola Tesla [12]. It is an unconventional bladeless turbomachinery, which uses the viscous stress acting on rotating disc walls to drag the rotor to rotate, as shown in Figure 1. The rotor of Tesla turbines is composed by several flat, parallel, rigid, co-rotating discs, which are placed closely and mounted on a central shaft. The narrow spaces between the adjacent discs are disc channels. In Tesla turbines, the working fluid accelerates in nozzles or volute and obtains its highest flow velocity at the stator outlet; then injects into the disc channels nearly tangentially; and finally flows spirally towards the disc holes or slots around the shaft.

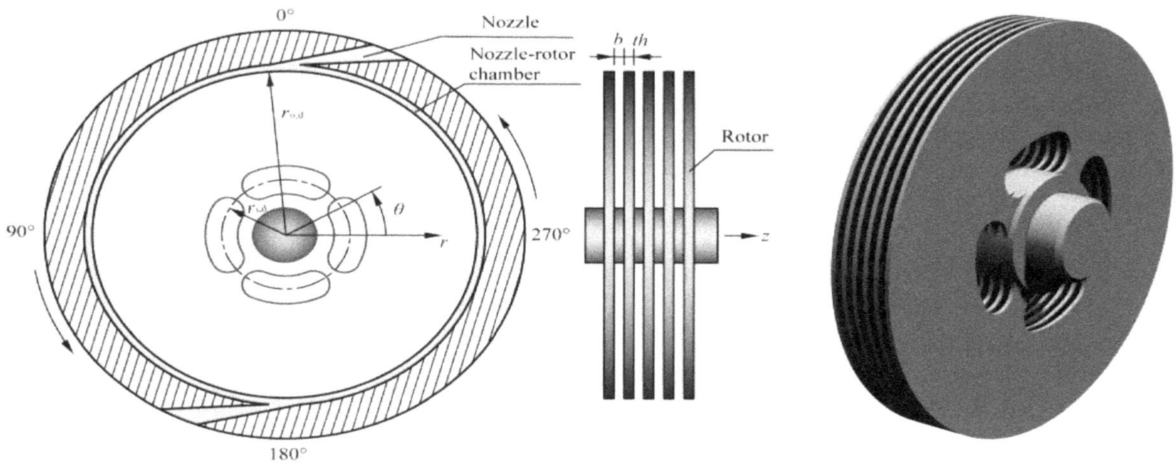

(a) (b)

Figure 1. Tesla turbine schematic: (**a**) 2-D sketch; and (**b**) 3-D rotor.

During 1913–1950, little research was conducted on Tesla turbines due to the invention of gas turbines and its high efficiency. Thereafter, the theoretical and experimental analysis on Tesla turbines has been restarted due to its advantages; for example, it is simple to manufacture and maintain with low cost and it can use kinds of working fluid [13–16]. However, the study progress of Tesla turbines was still slow. In the last two decades, much research on Tesla turbines has been reported using not only theoretical and experimental methods but also numerical simulations due to the rapid development of computational fluid dynamics (CFD) and advanced computer techniques.

According to the inlet geometry, Tesla turbines are classified into two types: voluted Tesla turbine and nozzled Tesla turbine. There are few reports on the voluted Tesla turbine. The flow characteristics and loss mechanism of voluted Tesla turbines were investigated using experimental and numerical methods [17]. The results show that the theoretical limit of the isentropic efficiency is about 40%, if the parasitic losses, mainly including bearing loss, viscous loss on end walls, and eddy loss in the volute, can be minimized.

The main focus of Tesla turbines is on investigating flow characteristics in nozzled Tesla turbines. A one-dimensional analysis method of flow field in the disc channel is proposed with some hypotheses, and the model agrees well with previous experimental performance data [18]. Sengupta and Guha formulated a mathematical theory by simplifying the Navier–Stokes equations using a magnitude analysis method, which can assess the turbine performance [19–21]. Talluri et al. developed a new method for designing of a Tesla turbine for Organic Rankine Cycle (ORC) applications, in which almost all losses are considered using real gas physical properties [22]. It can calculate the aerodynamic performance of the multichannel Tesla turbine; however, the results have not been validated by numerical and experimental analysis. To sum up, most theoretical analysis can only be applied to one channel Tesla turbines, in which the influence of disc thickness has not been taken into account yet.

Some researchers investigated the total aerodynamic performance and detailed flow fields using experimental method. The research group of Guha set up a test rig of Tesla turbines, and put forward several methods for measuring power [23]. In addition, they improved the inlet and nozzle to enhance the turbine efficiency [24]. Schosser et al. designed and set up a test rig to reveal the detailed flow field in the disc spacing of one channel Tesla turbine using 3D tomographic Particle Image Velocimetry (PIV) and Particle Tracking Velocimetry (PTV) measurements, however its rotational speed is restricted because of the mechanical stress [25]. From the above research, the detailed flow field in multichannel Tesla turbines cannot be investigated experimentally.

The complicated turbine model cannot be analyzed theoretically, although it can be simulated by CFD method to reveal the detailed flow characteristics. The effects of some parameters, such as the rotational speed and nozzle number, have been studied for a Tesla turbine with a low-boiling medium [11,26]. The effects of disc spacing distance and rotational speed on a Tesla turbine were analyzed numerically by our research group, and the results show that an optimal value of both parameters exists for the turbine to obtain its highest efficiency, respectively [27–29]. The flow fields in the disc channel of one channel Tesla turbine (no stator) with different inlet conditions, including flow coefficient and inlet geometries, were simulated numerically; the efficiency decreases dramatically with increasing flow coefficient, and the inlet non-uniformity is also a factor that makes turbine efficiency decrease [30]. Sengupta et al. studied the influence of four parameters, namely the nozzle number, disc thickness, rotational speed and nozzle-rotor radial clearance, on the performance of a Tesla turbine, and the results show that in the turbine design, thin discs with flat disc tip edge, more nuzzles and an optimal radial clearance are suggested [31].

In the authors' previous study, the nozzled multichannel Tesla turbine is classified into two categories according to nozzle geometry: one nozzle channel to one disc channel (one-to-one Tesla turbines) and one nozzle channel to several disc channels (one-to-many Tesla turbines), as shown in Figure 2 [29]. The one-to-one Tesla turbine and the one-to-many turbine with the same geometry operating under the same conditions were studied numerically and the only difference between two kinds of Tesla turbines is their nozzle geometries. The objective was to reveal and compare their fluid dynamic, and the results show that the flow mechanism of the two Tesla turbines are totally different.

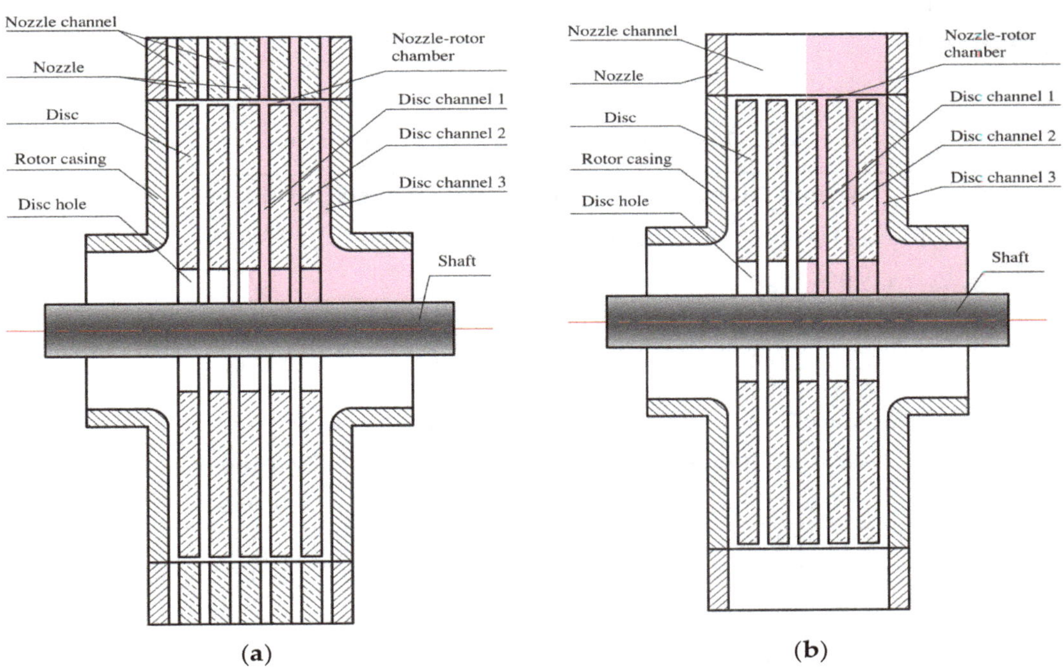

Figure 2. Sketch maps of multichannel Tesla turbines: (**a**) one-to-one Tesla turbine; and (**b**) one-to-many Tesla turbine.

For multichannel Tesla turbines, the affecting factors include the geometrical parameters (such as nozzle number, nozzle geometry, disc outer diameter, disc spacing distance and disc thickness) and operating parameters (such as rotational speed mass flow rate and turbine pressure ratio). The above reference review shows that the influence of most affecting parameters has been investigated, although some research is on one channel Tesla turbine. However, up to now, few studies have been conducted on the impacts of disc thickness and disc spacing distance on the performance of multichannel Tesla turbines. The effect of disc thickness is studied in Ref. [31], but only the flow in the nozzle-rotor chamber and the rotor are included in numerical calculations. In addition, it has been analyzed experimentally for the one-to-many turbine, while, for the one-to-one turbine, its influence and the fluid mechanism have not been investigated [32].

The impacts of disc spacing distance on the one channel Tesla turbine were analyzed by our group, and the results show that an optimal disc spacing distance exists to enable the Tesla turbine to obtain its best performance [28]. In the turbine with narrower disc spacing distance, two boundary layers on two disc walls overlap, resulting in a decrease in frictional force and torque; in the turbine with wider disc spacing distance, more working fluid outside the boundary layer flows out of the disc channels resulting in less momentum exchange to drive a rotor. Because disc thickness has to be taken into consideration in multichannel Tesla turbines, the influence of disc spacing distance on flow fields must be different from that in one channel Tesla turbines.

In this paper, two kinds of multichannel Tesla turbines (one-to-one turbines and one-to-many turbines) with different disc spacing distance and disc thickness are simulated numerically to investigate their influence on the aerodynamic performance and flow characteristics. This paper provides theoretical and engineering reference for designing a multichannel Tesla turbine.

2. Numerical Approach

2.1. Geometry Model and Boundary Conditions

In this study, two groups of multichannel Tesla turbines were calculated numerically. One group is the one-to-one Tesla turbine, and the other is the one-to-many Tesla turbine. Each group has six cases of Tesla turbines, and the parameters of disc spacing distance and disc thickness are given in Table 1, in which the case is named as "disc thickness-disc spacing distance". In this research, the working fluid is compressed air.

Table 1. Geometrical parameters of each case.

Group 1 (One-To-One Multichannel Tesla Turbines)			
Case	***th* (mm)**	***b* (mm)**	***b*/*th* (-)**
Case 1-0.3	1	0.3	0.3
Case 1-0.5	1	0.5	0.5
Case 1-1	1	1	1
Case 2-0.3	2	0.3	0.15
Case 2-0.5	2	0.5	0.25
Case 2-1	2	1	0.5
Group 2 (One-To-Many Multichannel Tesla Turbines)			
Case	***th* (mm)**	***b* (mm)**	***b*/*th* (-)**
Case 1-0.3	1	0.3	0.3
Case 1-0.5	1	0.5	0.5
Case 1-1	1	1	1
Case 2-0.5	2	0.5	0.25
Case 2-1	2	1	0.5
Case 2-2	2	2	1

Each turbine case consists of five discs and six disc channels. The other geometrical parameters and aerodynamic parameters are all the same for the two groups of the Tesla turbine, which are the same as those in the authors' previous study [29] (Table 2).

To save computing time and resources, the calculation domain was reduced based on a rotational symmetry and a symmetry about a middle plane that is normal to the axis, which finally became a quarter of the whole turbine. It is shown in Figure 2 with red color.

Table 2. Geometrical and aerodynamic parameters.

Symbol	Unit	Value
N_n	(-)	2
$d_{o,d}$	(mm)	100
$d_{i,d}$	(mm)	38.4
c	(mm)	0.25
N_d	(-)	5
N_{dc}	(-)	6
α	(°)	10
p_{nt}/p_i	(-)	3.42
T_{nt}	(K)	373

The boundary conditions are given in Table 2. In addition, because of the calculation domain reduction, the symmetry and rotational periodicity boundary conditions were set up at corresponding surfaces. The adiabatic and no-slip boundary condition was adopted for all walls. The frozen rotor method was applied to deal with the rotor-stator interface.

2.2. Numerical Solver and Mesh Sensitivity

In this research, all numerical calculations were conducted using commercial software ANSYS CFX (ANSYS Inc., Canonsburg, PA, USA), in which the Reynolds Averaged Navier–Stokes (RANS) equations in the calculation domain were solved for turbulent flow. The RANS equation groups are as follows.

Continuity equation:

$$\frac{\partial \rho}{\partial t} + \frac{\partial}{\partial x_j}(\rho U_j) = 0 \tag{1}$$

Momentum equations:

$$\frac{\partial \rho U_i}{\partial t} + \frac{\partial}{\partial x_j}(\rho U_i U_j) = -\frac{\partial p}{\partial x_i} + \frac{\partial}{\partial x_j}(\tau_{ij} - \rho\overline{u_i u_j}) + S_M \tag{2}$$

Energy equations:

$$\frac{\partial(\rho h_{tot})}{\partial t} - \frac{\partial p}{\partial t} + \frac{\partial}{\partial x_j}(\rho U_j h_{tot}) = \frac{\partial}{\partial x_j}(\lambda\frac{\partial T}{\partial x_j} - \rho\overline{u_j h}) + \frac{\partial}{\partial x_j}[U_i(\tau_{ij} - \rho\overline{u_i u_j})] + S_E \tag{3}$$

in which x_j are coordinates in three directions, U_j are average velocity in three directions, p is the static pressure, T is the temperature, ρ is the density, and τ is the molecular stress tensor.

In Equations (2) and (3), S_M and S_E are source terms, $\rho\overline{u_i u_j}$ are the Reynolds stresses, $\rho\overline{u_j h}$ is an additional turbulence flux term, $\frac{\partial}{\partial x_j}[U_i(\tau_{ij} - \rho\overline{u_i u_j})]$ represents the viscous work term. h_{tot} is the mean total enthalpy, h is the static enthalpy, and $h_{tot} = h + U_i U_j/2 + k$, where k is the turbulent kinetic energy and $k = \overline{u_i^2}/2$.

In addition, the density can be solved according to the state equation for ideal air,

$$\rho = p_{\mathrm{abs}}/RT \tag{4}$$

where p_{abs} is the absolute pressure and R is the specific gas constant of air.

According to the above discussion, it can be found that RANS equations are not closed due to two additional terms, which are Reynolds stress and Reynolds flux. Therefore, some turbulence models in ANSYS CFX are provided to calculate turbulent flow.

For Tesla turbines, the detailed experimental data cannot be found in public references, therefore the flow model verification could not be conducted. The critical Reynolds number for the plane Poiseuille flow is about 1000 [33], and is defined by the half disc spacing distance and the average velocity in Tesla turbines. In most Tesla turbines, the Reynolds number is always above the critical values, and the minimum Reynolds number was about 1500 in our computational cases. Therefore, the Shear Stress Transport (SST) turbulence model was adopted in this research, which is also used in numerical research by other researchers [11,25,26].

SST turbulence model is a kind of eddy viscosity model, and, according to eddy viscosity hypothesis, the Reynolds stresses are considered to be proportional to mean velocity gradients. They can be expressed as follows.

$$-\rho\overline{u_i u_j} = \mu_t(\frac{\partial U_i}{\partial x_j} + \frac{\partial U_j}{\partial x_i}) - \frac{2}{3}\delta_{ij}(\rho k + \mu_t\frac{\partial U_k}{\partial x_k}) \tag{5}$$

where μ_t is the eddy viscosity.

In addition, the Reynolds fluxes of a scalar is stated as a linear relationship to the mean scalar gradient based on the eddy diffusivity hypothesis:

$$-\rho\overline{u_i\phi} = \Gamma_t\frac{\partial\Phi}{\partial x_i} \tag{6}$$

where $\rho\overline{u_i\phi}$ is the Reynolds flux of a scalar Φ, Γ_t is the eddy diffusivity, written as $\Gamma_t = \mu_t/\mathrm{Pr}_t$, and Pr_t is the turbulent Prandtl number.

Equations (5) and (6) give expressions of turbulent fluctuations as functions of the mean variables when the turbulent viscosity is known. In ANSYS CFX solver, SST turbulence model, which derives from the $k-\varepsilon$ and $k-\omega$ turbulence models, uses this variable and expresses the turbulent viscosity as,

$$\mu_t = \rho\frac{k}{\omega} \tag{7}$$

where ω is turbulent frequency.

The Reynolds averaged transport equations for k and ω in SST turbulence model are as follows.

$$\frac{\partial(\rho\omega)}{\partial t} + \frac{\partial}{\partial x_j}(\rho U_j\omega) = \frac{\partial}{\partial x_j}\left[\left(\mu + \frac{\mu_t}{\sigma_{\omega 3}}\right)\frac{\partial\omega}{\partial x_j}\right] + (1-F1)2\rho\frac{1}{\sigma_{\omega 2}\omega}\frac{\partial k}{\partial x_j}\frac{\partial\omega}{\partial x_j} + \alpha_3\frac{\omega}{k}P_k - \beta_3\rho\omega^2 + P_{kb} \tag{8}$$

$$\frac{\partial(\rho k)}{\partial t} + \frac{\partial}{\partial x_j}(\rho U_j\, k) = \frac{\partial}{\partial x_j}\left[\left(\mu + \frac{\mu_t}{\sigma_{k2}}\right)\frac{\partial k}{\partial x_j}\right] + P_k - \beta'\rho k\omega + P_{kb} \tag{9}$$

For the detailed coefficients in Equations (8) and (9), refer to Ref. [34]. Up to now, the RANS equations are closed, and can be solved to obtain the turbulent flow fields.

For all numerical simulations in this research, spatial discretization for the above flow governing equations was conducted with high-resolution second-order central difference scheme and time discretization was with second-order backward Euler scheme. In addition, an empirical automatic wall function was adopted to guarantee the solution accuracy. The accuracy of second order was obtained in all numerical simulations. In addition, the y+ should be less than 2 for all walls as the SST turbulence model requires.

In this study, a structured grid was generated for fluid domain, using ANSYS mesh generation software ICEM CFD (ANSYS Inc., Canonsburg, PA, USA). The mesh independence analysis for the one-to-many turbine Case 1-0.5 was conducted to ensure the accuracy of the results.

Three sets of mesh for this turbine were generated and the detailed information are given in Table 3. The N-R chamber stands for the chamber between the nozzle and the rotor.

Table 3. Mesh information.

Case No.	Stator (Nozzle/N-R Chamber)		Rotor (Each Disc Channel)	
	Number of Nodes (r,θ,z Directions)	Total Node Number	Number of Nodes (r,θ,z Directions)	Total Node Number
Case 1	$(55/13) \times (36/269) \times 99$	526,516	$65 \times 288 \times 23$	400,660
Case 2	$(67/17) \times (45/335) \times 107$	923,517	$81 \times 333 \times 29$	782,217
Case 3	$(87/21) \times (57/417) \times 135$	1,830,306	$102 \times 417 \times 37$	1,581,306

The aerodynamic performance parameters and their relative variation values (based on the results of Case 3) are given in Table 4. The isentropic efficiency η is defined as the actual shaft power P divided by the isentropic power across the whole turbine $m\Delta h_{\text{isen}}$, as follows:

$$\eta = \frac{P}{m\Delta h_{\text{isen}}} = \frac{M\Omega}{mc_p T_{\text{nt}}\left[1-(p_{\text{i}}/p_{\text{nt}})^{(\gamma-1)/\gamma}\right]} \tag{10}$$

where P equals the whole torque M multiplied by the angular speed of the rotor Ω. c_p and γ are the physical properties of the working fluid, which are specific heat at constant pressure and the specific heat ratio. The subscript "nt" represent the total parameters at the turbine inlet, while "i" indicates the static parameter at the turbine outlet.

Figure 3 presents the Mach number contours on the mid-section of DC1 ("DC" is short for "disc channel") for the one-to-many turbine Case 1-0.5 with three different node cases. The detailed locations of the three DCs are indicated in Figure 2. Note that DC1 and DC2 each consists of two rotating disc walls, and DC3 has one rotating disc wall and one motionless casing wall.

Table 4 indicates that the variation value of each aerodynamic performance decreases with node number. Figure 3 shows that the Mach number distribution for each case is similar, but the difference between Cases 2 and 3 is much less than that between Cases 1 and 2. To sum up, the results of Case 2 fulfill the requirements for mesh independence, therefore the mesh of Case 2 was applied for the this turbine in numerical simulations. Moreover, for other turbine cases with different geometry, the grid distribution changes accordingly to get a reliable result.

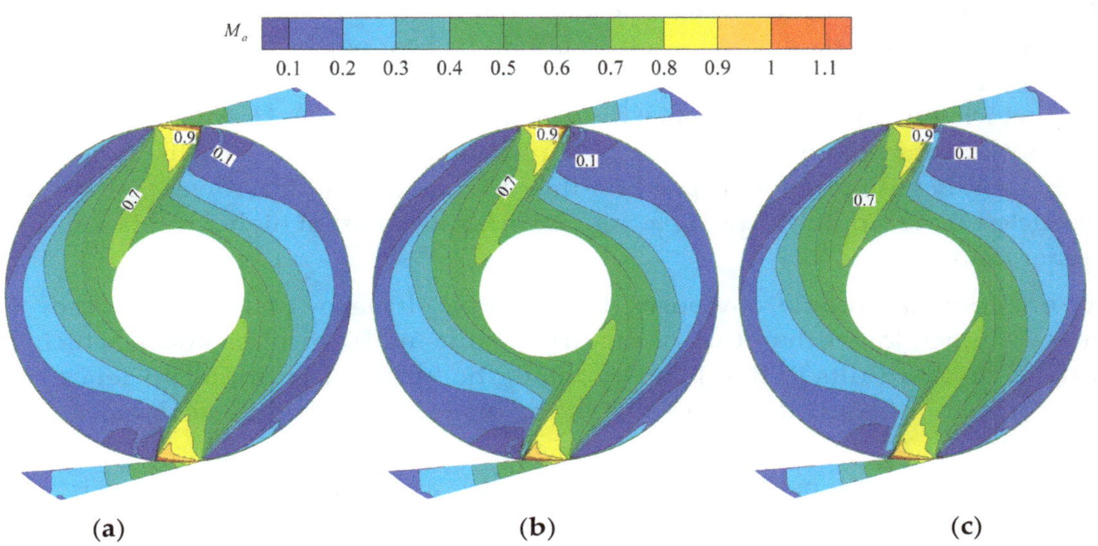

Figure 3. Mach number contours on the mid-sections of DC1 for the one-to-many turbine Case 1-0.5 with three different node cases: (**a**) Case 1; (**b**) Case 2; and (**c**) Case 3.

Table 4. Mesh independence.

Case No.	Node Number (million)	m (kg/s)	δm(%)	P (kW)	δP (%)	η (-)	$\delta\eta$ (%)
Case 1	1.72	0.03576	0.619	0.5960	1.568	0.1504	0.940
Case 2	3.27	0.03562	0.225	0.5886	0.307	0.1491	0.067
Case 3	6.57	0.03554	0	0.5868	0	0.1490	0

3. Results and Discussions

3.1. Total Aerodynamic Performance of Two Kinds of Multichannel Tesla Turbines

Two kinds of multichannel Tesla turbines with disc thickness of 1 mm and 2 mm were calculated numerically, respectively. It is obvious that, if the disc thickness is too large, the air flowing into disc channels will be more difficult, therefore, the aerodynamic performance becomes worse. However, the disc thickness should not be too small, which is mainly confined to its mechanical stress and processing problem. As disc thickness decreases, the stiffness of the discs decreases rapidly, and accordingly the mechanical stress increases, which should be lower than material allowable stress. Previous study on the computational solid mechanical (CSM) analysis indicates that the disc thickness should not be less than 1 mm for the Tesla turbine with disc outer diameter of 100 mm.

3.1.1. Isentropic Efficiency

Figure 4 shows the variation curves of the isentropic efficiency for each turbine case with the rotational speed for the two kinds of multichannel Tesla turbines. In Figure 4, note that the negative isentropic efficiency has no physical meaning and only indicates that under this condition the Tesla turbine consumes external energy without power outputting. It can be seen that the isentropic efficiency of the one-to-one turbine is much higher than that of the one-to-many Tesla turbine, and it changes much faster with rotational speed.

For the one-to-one turbine, the isentropic efficiency of the turbine cases with same disc spacing distance but different disc thickness is almost equal, although the turbine cases with smaller disc thickness are slightly more efficient, which is more obvious at higher rotational speeds. For example, the isentropic efficiency of Case 2-0.5 has a less than 7% reduction of that of Case 1-0.5 at all rotational speeds lower than 45,000 r/min. The turbine case with disc spacing distance of 0.5 mm obtains the highest efficiency at lower rotational speeds. That is, an optimal disc spacing distance for the one-to-one multichannel Tesla turbine exists to obtain its best performance, which is in accordance with that for one channel Tesla turbines [28]. In fact, a high efficiency range of disc spacing distance exists, among which the isentropic efficiency can remain at a high level. However, the turbine with disc spacing distance of 1 mm performs best at higher rotational speeds, because the isentropic efficiency changes more rapidly with rotational speed for the turbine case with narrower disc spacing distance. It can be predicted that an optimal disc spacing distance must exist at higher rotational speeds, which must be higher than 0.5 mm.

For the one-to-many turbine, obviously, the isentropic efficiency of the turbine cases with greater disc thickness is much lower than that of the cases with smaller disc thickness, which agrees well with the experimental results in Ref. [32]. In detail, the turbine isentropic efficiency of Case 2-0.5 is about 55% of that of Case 1-0.5. Meanwhile, the one-to-many turbine also has an optimal disc spacing distance.

To sum up, for designing multichannel Tesla turbines, a one-to-one Tesla turbine is recommended. Moreover, the disc spacing distance should be in its high efficiency range, and the disc thickness should be as small as possible while satisfying the requirement of allowable material stress. Usually, the disc thickness can be determined by the results of CSM, and it should be larger than 1 mm for discs of 100 mm diameter.

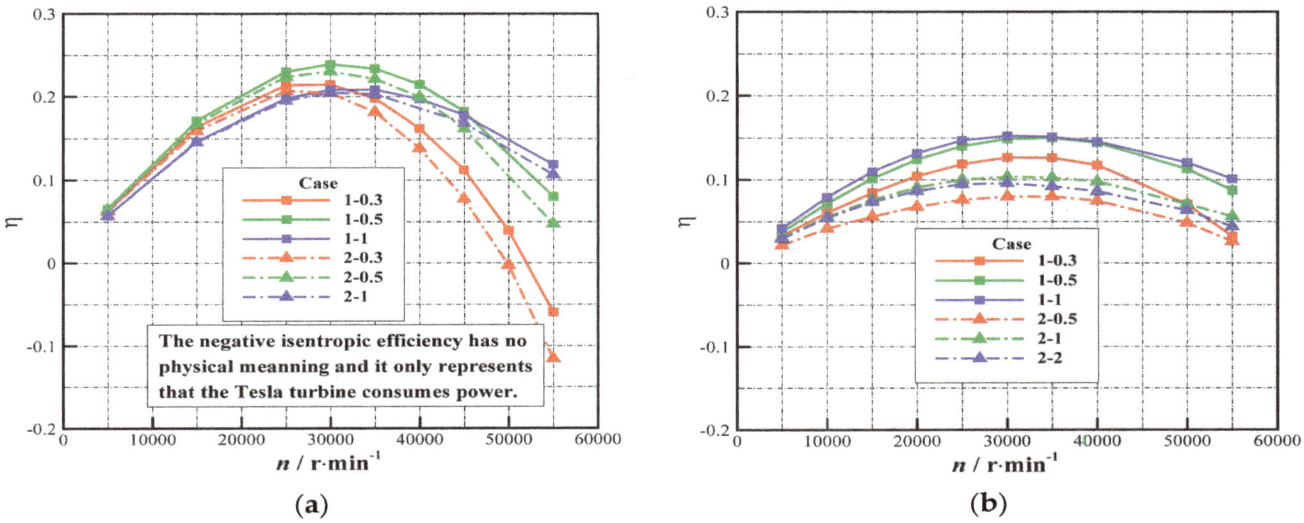

Figure 4. Isentropic efficiency versus rotational speed of each turbine case for two kinds of multichannel Tesla turbines: (**a**) one-to-one Tesla turbine; and (**b**) one-to-many Tesla turbine.

3.1.2. Flow Coefficient

Figure 5 presents the curves of the flow coefficient for each turbine case versus the rotational speed. The flow coefficient C_m is defined as:

$$C_m = \frac{v_{\text{o,d}}}{\Omega r_{\text{o,d}}} = \frac{m}{2\pi\rho_{\text{o,d}}\Omega b r_{\text{o,d}}^2} \tag{11}$$

where $v_{\text{o,d}}$ and $\omega r_{\text{o,d}}$ represent the average radial velocity and the disc rotational linear velocity at the rotor inlet, respectively.

The flow coefficient of the one-to-one Tesla turbine is much less than that of the one-to-many turbine with same disc spacing distance and disc thickness (Figure 5). In detail, for turbine Case 1-0.5, that of the one-to-many is nearly double that of the one-to-one turbine. Compared to the one-to-one turbine, when the operating parameters and other geometrical parameters are given, the mass flow rate of the one-to-many turbine is much higher due to much larger nozzle throat area (for Case 1-0.5, the one-to-many turbine is about 2.6 times); the density at the rotor inlet changes much less with the multichannel Tesla turbine type (for Case 1-0.5, the one-to-many turbine is about 1.3 times) than the mass flow rate; therefore, the flow coefficient of the one-to-many turbine is much higher.

For the one-to-one turbine, the flow coefficient of the turbine cases with same disc spacing distance is almost the same (the relative variation between Cases 1-0.5 and 2-0.5 is less than 2%), although it has a little bit increase with disc thickness, especially at higher rotational speeds. Meanwhile, this coefficient of the turbine cases with disc spacing distance of 0.3 mm and 0.5 mm is almost the same, and clearly higher than that of 1 mm, which is because the air density at the rotor inlet goes up with disc spacing distance.

For the one-to-many turbine, the variation relationships of the flow coefficient with disc thickness and spacing distance become complicated, and these two geometrical parameters affect significantly on this coefficient. As shown in Figure 2, the axial length ratio of nozzle channel to disc channels of the one-to-many Tesla turbine is larger than 1 and it goes up with increasing disc thickness or decreasing disc spacing distance, thus the flow coefficient increases as disc thickness goes up and disc spacing distance goes down, respectively. Same with the one-to-one turbine, the air density at the rotor inlet mainly depends on disc spacing distance, and increases with it, which also makes the coefficient decrease as disc spacing distance increases.

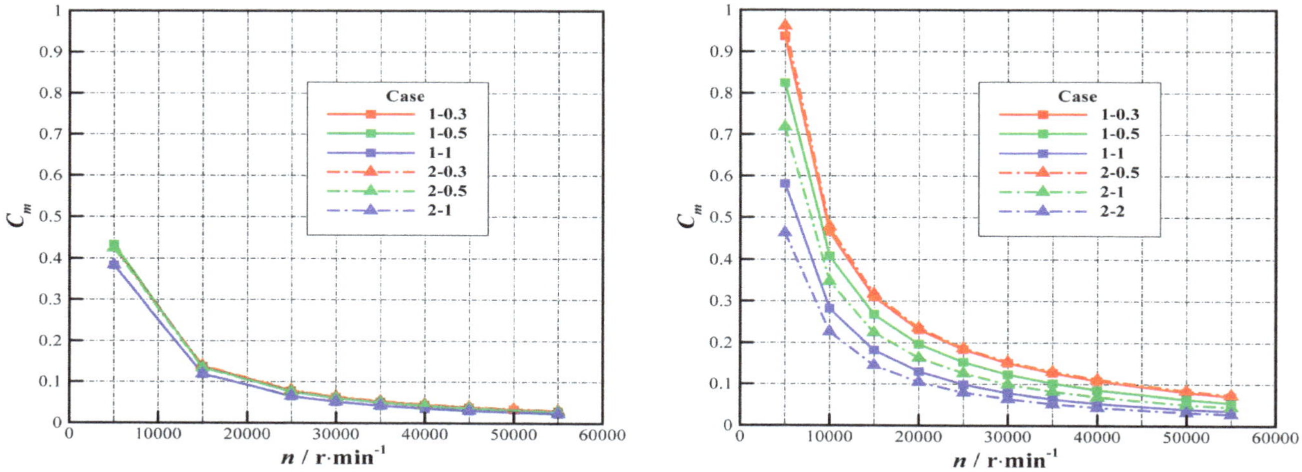

(a) (b)

Figure 5. Flow coefficient versus rotational speed of each case for two kinds of multichannel Tesla turbines: (**a**) one-to-one Tesla turbine; and (**b**) one-to-many Tesla turbine.

3.1.3. Percentages of Mass Flow Rate and Torque in Disc Channels

Figure 6 shows the mass flow rate and torque percentages in three DCs for all the turbine cases at 30,000 r/min. In Figure 6, both percentages in DCs 1 and 2 are nearly equal for both kinds of turbines, which indicates that the flow fields in DCs 1 and 2 are similar. In addition, DCs 1 and 2 have more air than DC3, while generating more torque.

For the one-to-one turbine, as disc thickness changes, both percentages in each disc channel are almost the same for turbine cases with same disc spacing distance, respectively. The torque percentage of DC3 for the turbine cases with 0.5 mm in disc spacing distance are almost the same, about 17%, while that for turbine with 0.3 mm in disc spacing distance decreases to 12%. All these flow phenomena are explained in the next sections.

For the one-to-many turbine, the variation relationship of mass flow rate percentage with disc spacing distance is not coincident. For the turbine cases with disc thickness of 1 mm, those in DCs 1 and 2 decrease as disc spacing distance increases, while for the turbine cases with disc thickness of 2 mm, they decrease first and then increase. In detail, those of Cases 1-1 and 2-1 reach the minimum values respectively. Moreover, disc thickness and spacing distance have few effects on the torque percentage in each disc channel.

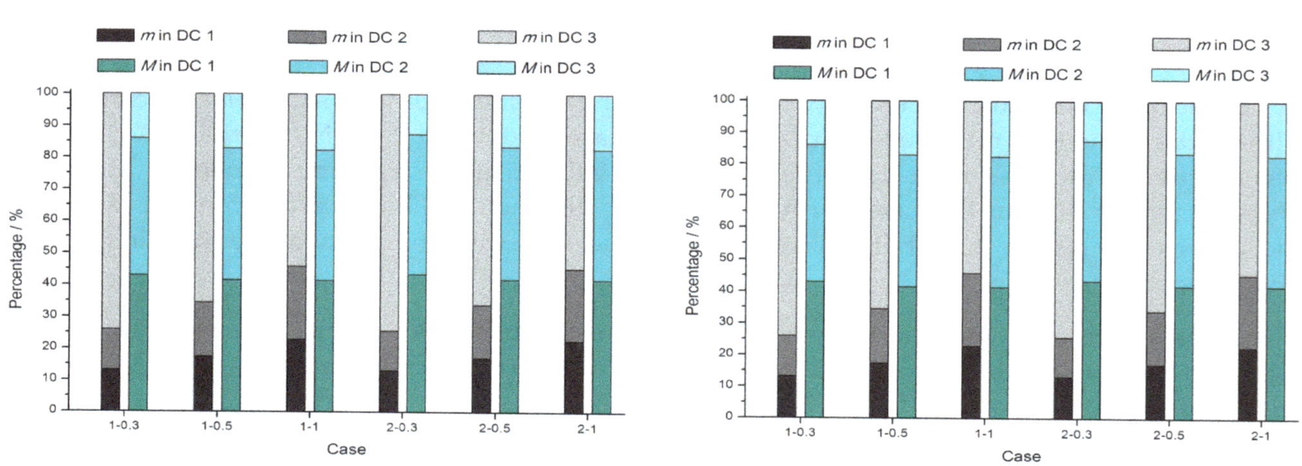

(a) (b)

Figure 6. Mass flow rate and torque percentages in three DCs for two kinds of Tesla turbines at 30,000 r/min: (**a**) one-to-one Tesla turbine; and (**b**) one-to-many Tesla turbine.

To be noted, the calculation domain has five disc rotating walls and the average torque percentage is 20%. However, for all the turbine cases, the torque percentage in DC3 is less than 20%. Meanwhile, DC3 flows more working fluid than DCs 1 and 2. That means DC3 uses much more working fluid and generates less torque and power. Thus, the disc spacing distance of DC3 should be smaller to get less working fluid. To sum up, disc thickness slightly influences the aerodynamic performance of the one-to-one turbine. The isentropic efficiency of the turbine cases with disc thickness of 2 mm is a little lower than that of 1 mm. Specifically, for turbine cases with 0.5 mm in disc spacing distance, it decreases by only 7% at most rotational speeds. Moreover, the disc spacing distance affects greatly the aerodynamic performance of the one-to-one turbine, and an optimal disc spacing distance exists to make the turbine obtain the highest isentropic efficiency.

Both the disc thickness and spacing distance have significant impacts on the aerodynamic performance of the one-to-many turbine. The one-to-many turbine also has an optimal disc spacing distance, which is larger than that for the one-to-one turbine (1 mm and 0.5 mm, respectively, for turbine with disc thickness of 1 mm). In addition, the performance becomes remarkably worse with an increase in disc thickness (a reduction of about 45%).

3.2. Flow Status of One-To-One Multichannel Tesla Turbines

From the above discussion, the isentropic efficiency of the one-to-one Tesla turbine drops a little as disc thickness increases. It increases first and then decreases with increasing disc spacing distance, and reaches its maximum value for the turbine case with disc spacing distance of 0.5 mm at lower rotational speeds including the optimal rotational speed of 30,000 r/min. To reveal the reasons that the disc thickness and spacing distance affect the turbine performance, the flow characteristics in the Tesla turbine were analyzed.

3.2.1. One-To-One Turbine with Different Disc Thickness

Figures 7 and 8 show Mach number contours and streamlines and on the mid-sections of three DCs for the one-to-one turbine Cases 1-0.5 and 2-0.5 at 30,000 r/min, respectively. The streamlines and Mach number contours of DCs 1 and 2 are almost the same, but different from that of DC3, which agrees well with the above analysis. Therefore, in the next analysis, only the flow fields of DCs 1 and 3 are given.

Obviously, in flow channels 1 and 2, part of the air that just injects nearly tangentially into the disc channels flows into the N-R chamber due to centrifugal force, and finally flows into DC3. In flow channel 3, the air from the nozzle together with that from the N-R chamber flows into DC3 at a larger flow angle (relative to the tangential direction). Thus, it is easy to know that for the one-to-one turbine less air flows through DCs 1 and 2 than that through DC3. In addition, it can be observed that the Mach number on the mid-sections of DCs 1 and 2 is much higher than that of DC3, and the flow angle is much less than that in DC3, which leads to much higher relative tangential velocity and longer path lines in DCs 1 and 2. Therefore, the torque percentage in DC3 is lower than its average value of 20%.

Comparing Figures 7 and 8, the flow fields in each disc channel of the turbine cases with same disc spacing distance but different disc thickness have little difference. This indicates that their aerodynamic performance also has little difference.

Figure 9 illustrates the contours of relative tangential velocity and vector distributions on the radial cross-section of 0°, whose location is marked in Figure 1. In this figure, only a small part of the disc channels is presented and the axial size is doubled to show the flow field clearly. The reference vector is given near the text of "rotor".

In the disc channels, the contour represents the relative tangential velocity, and the vector indicates the radial and axial velocities. In the stator, the contour shows the velocity normal to this section, and the vector describes the axial velocity and another velocity on this section. In the disc channels, the positive relative tangential velocity indicates that the air at this region generates usable torque; otherwise, it consumes external mechanical energy, most of which appears near the casing wall.

Figure 9 indicates that on this section the axial velocity is close to zero, and the radial velocity is much lower than the relative tangential velocity. In addition, the air with high velocity flows from the nozzle through the N-R chamber and finally into the disc channels rapidly with little flow deflection to the axial direction. Some vortex generates at some region of the N-R chamber.

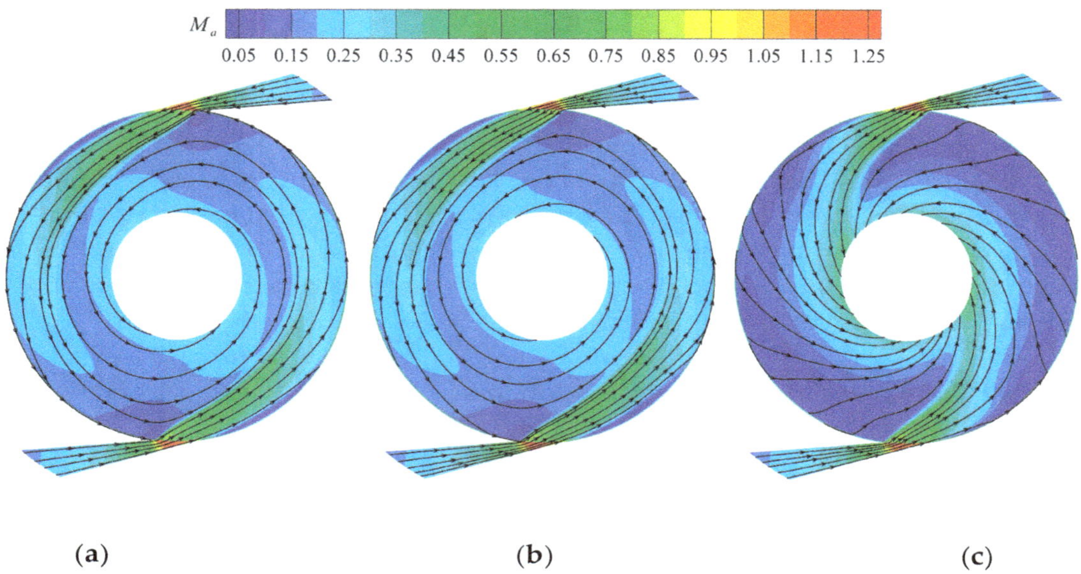

Figure 7. Mach number contours and streamlines on the mid-sections of three DCs for the one-to-one turbine Case 1-0.5 at 30,000 r/min: (**a**) DC1; (**b**) DC2; and (**c**) DC3.

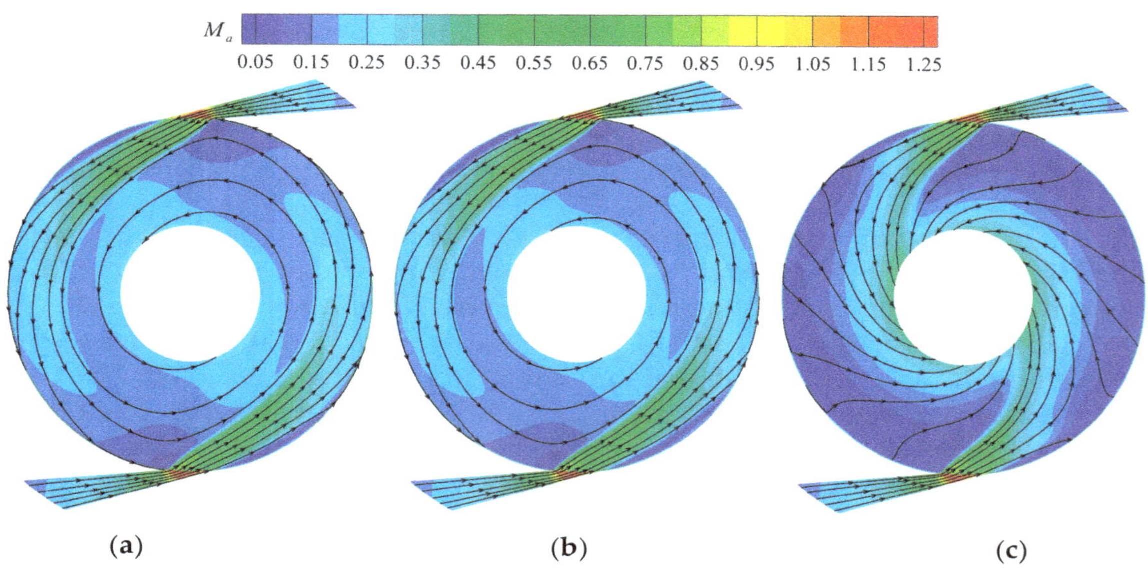

Figure 8. Mach number contours and streamlines on the mid-sections of three DCs for the one-to-one turbine Case 2-0.5 at 30,000 r/min: (**a**) DC1; (**b**) DC2; and (**c**) DC3.

The flow fields for turbine Cases 1-0.5 and 2-0.5 are almost the same in most regions on the radial cross-section. However, the flow velocity of the working fluid in the N-R chamber for turbine Case 2-0.5 is much lower than that for Case 1-0.5, which consumes more extra energy to move them. Moreover, it can be observed from the relative tangential velocity contours that the tangential velocity gradient near the rotating disc walls for Case 2-0.5 is a little less than that for Case 1-0.5, which leads to lower torque. All these factors lead to the isentropic efficiency of Case 2-0.5 a little lower than that of 1-0.5 (23.92% and 23.04%, respectively).

To sum up, the flow field and aerodynamic performance of the one-to-one Tesla turbine changes a little with disc thickness. In detail, the isentropic efficiency of the turbine with the thicker disc is little lower, which results from lower velocity in the N-R chamber and a little lower tangential velocity gradient close to the disc walls.

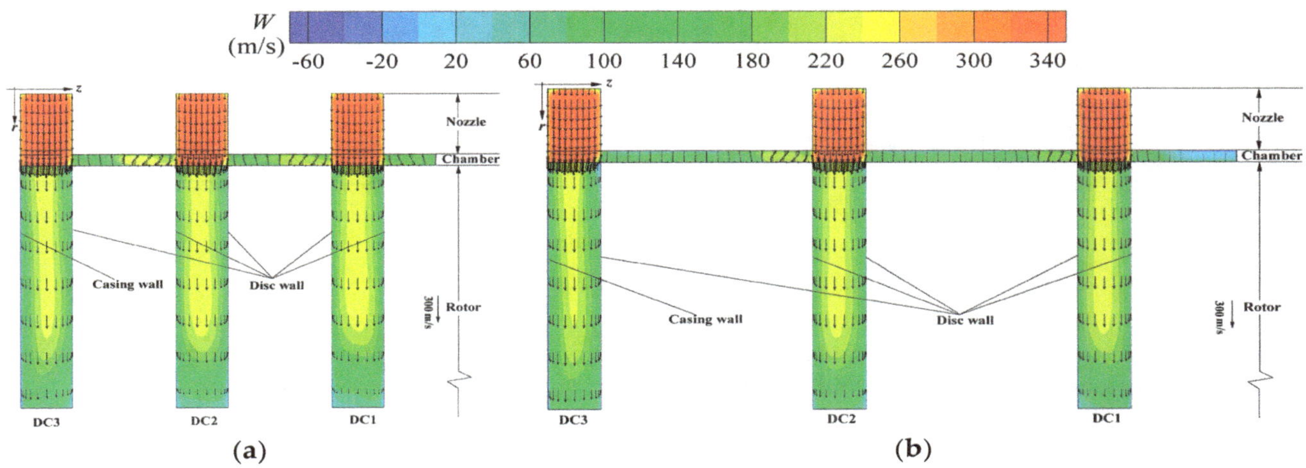

Figure 9. Contours of relative tangential velocity and vector distributions on radial cross-sections of 0° for the one-to-one Tesla turbine cases at 30,000 r/min: (**a**) Case 1-0.5; and (**b**) Case 2-0.5.

3.2.2. One-To-One Turbine with Different Disc Spacing Distance

To reveal the detailed flow characteristics in the disc channels of the one-to-one turbine with different disc spacing distance, Figure 10 shows the variation curves of circumferential mass flow average velocity, including relative tangential velocity and radial velocity, versus the radius ratio (local radius to rotor inlet radius) for different one-to-one turbine cases. In this figure, the abscissa values of 1 and 0.384 represent the rotor inlet and outlet, respectively.

As shown in Figure 10a, for all turbine cases, the average relative tangential velocity in DC1 obtains the highest value at the rotor inlet, and decreases with radius ratio due to the contribution to the output power. For turbine Cases 1-0.5 and 1-1, it then increases as radius ratio continues to decrease (the air flows towards the turbine outlet), and the stationary point of velocity occurs much earlier with increasing disc spacing distance. In addition, the average relative tangential velocity decreases at the region near the rotor inlet and increases at the region near the rotor outlet with increasing disc spacing distance. Similarly, the average relative tangential velocity in DC3 for all turbine cases decreases first and then increases as the air flows towards the turbine outlet. However, it is much higher for the turbine cases with larger disc spacing distance. In general, the average relative tangential velocity difference between DCs 1 and 3 decreases with increasing disc spacing distance.

As shown in Figure 10b, for all turbine cases, the average radial velocity in DC1 decreases with radius ratio, which seems to violate the continuity equation of working fluid. The flow area (cylindrical surface of the disc channel) decreases as the air flows through the disc channels to the rotor outlet (radius ratio decreases), while the density also decreases, thus the radial velocity should increase based on the continuity equation. However, it is completely opposite to the results in Figure 10, which is due to the partial admission of the nozzled Tesla turbine. These discrete nozzles are installed symmetrically at the disc tip, and most air is injected into DCs 1 and 2 by the nozzles. Although the flow area decreases as radius ratio decreases, the effective flow area increases due to the air diffusion when the air flows from the rotor inlet to outlet, which leads to a decrease in average radial velocity in DC1.

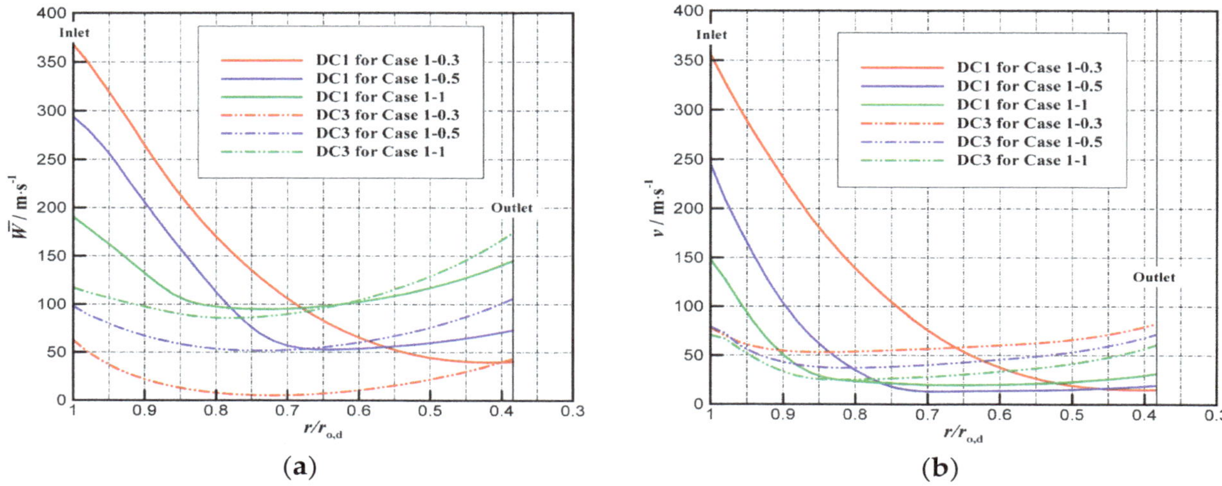

Figure 10. Variation curves of circumferential mass flow average velocity versus the radius ratio for three one-to-one turbine cases: (**a**) relative tangential velocity; and (**b**) radial velocity.

In addition, the average radial velocity in DC3 decreases first and then increases; however, it has much less variation than that in DC1. Some air from the N-R chamber flows into DC3, and the effective flow area of DC3 is much larger than that of DC1. However, the flow velocity at the rotor inlet from the N-R chamber is much less, and the average radial velocity still decreases slightly due to air diffusion. As the air flows towards the rotor outlet, the average radial velocity has a slight increase due to the pressure drop the rotor and the decrease in effective flow area. Moreover, the average radial velocity in DCs 1 and 3 both decreases with increasing disc spacing distance, and the average radial velocity at the rotor inlet in DC1 is much higher than that in DC3.

In summary, the relative tangential velocity, determining the torque and power, decreases with radius ratio due to outputting power, and then increases slightly, resulting from the decrease in disc rotational linear velocity and the pressure in the drop. Furthermore, its difference between DCs 1 and 3 decreases with increasing disc spacing distance. Moreover, the radial velocity, depending on the mass flow rate, has a strong relationship with the effective flow area, and it decreases with increasing disc spacing distance.

Figure 11 shows Mach number contours and streamlines and on the mid-sections of DCs 1 and 3 for the one-to-one turbine cases with different disc spacing distance at 30,000 r/min.

As shown in Figure 11, the Mach number at the nozzle outlet decreases as disc spacing distance increases because of the higher pressure drop in the disc channels of the turbine cases with wider disc spacing. With increasing flow velocity at the nozzle outlet, the flow angle at the rotor inlet should decrease if the nozzle outlet flow angle of different turbine cases is the same. However, it is totally opposite to the flow fields in Figure 11, which is because the nozzle outlet flow angle changes with disc spacing distance despite their same nozzle outlet geometrical angles. The air in the scarfed part of the nozzle tends to speed up and changes its direction to the side without the nozzle wall, when the pressure ratio of the nozzle (nozzle outlet pressure to nozzle inlet pressure) is under the critical pressure ratio of air. The deflection angle goes up with an increase in difference value between the nozzle pressure ratio and the critical pressure ratio, as shown in turbine Cases 1-0.3 and 1-0.5 in Figure 11. When the nozzle pressure ratio is over the critical pressure ratio, the flow will not deflect in the scarfed part and the flow angle at the rotor inlet is almost equal to the nozzle outlet geometrical angle, shown in turbine Case 1-1.

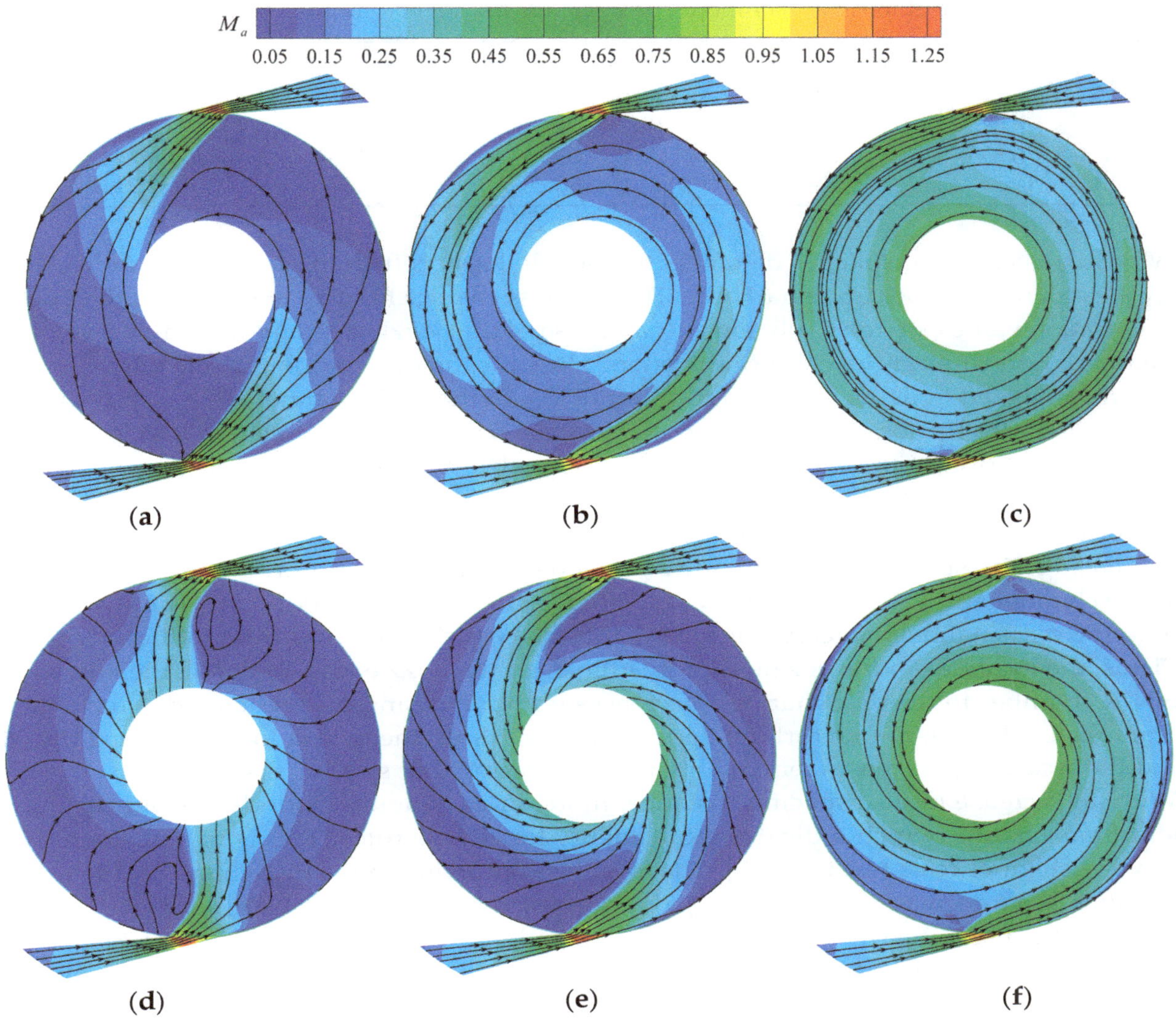

Figure 11. Mach number contours and streamlines on the mid-sections of DCs 1 and 3 for the one-to-one turbine cases with different disc spacing distance at 30,000 r/min: (**a**) DC1 for Case 1-0.3; (**b**) DC1 for Case 1-0.5; (**c**) DC1 for Case 1-1; (**d**) DC3 for Case 1-0.3; (**e**) DC3 for Case 1-0.5; and (**f**) DC3 for Case 1-1.

Comparing the flow fields in DCs 1 and 3, it can be found that more air flows into DC3 from both the nozzle and the N-R chamber, and the velocity of the working fluid from the N-R chamber is much lower. Therefore, the average relative tangential velocity and radial velocity at the rotor inlet of DC3 are under those of DC1. Apart from DC1 for turbine Case 1-0.3, the flow velocity in other disc channels decreases first and then increases as the radius decreases. Moreover, the flow field difference between DCs 1 and 3 decreases with increasing disc spacing distance. All the above flow phenomena are the same as those in Figure 10.

Based on the above discussion, the isentropic efficiency of the one-to-one turbine goes up first and then down with increasing disc spacing distance, that is to say, an optimal disc spacing distance exists to allow a Tesla turbine to obtain the highest isentropic efficiency. To explain this phenomenon, Table 5 gives the coefficients of component energy loss for the one-to-one turbine cases at 30,000 r/min, including the nozzle loss, the disc loss and the leaving-velocity loss, which are defined as each component energy loss divided by the isentropic enthalpy drop across the whole turbine.

Table 5. Coefficients of component energy loss for one-to-one Tesla turbine cases.

Case Name	Nozzle Loss Coefficient	Disc Loss Coefficient	Leaving-Velocity Loss Coefficient	Isentropic Efficiency
Case 1-0.3	0.2857	0.4196	0.0760	0.2187
Case 1-0.5	0.1849	0.4405	0.1303	0.2443
Case 1-1	0.1085	0.4358	0.2440	0.2117

With increasing disc spacing distance, the nozzle loss coefficient goes down and the leaving velocity loss coefficient goes up remarkably. The disc loss coefficient increases first and then decreases, and its variation values are relatively lower than those of the nozzle loss and the leaving-velocity loss.

The variation rules of the component energy loss for the one-to-one turbine with disc spacing distance are analyzed in detail based on the flow fields. In the multichannel Tesla turbine, the nozzle energy loss includes the loss resulting from the viscous friction (called "friction loss in the nozzle"), and the loss occurring in the nozzle and the N-R chamber caused by the sudden variation of flow area, when the air flows through the nozzle and the N-R chamber and finally to the disc channels (called "local loss in the nozzle"). For the one-to-one turbine, the friction loss in the nozzle decreases with increasing disc spacing distance, caused by a decrease in proportion of the boundary layer in the nozzle channel and a decrease in flow velocity (Figure 11). In addition, most air flows out of the nozzle to the disc channels quickly (Figure 9), and the N-R chamber has little influence on the flow field. Thus, the local loss in the nozzle slightly influences the nozzle loss. In detail, the flow area ratio of the N-R chamber to the disc channels decreases with increasing disc spacing distance, therefore, the local loss for the one-to-one turbine decreases. As a result, the nozzle loss, including the friction loss and the local loss, decreases remarkably with increasing disc spacing distance.

With an increase in disc spacing distance, more air outside the boundary layer has no use in momentum exchange, which indicates more energy loss in the rotor. This is in agreement with the result in one channel Tesla turbines [28]. The local loss occurring in the disc channels due to the flow area variation has a decrease with increasing disc spacing distance, although the local loss also has little influence on the disc loss. Therefore, the combined actions of the two factors lead to increasing first and then decreasing disc loss as disc spacing distance increases. The Mach number at the turbine outlet rises with disc spacing distance, resulting in increasing leaving-velocity loss coefficient, as shown in Figure 11. These results are the same as those in Table 5.

3.3. Flow Status of One-To-Many Multichannel Tesla Turbines

As discussed in the above section, the isentropic efficiency of the one-to-many turbine goes up first and then down with increasing disc spacing distance. It decreases remarkably as disc thickness increases. Table 6 gives the coefficients of component energy loss for the one-to-many turbine cases at 30,000 r/min. With increasing disc spacing distance, the nozzle loss coefficient and disc loss coefficient decrease, and the leaving-velocity loss coefficient rises. With increasing disc thickness, the nozzle loss has a slight increase, the disc loss decreases, and the leaving-velocity loss increases remarkably. The variation rules of these coefficients are explained in the following text.

Table 6. Coefficients of component energy loss for one-to-many Tesla turbine cases.

Case Name	Nozzle Loss Coefficient	Disc Loss Coefficient	Leaving-Velocity Loss Coefficient	Isentropic Efficiency
Case 1-0.3	0.0724	0.5539	0.2387	0.1350
Case 1-0.5	0.0578	0.4574	0.3310	0.1538
Case 1-1	0.0417	0.3936	0.4055	0.1592
Case 2-0.5	0.0699	0.4324	0.4040	0.0937
Case 2-1	0.0471	0.3504	0.4925	0.1100
Case 2-2	0.0375	0.3083	0.5561	0.0981

3.3.1. One-To-Many Turbine with Different Disc Thickness

To explain the variation rules of the aerodynamic performance of the one-to-many turbine with disc thickness, Figure 12 presents Mach number contours and streamlines on the mid-sections of DCs 1 and 3 for turbine Cases 1-0.5 and 2-0.5.

The Mach number at the nozzle outlet for Case 1-0.5 is slightly higher than that for Case 2-0.5. Therefore, the flow angle at the rotor inlet is a little lower. Moreover, the flow angle in most regions of the disc channels for Case 1-0.5 is lower than that for Case 2-0.5, leading to longer path lines and more momentum exchange. Different from the one-to-one turbine, Mach number increases suddenly at the rotor inlet directly facing the nozzle outlet, which is due to a sudden increase in flow area when the air passes the N-R chamber to the disc channels. The flow velocity at that place increases more significantly for Case 2-0.5 than that for Case 1-0.5.

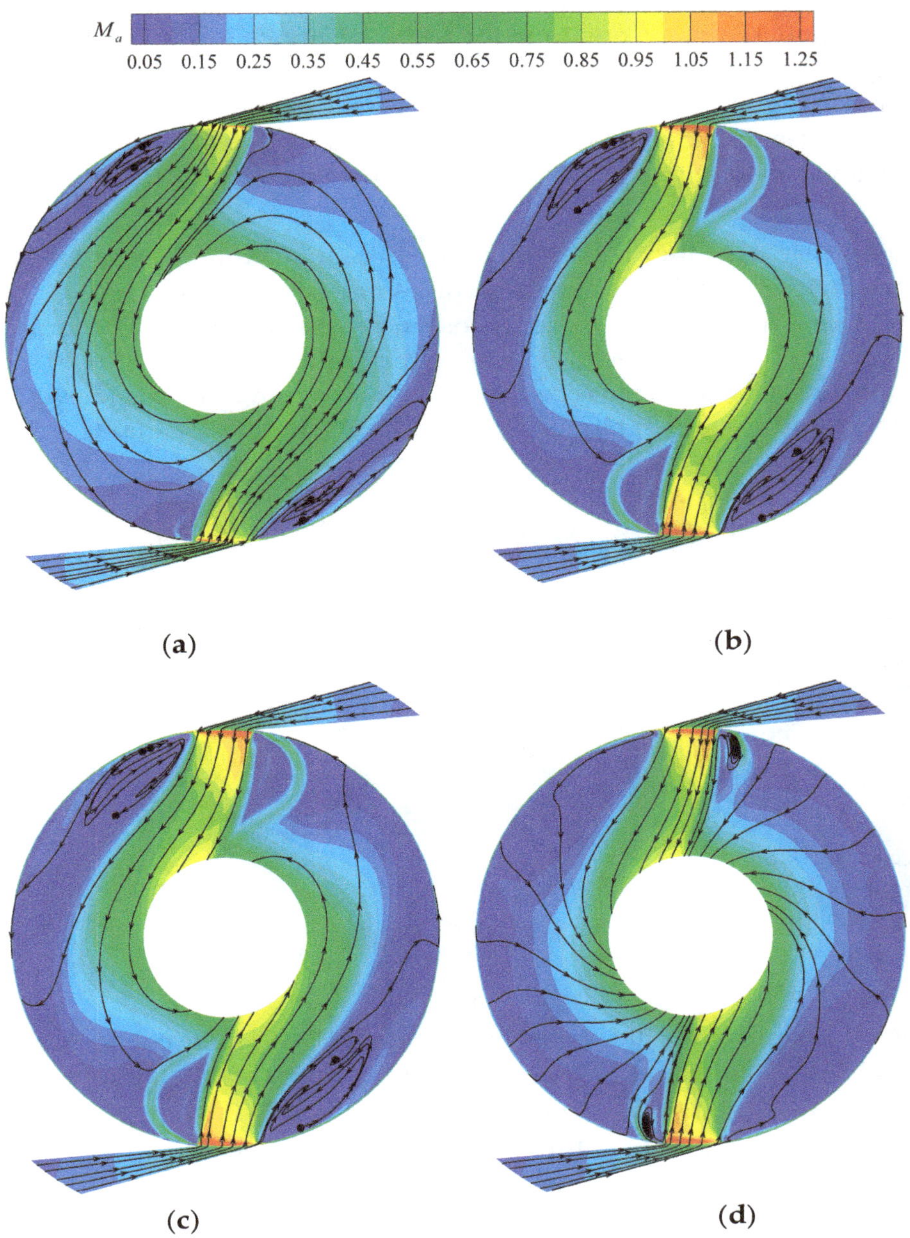

Figure 12. Mach number contours and streamlines on the mid-sections of DCs 1 and 3 for the one-to-many turbine cases with different disc thickness at 30,000 r/min: (**a**) DC1 for Case 1-0.5; (**b**) DC1 for Case 2-0.5; (**c**) DC3 for Case 1-0.5; and (**d**) DC3 for Case 2-0.5.

Figure 13 presents the contours of the pressure ratio on the mid-sections of DC1 for the one-to-many turbine cases, and the pressure ratio equals the ratio of the local pressure to the turbine inlet total pressure. The pressure ratio at the nozzle outlet for Case 1-0.5 is lower than that for Case 2-0.5, due to lower flow area ratio of the nozzle channel to the disc channels. It indicates a higher pressure drop in the nozzle and higher flow velocity at the nozzle outlet for Case 1-0.5 (Figures 12 and 13). A sudden decrease in pressure ratio also occurs at the same region of the sudden increase in Mach number. The pressure ratio decreases more remarkably for Case 2-0.5 than that for Case 1-0.5, because of much more decrease in flow area when the air flows through the N-R chamber to the disc channels.

The following is the detailed analysis of the component energy loss for the one-to-many turbine cases with different disc thickness based on the flow fields. With increasing disc thickness, the proportion of the boundary layer in the nozzle channel decreases and the flow velocity also decreases, which leads to lower friction loss in the nozzle. Actually, compared with the one-to-one turbine, the nozzle friction loss of the one-to-many turbine is much lower due to much less proportion of the boundary layer in the nozzle channel. It should be quite little and has a little influence on the nozzle loss. The local loss in the nozzle for the turbine case with thicker discs is much higher than that with thinner discs because the flow area reduces more when the air flows out of the N-R chamber to the disc channels They lead to a little increase in nozzle loss coefficient with increasing disc thickness (see Table 6).

As disc thickness increases, the working fluid becomes much more due to the larger nozzle throat area, and the flow velocity in the disc channels becomes much higher (see Figure 12). This leads to less boundary layer thickness, leading to less energy loss in the disc channels. The local loss at the inlet of the disc channels goes up with disc thickness, due to the higher flow area ratio of the N-R chamber to the disc channels. As a result, the disc loss decreases with increasing disc thickness (Table 6). The Mach number at the turbine outlet is much higher for Case 2-0.5 than that for Case 1-0.5 (Figure 12), which leads to higher leaving-velocity loss.

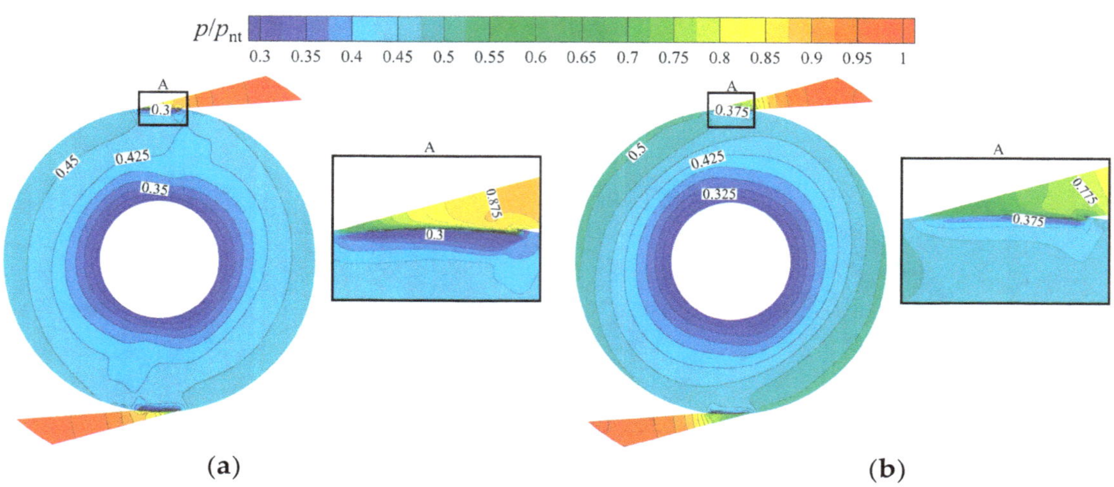

Figure 13. Contours of pressure ratio on the mid-sections of DC1 for the one-to-many Tesla turbine cases with different disc thickness at 30,000 r/min: (**a**) Case 1-0.5; and (**b**) Case 2-0.5.

Figure 14 gives contours of relative tangential velocity and vector distributions on the radial cross-sections of 0° for the one-to-many Tesla turbine cases. Unlike the one-to-one turbine, the air at the nozzle outlet in the one-to-many turbine bends to the axial direction to flow into the disc channels, especially at the nozzle of which downstream is a wall. The vortex occurs on the downstream wall of the N-R chamber and the inlet of the disc channels, leading to the energy loss at those places. The energy loss on the downstream wall of the N-R chamber wall is the main source of the local loss in the nozzle. That at the inlet of the disc channels is a part of the disc loss, which also includes the friction loss.

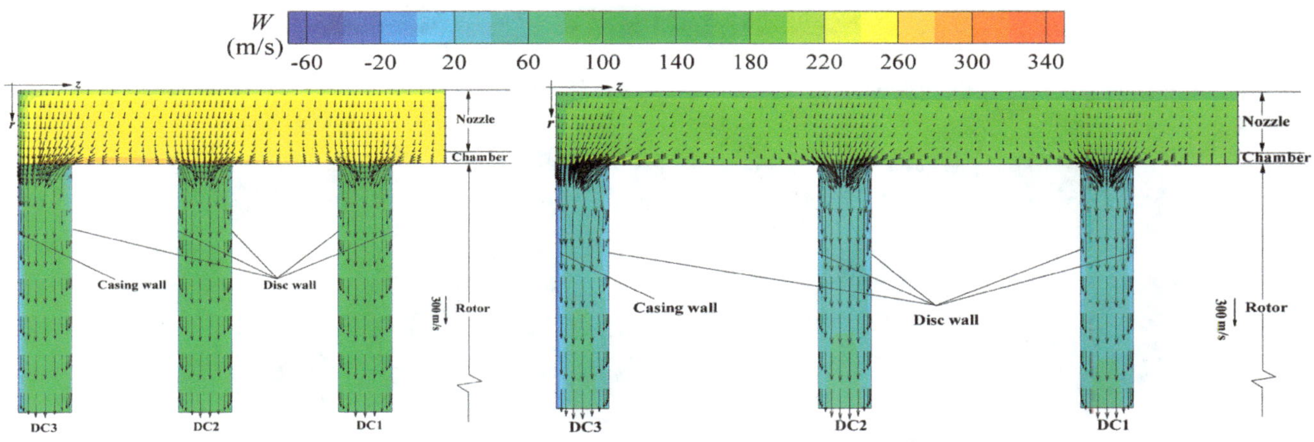

Figure 14. Contours of relative tangential velocity and vector distributions on radial cross-sections of 0° for the one-to-many Tesla turbine cases with different disc thickness: (**a**) Case 1-0.5; and (**b**) Case 2-0.5.

Compared with turbine Case 1-0.5, the flow velocity at the nozzle outlet for turbine Case 2-0.5 is lower, thus leading to a larger flow angle at the rotor inlet. The radial velocity at the rotor inlet is higher and the relative tangential velocity is lower. In addition, the vortex on the downstream wall of the N-R chamber and at the inlet of the disc channels for turbine Case 2-0.5 is much larger than that for turbine Case 1-0.5, therefore the energy loss caused by the variation of the flow area for Case 2-0.5 is much higher.

Combining Figures 12 and 14, the relationship of the mass flow rate percentages in each disc channel of the one-to-many turbine with disc thickness is analyzed. Compared with Case 1-0.5, the flow angle in DC1 for Case 2-0.5 is much larger, causing more air to flow through DC1 for Case 2-0.5. The mass flow rate of the air flowing into each disc channel of the one-to-many turbine has a significant difference from that of the one-to-one turbine. Figure 14 shows that the nozzle outlet area of the air flowing into DCs 1 and 2 is the same, and equals to the corresponding area of one disc spacing distance and one disc thickness. It is larger than that flowing into DC3, which equals to the corresponding area of one disc spacing distance and half of disc thickness. The area ratio of the working fluid flowing into DC1 to DC3 increases with disc thickness, leading to increasing mass flow rate percentage in DC1. Based on the two factors, the mass flow rate percentage in DC1 for Case 2-0.5 is higher than that for Case 1-0.5, which agrees with the results in Figure 6.

3.3.2. One-To-Many Turbine with Different Disc Spacing Distance

Mach number contours and streamlines on the mid-sections of DCs 1 and 3 for the one-to-many turbine cases with different disc spacing distance are presented in Figure 15. With increasing disc spacing distance, the Mach number increases at the nozzle outlet, which leads to lower flow angle at the rotor inlet. Thus, the turbine cases with larger disc spacing have longer path lines. Moreover, the sudden increase in Mach number at the rotor inlet decreases significantly with increasing disc spacing distance.

As discussed above, some air in DC1 passes through the N-R chamber and finally flows into DC3, which leads to more air in DC3 than that in DC1. For the turbine cases with disc thickness of 1 mm, with increasing disc spacing distance, the working fluid in DC1 flowing into the N-R chamber becomes much less due to less flow angle; with increasing disc spacing distance, the nozzle outlet area ratio of the air flowing into DC1 to that into DC3 decreases, causing less air in DC1. Under the combined actions of the two factors, the air in DC1 becomes less with increasing disc spacing distance.

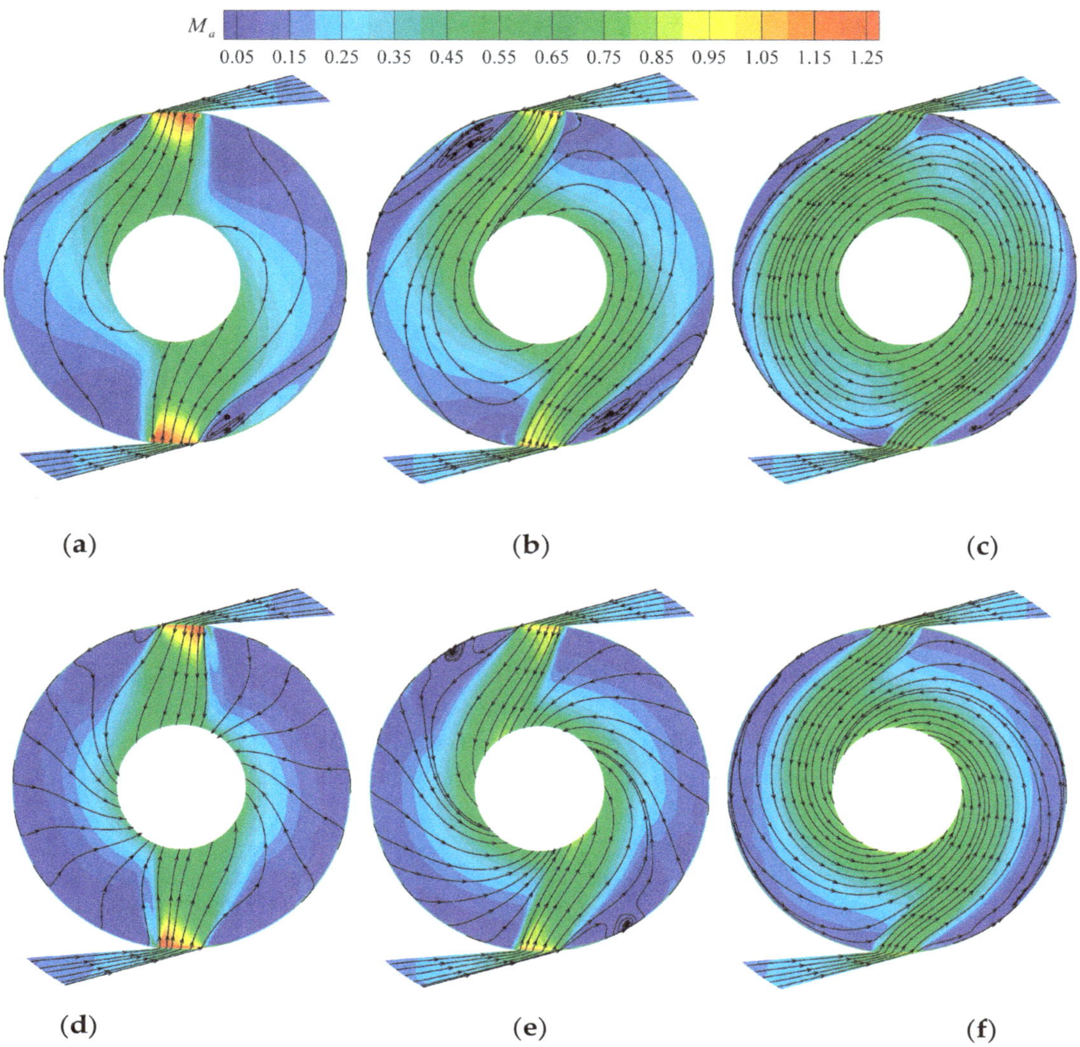

Figure 15. Mach number contours and streamlines on the mid-sections of DCs 1 and 3 for the one-to-many turbine cases with different disc spacing distance at 30,000 r/min: (**a**) DC1 for Case 1-0.3; (**b**) DC1 for Case 1-0.5; (**c**) DC1 for Case 1-1; (**d**) DC3 for Case 1-0.3; (**e**) DC3 for Case 1-0.5; and (**f**) DC3 for Case 1-1.

With increasing disc spacing distance, the local loss in the N-R chamber decreases, resulting from a decrease in flow area ratio of the N-R chamber to the disc channels. Based on the above analysis, the friction loss in the nozzle of the one-to-many turbine is quite little and has few contributions to the nozzle loss. Thus, the nozzle loss decreases with increasing disc spacing distance (Table 6).

For the one-to-many turbine, the energy loss in the disc channels also increases with disc spacing distance, which is the same as the one-to-one turbine. In addition, the local loss occurring at the inlet of the disc channels decreases with increasing disc spacing distance significantly. Thus, the disc loss of the one-to-many turbine decreases significantly with increasing disc spacing distance. Moreover, the leaving-velocity loss goes up with disc spacing distance, resulting from increasing Mach number at the turbine outlet (Figure 15).

Figure 16 shows the contours of pressure ratio on the mid-sections of DC1 for the one-to-many turbine cases with different disc spacing distance. The pressure ratio variation with increasing disc spacing distance is same as that with decreasing disc thickness, which is because both decreasing disc thickness and increasing disc spacing distance lead to a decrease in flow area ratio of the nozzle channel to the disc channels for the one-to-many turbine. In general, an increase in flow area ratio of the nozzle to disc channels leads to an increase in pressure at the nozzle outlet and a significant decrease in pressure at the rotor inlet facing the nozzle outlet. Therefore, the flow velocity at the nozzle outlet decreases and that at the rotor inlet increases much more.

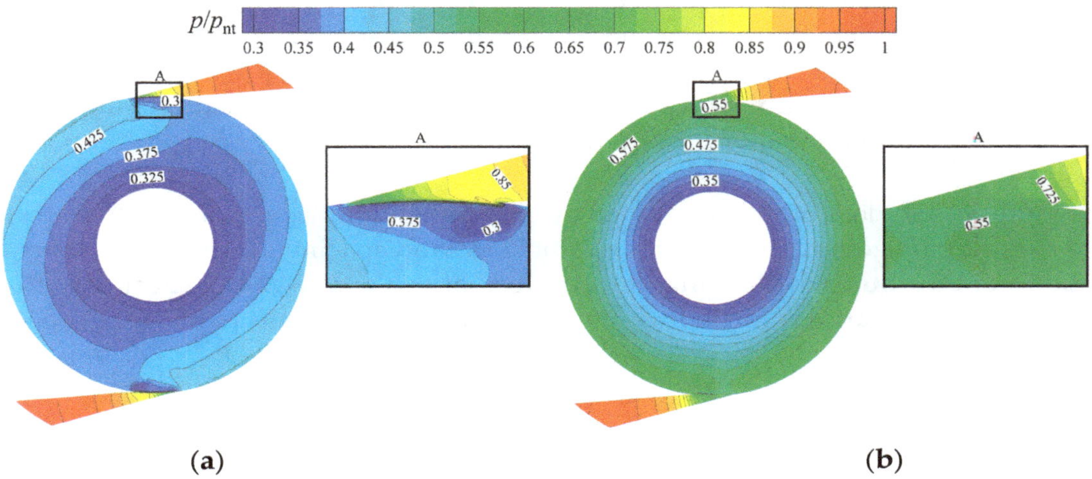

Figure 16. Contours of pressure ratio on the mid-sections of DC1 for one-to-many Tesla turbine cases with different disc spacing distance at 30,000 r/min: (**a**) Case 1-0.3; and (**b**) Case 1-1.

3.4. Influence of Disc Thickness and Spacing Distance on Energy Loss

As discussed above, the disc thickness has a little influence on the aerodynamic performance of the one-to-one Tesla turbine. In detail, the isentropic efficiency has a slight decrease with increasing disc thickness (less than 7% reduction for the turbine with 0.5 mm in disc spacing distance). That of the one-to-many turbine decreases significantly with disc thickness (decrease of about 45% for the turbine with 0.5 mm in disc spacing distance), which is due to higher leaving-velocity. For both the one-to-one turbine and the one-to-many turbine, the isentropic efficiency goes up first and then down with increasing disc spacing distance.

For the one-to-one turbine, the flow coefficient remains unchanged with the variation of disc thickness, and it increases with disc thickness for the one-to-many turbine. In addition, it increases with decreasing disc spacing distance for both two kinds of turbines. This coefficient of the one-to-one turbine is much lower than that of the one-to-many turbine with same disc spacing distance and disc thickness (The former is about half of the latter for turbine Case 1-0.5).

The nozzle loss of the multichannel Tesla turbine includes the friction loss and the local loss. That of the one-to-one turbine decreases with increasing disc spacing distance; that of the one-to-many turbine decreases with increasing disc spacing distance and increases slightly with increasing disc thickness. In general, the one-to-one turbine has higher nozzle loss than the one-to-many turbine (18.49% and 5.78%, respectively, for Case 1-0.5) due to much higher friction loss of the one-to-one turbine, although the local loss of the one-to-one is lower than that of the one-to-many turbine. Combining the discussion in the previous sections, it can be concluded that most of the nozzle loss is the friction loss for the one-to-one turbine, while, for the one-to-many turbine, the local loss in the nozzle is competitive with the friction loss in the nozzle. The friction loss decreases rapidly with a decrease in proportion ratio of the boundary layer in the nozzle channel. In detail, it drops with increasing disc spacing distance for the one-to-one turbine and increasing disc spacing distance or disc thickness for the one-to-many turbine, and the friction loss of the one-to-many turbine is quite low. The local loss in the nozzle decreases with decreasing flow area ratio of the N-R chamber to the disc channels, which is increasing disc spacing distance or decreasing disc thickness.

For the one-to-one turbine, the disc loss goes up first and then down with increasing disc spacing distance, and its variation range is much less (about 2%), while, for the one-to-many turbine, it decreases significantly with increasing disc spacing distance and decreases a little with increasing disc thickness (the maximum variation is 24.56%). For both the one-to-one turbine and the one-to-many turbine, the disc energy loss in the disc channels rises with disc spacing distance. In addition, the local loss at the inlet of the disc channels decreases with increasing disc spacing distance. According to the variation rule of the local loss with disc spacing distance, it can be concluded that the local loss of the one-to-one turbine is less than that of the one-to-many turbine.

The leaving-velocity loss of the one-to-one turbine rises with disc spacing distance, while that of the one-to-many turbine goes up with both disc thickness and spacing distance. Moreover, compared with the one-to-one turbine, the leaving-velocity loss of the one-to-many turbine with same disc thickness and spacing distance is much higher (33.1% and 13.03%, respectively, for Case 1-0.5).

4. Conclusions

As one of the best alternative options for small-scale conventional bladed turbines, Tesla turbines can be comparable with conventional turbines when they are well designed. Two kinds of multichannel Tesla turbines (one-to-one Tesla turbine and one-to-many Tesla turbine) with different disc spacing distance and disc thickness were studied numerically to analyze the influence of the two rotor geometrical parameters on their aerodynamic performance and flow characteristics, which offers a theoretical guidance in the design method. The main conclusions are summarized as follows.

(1) For both the one-to-one turbine and the one-to-many turbine, the isentropic efficiency goes up first and then down with increasing disc spacing distance, thus an optimal disc spacing distance exists to allow the multichannel Tesla turbine to obtain the highest efficiency. With increasing disc thickness, the isentropic efficiency of the one-to-many turbine decreases dramatically, while that of the one-to-one turbine decreases slightly. For turbine cases with disc spacing distance of 0.5 mm, as disc thickness increases from 1 mm to 2 mm, the isentropic efficiency of the one-to-one turbine was lowered by less than 7% (relative variation), while that of the one-to-many turbine drops by about 45% (relative variation). In addition, the isentropic efficiency of the one-to-one turbine is more sensitive to rotational speed than the one-to-many turbine.

(2) For the one-to-one turbine, with increasing disc thickness, the flow field is similar for turbine cases with same disc spacing distance; the flow velocity in the N-R chamber and the relative tangential velocity gradient close to the disc walls decrease slightly, causing the isentropic efficiency to decrease slightly. With increasing disc spacing distance, the flow velocity at the nozzle outlet decreases, and the flow angle at the rotor inlet increases, leading to an increase in mass flow rate percentage of DCs 1 and 2.

(3) For the one-to-many turbine, with increasing disc thickness, the flow velocity at the nozzle outlet decreases and its sudden increase at the rotor inlet facing the nozzle outlet is greater; the flow angle in the disc channels increases slightly and the flow velocity at the turbine outlet also increases. With increasing disc spacing distance, the flow velocity at the nozzle outlet goes up and its sudden increase becomes less; the flow angle in the disc channels decreases and the velocity at the rotor outlet increases.

Author Contributions: Conceptualization, W.Q. and Q.D.; Methodology, W.Q. and Q.D.; Investigation, W.Q. and Y.J.; Data Curation, W.Q. and Y.J.; Writing—Original Draft Preparation, W.Q.; Writing—Review and Editing, Q.D., Q.Y. and Z.F.; Visualization, W.Q. and Y.J.; and Supervision, Q.Y. and Z.F.

Nomenclature

b	disc spacing distance, mm
c	radial clearance of nozzle-rotor chamber, mm
c_p	specific heat at constant pressure, J/kg·K
C_m	flow coefficient
d	diameter, mm
h	static enthalpy, m^2/s^2
h_{tot}	mean total enthalpy, m^2/s^2
k	turbulent kinetic energy,
m	mass flow rate, kg/s
M	torque, N·m
M_a	Mach number
n	rotational speed of the rotor, r/min
N	number
p_{abs}	absolute pressure
$p_{\mathrm{nt}}/p_{\mathrm{i}}$	ratio of total pressure at the nozzle inlet to pressure at the turbine outlet
P	power, kW
Pr_t	turbulent Prandtl number
r	radial coordinate or radius, mm
R	specific gas constant of air,
S_E	energy source, $kg/(m \cdot s^3)$
S_M	momentum source, $kg/(m^2 \cdot s^2)$
t	time, s
th	disc thickness, mm
T	temperature, K
T_{nt}	total temperature at the nozzle inlet, K
U_j	averaged velocity in three directions, m/s
v	average radial velocity, m/s
W	relative tangential velocity, m/s
$\overline{W}$	average relative tangential velocity, m/s
x_j	coordinates in three directions, m
z	axial coordinate, m
α	nozzle exit geometrical angle (relative to the tangential direction), °
γ	specific heat ratio
Γ_t	eddy diffusivity, $kg/(m \cdot s)$
δ	relative variation of parameters
Δh_{isen}	isentropic enthalpy drop of the whole turbine, J/kg
η	isentropic efficiency
θ	circumferential coordinate, rad
μ_t	eddy viscosity or turbulent viscosity, $kg/(m \cdot s)$
ρ	density, kg/m^3
$\rho\overline{u_i u_j}$	Reynolds stresses
$\rho\overline{u_j h}$	turbulence flux
$\rho\overline{u_i \phi}$	Reynolds flux
τ	molecular stress tensor, $kg/(m \cdot s^2)$
ω	turbulent frequency, s^{-1}
Ω	rotational angular speed, rad/s

Subscripts

d	disc
dc	disc channel
i	inner
n	nozzle
o	outer

References

1. Gerendas, M.; Pfister, R. Development of a very small aero-engine. In Proceedings of the ASME Turbo Expo 2000, Munich, Germany, 8–11 May 2000.
2. Capata, R.; Sciubba, E. Experimental fitting of the re-scaled Balje maps for low-Reynolds radial turbomachinary. *Energies* **2015**, *8*, 7686–8000. [CrossRef]
3. Capata, R.; Saracchini, M. Experimental campaign tests on ultra micro gas turbines, fuel supply comparison and optimization. *Energies* **2018**, *11*, 799. [CrossRef]
4. Epstein, A.H. Millimeter-scale, MEMS gas turbine engines. *J. Eng. Gas Turbines Power* **2015**, *126*, 205–226. [CrossRef]
5. Fu, L.; Feng, Z.P.; Li, G.J. Experimental investigation on overall performance of a millimeter-scale radial turbine for micro gas turbine. *Energy* **2017**, *134*, 1–9. [CrossRef]
6. Fu, L.; Feng, Z.P.; Li, G.J. Investigation on design flow of a millimeter-scale radial turbine for micro gas turbine. *Microsyst. Technol.* **2018**, *24*, 2333–2347. [CrossRef]
7. Shan, X.C.; Zhang, Q.D.; Sun, Y.F.; Wang, Z.F. Design, fabrication and characterization of an air-driven micro turbine device. *J. Phys.* **2006**, *34*, 316–321. [CrossRef]
8. Isomura, K.; Murayama, M.; Teramoto, S.; Hikichi, K.; Endo, Y.; Togo, S.; Tanaka, S. Experiment verification of the feasibility of a 100 W class micro-scale gas turbine at an impeller diameter of 10 mm. *J. Micromech. Microeng.* **2006**, *16*, s254–s261. [CrossRef]
9. Kim, M.J.; Kim, J.H.; Kim, T.S. The effects of internal leakage on the performance of a micro gas turbine. *Appl. Energy* **2018**, *212*, 175–184. [CrossRef]
10. Deam, R.T.; Lemma, E.; Mace, B.; Collins, R. On scaling down turbines to millimeter size. *J. Eng. Gas Turbines Power* **2008**, *130*, 052301. [CrossRef]
11. Lampart, P.; Kosowski, K.; Piwowarski, M.; Jedrzejewski, L. Design analysis of Tesla micro-turbine operating on a low-boiling medium. *Pol. Marit. Res.* **2009**, *16*, 28–33. [CrossRef]
12. Tesla, N. Turbine. U.S. Patent 1,061,206, 6 May 1913.
13. Beans, E.W. Investigation into the performance characteristics of a friction turbine. *J. Spacecr.* **1966**, *3*, 131–134. [CrossRef]
14. Breiter, M.C.; Pohlhausen, K. Laminar Flow between Two Parallel Rotating Disks. Available online: https://apps.dtic.mil/dtic/tr/fulltext/u2/275562.pdf (accessed on 8 April 2016).
15. Rice, W. An analytical and experimental investigation of multiple-disk turbines. *J. Eng. Power* **1965**, *87*, 29–36. [CrossRef]
16. Adams, R.; Rice, W. Experimental investigation of the flow between corotating disks. *J. Appl. Mech.* **1970**, *37*, 844–849. [CrossRef]
17. Lemma, E.; Deam, R.T.; Toncich, D.; Collins, R. Characterisation of a small viscous flow turbine. *Exp. Therm. Fluid Sci.* **2008**, *33*, 96–105. [CrossRef]
18. Carey, V.P. Assessment of Tesla turbine performance for small scale Rankine combined heat and power systems. *J. Eng. Gas Turbines Power* **2010**, *132*, 122301. [CrossRef]
19. Sengupta, S.; Guha, A. A theory of Tesla disc turbines. *Proc. Inst. Mech. Eng. Part A J. Power Energy* **2012**, *226*, 650–663. [CrossRef]
20. Guha, A.; Sengupta, S. The fluid dynamics of the rotating flow in a Tesla disc turbine. *Eur. J. Mech. B/Fluid* **2013**, *37*, 112–123. [CrossRef]
21. Sengupta, S.; Guha, A. Analytical and computational solutions for three-dimensional flow-field and relative pathlines for the rotating flow in a Tesla disc turbine. *Comput. Fluids* **2013**, *88*, 344–353. [CrossRef]
22. Talluri, L.; Fiaschi, D.; Neri, G.; Ciappi, L. Design and optimization of a Tesla turbine for ORC applications. *Appl. Energy* **2018**, *226*, 300–319. [CrossRef]
23. Hoya, G.P.; Guha, A. The design of a test rig and study of the performance and efficiency of a Tesla disc turbine. *Proc. Inst. Mech. Eng. Part A J. Power Energy* **2009**, *223*, 451–465. [CrossRef]
24. Guha, A.; Smiley, B. Experiment and analysis for an improved design of the inlet and nozzle in Tesla disc turbines. *Proc. Inst. Mech. Eng. Part A J. Power Energy* **2010**, *224*, 261–277. [CrossRef]
25. Schosser, C.; Lecheler, S.; Pfitzner, M. A test rig for the investigation of the performance and flow field of Tesla friction turbines. In Proceedings of the ASME Turbo Expo 2014, Düsseldorf, Germany, 16–20 June 2014.
26. Lampart, P.; Jedrzejewski, L. Investigations of aerodynamics of Tesla bladeless microturbines. *J. Theor. Appl. Mech.* **2011**, *49*, 477–499.

27. Deng, Q.H.; Qi, W.J.; Feng, Z.P. Improvement of a theoretical analysis method for Tesla turbines. In Proceedings of the ASME Turbo Expo 2013, San Antonio, TX, USA, 3–7 June 2013.
28. Qi, W.J.; Deng, Q.H.; Feng, Z.P.; Yuan, Q. Influence of disc spacing distance on the aerodynamic performance and flow field of Tesla turbines. In Proceedings of the ASME Turbo Expo 2016, Seoul, Korea, 13–17 June 2016.
29. Qi, W.J.; Deng, Q.H.; Jiang, Y.; Feng, Z.P.; Yuan, Q. Aerodynamic performance and flow characteristics analysis of Tesla turbines with different nozzle and outlet geometries. *Proc. Inst. Mech. Eng. Part A J. Power Energy* **2018**, accepted. [CrossRef]
30. Hidema, T.; Okamoto, K.; Teramoto, S.; Nagashima, T. Numerical investigation of inlet effects on Tesla turbine performance. In Proceedings of the AJCPP, Miyazaki, Japan, 4–6 March 2010.
31. Sengupta, S.; Guha, A. Inflow-rotor interaction in Tesla disc turbines: Effects of discrete inflows, finite disc thickness, and radial clearance on the fluid dynamics and performance of the turbine. *Proc. Inst. Mech. Eng. Part A J. Power Energy* **2018**. accepted. [CrossRef]
32. Okamoto, K.; Goto, K.; Teramoto, S.; Yamaguchi, K. Experimental investigation of inflow condition effects on Tesla turbine performance. In Proceedings of the ISABE Conference 2017, Manchester, UK, 3–8 September 2017.
33. Carlson, D.R.; Widnall, S.E.; Peeters, M.F. A flow-visualization study of transition in plane Poiseuille flow. *J. Fluid Mech.* **1982**, *121*, 487–505. [CrossRef]
34. ANSYS. *CFX-Solver Theory Guide*; Release 12.1; ANSYS Inc.: Canonsburg, PA, USA, 2009.

12

Numerical and Experimental Investigation of Equivalence Ratio (ER) and Feedstock Particle Size on Birchwood Gasification

Rukshan Jayathilake and Souman Rudra *

Department of Engineering and Sciences, University of Agder, Grimstad 4879, Norway; rukshanma@gmail.com
* Correspondence: souman.rudra@uia.no.

Abstract: This paper discusses the characteristics of Birchwood gasification using the simulated results of a Computational Fluid Dynamics (CFD) model. The CFD model is developed and validated with the experimental results obtained with the fixed bed downdraft gasifier available at the University of Agder (UIA), Norway. In this work, several parameters are examined and given importance, such as producer gas yield, syngas composition, lower heating value (LHV), and cold gas efficiency (CGE) of the syngas. The behavior of the parameters mentioned above is examined by varying the biomass particle size. The diameters of the two biomass particles are 11.5 mm and 9.18 mm. All the parameters investigate within the Equivalences Ratio (ER) range from 0.2 to 0.5. In the simulations, a variable air inflow rate is used to achieve different ER values. For the different biomass particle sizes, CO, CO_2, CH_4, and H_2 mass fractions of the syngas are analyzed along with syngas yield, LHV, and CGE. At an ER value of 0.35, 9.18 mm diameter particle shows average maximum values of 60% of CGE and 2.79 Nm^3/h of syngas yield, in turn showing 3.4% and 0.09 Nm^3/h improvement in the respective parameters over the 11.5 mm diameter biomass particle.

Keywords: Birchwood gasification; computational fluid dynamics; equivalence ratio; cold gas efficiency; syngas

---

1. Introduction

Among the available energy sources, biomass is envisaged to play a major role in the future energy supplement. It produces no net carbon emission, while being the fourth largest energy source available in the world [1–4]. Therefore, biomass has high potential in contributing to satisfy the future energy demands of the world. Further, it is seen that for countries where the economy is mostly based on agriculture, they can utilize the potential of biomass efficiently. To recover energy from biomass, either thermochemical or biochemical process can be used [5–7]. The gasification process is a thermochemical process which gives a set of gases as output, consisting of CO, CO_2, H_2, CH_4, and N_2 by converting organic or carbonaceous materials like coal or biomass [8–10]. In gasification, several types of reactors used, such as entrained flow, fixed bed, fluidized bed, and moving bed gasifiers [7,11–13].

The producer gas is a result of a set of endothermic and exothermic reactions. The required heat for the endothermic reactions is provided by the exothermic reactions [11,14]. Once the steady state is achieved, the gasifier could work in a certain temperature range. Then, the producer gas can be obtained, as long as the fuel is being fed [14,15].

It was proven that producer gas was more versatile and useful than the biomass [16]. The quality of the producer gas is established to be one of the most important aspects of the gasification which should be enhanced, as it is used to generate power [11,16]. Chemical reactions inside the gasifier consist of both homogeneous and heterogeneous reactions. There are some factors which are important for the improvement of the gasification process such as reactor temperature, reactor pressure, the flow rate of the gasifying agent, and inflow rate of the biomass to the gasifier [11,17]. To evaluate gasification characteristics of biomass, a thorough evaluation is required. With the recent development of the computational fluid dynamics (CFD) and numerical simulations, sophisticated and robust models can be developed to give more qualitative information on biomass gasification [17,18]. In addition, CFD can produce important data with a relatively low cost [19]. Thus, CFD has become popular and an often-used tool to examine gasification characteristics [18,20,21].

Previous researches on gasification process have been performed on a regular basis investigating a broad range of parameters and variables. Most of them focused on the thermodynamics of the process, syngas composition, energy and exergy output, effect of temperature, the efficiency of the process, etc. [22,23]. Moreover, numerous numerical simulations have been done on the flow patterns and turbulence, as well as the different mathematical approaches. Ali et al. [24] developed a simulation model to discuss the co-gasification of coal and rice straw blend to investigate the syngas and cold gas efficiency (CGE), noting an 87.5% CGE value during the work. Slezak et al. [25] discussed an entrained flow gasifier using a CFD model to study the effects of coal particle density and size variations, where two devolatilization models were used in their work. It concluded that a higher fixed carbon conversion and H_2 could be achieved with a mix of different partitioned coal. Rogel et al. [26] presented a detailed investigation on the use of Eulerian approaches on 1-D and 2-D CFD models to examine pine wood gasification in a downdraft gasifier. Syngas lower heating value, syngas gas composition, temperature profiles, and carbon conversion efficiency were investigated [26] for both models to conclude that these methods could be used effectively in determining the above-mentioned parameters. Sharma [17] developed a CFD model of a downdraft biomass gasifier to investigate the thermodynamic and kinetic modeling of char reduction reactions. Further, a conclusion has been made that char bed length is less sensitive to equilibrium predictions, while CO and H_2 component, the calorific value of product gas and the endothermic heat absorption rate in reduction zone are found to be sensitive to the reaction temperature. Wu et al. [27] developed a 2-D CFD model of a downdraft gasifier to study the high-temperature agent biomass gasification, and observed a syngas with a high concentration of H_2 with a limited need of combustion inside the gasifier. Janejreh et al. [28] discussed the evaluation of species and temperature distribution using a CFD model with k-ε turbulence model and Lagrangian particle models. The gasification process was assessed using the CGE. Ismail et al. [29] developed a 2-D CFD model by using the Eulerian-Eulerian approach. Coffee husks are used as the feedstock to model the gasification process of a fluidized bed reactor. The effect of Equivalence Ratio (ER) and moisture content on gasification temperature, syngas composition, CGE, and HHV were discussed. After the analysis, high moisture content of coffee husks was found as a negative impact on the CGE and Higher Heating Value (HHV) as the ER increased. Monteiro et al. [30] developed a comprehensive 2-D CFD model to examine the potential of syngas from gasification of Portuguese Miscanthus. The Eulerian-Eulerian approach was used to model the exchange of mass, energy, and momentum. For the work, the effect of ER, steam to biomass ratio and temperature on syngas quality assessed. ER was proven to be a positive impact on the Carbon Conversion Rate (CCE), where an adverse impact was shown on the syngas quality and LHV. Further, the syngas quality and gasification efficiency was found to be increased by the increasing temperature. In order to describe the gasification process of three Portuguese biomasses, Silva et al. [31] developed a 2-D CFD model. A k-ε model was used to model the gas phase, while an Eulerian-Eulerian approach was used to model the transport of mass, energy, and momentum. After the simulations, the highest CGE value was shown by vine-pruning residues, while a higher H_2/CO ratio was shown by both coffee husks and vine pruning residues over the forest residues. Couto et al. [32] developed a 2-D CFD model using the Eulerian-Eulerian approach to evaluate the gasification of municipal solid waste. The gasification temperature was defined to be vastly important in syngas heating value in conclusion. In comparison to the above research works, the novelty of this study lies in the field of the Birchwood gasification. In addition, the literature lacks the use of 3-D CFD models and experiments on the effect of wood particle size on CGE. In this article, a 3-D CFD model is developed to investigate the Birchwood gasification in a fixed bed downdraft gasifier. Both heterogeneous and homogeneous reactions are considered and variation of syngas composition, syngas yield, LHV of the syngas, and CGE are measured for two sizes of feedstock, as well as for ER. Fixed bed downdraft gasifiers have demonstrated to perform well in the ER range from 0.25 to 0.43. Therefore, in the present work, the variations of the gasification parameters are considered in the ER range from 0.2 to 0.5 [14,15]. For the experimental work, the downdraft gasifier is used, which is available at the University of Agder (UIA), Norway. In addition, the CFD model is developed as the same geometry of the gasifier used for the experiments. For the validation purposes, experiments are done with the same gasifier with one of the wood particle sizes (11.5 mm woodchip) as it is the only biomass particle size available for the experiments. Then, with the developed simulation model, both biomass particle sizes are simulated.

2. Feedstock Characteristics

For the study, Birchwood is used as the feedstock. The ultimate and proximate analysis results for the Birchwood are listed down in Table 1. The data is extracted from a previous study [33], which is done on Birchwood in the same lab and under the same conditions.

Table 1. Proximate and ultimate analysis results of Birchwood [33].

Type of Analysis	Physical or Chemical Property	Value
Proximate Analysis (Dry Basis)	Moisture %	7
	Volatiles %	82.2
	Fixed Carbon %	10.45
	Ash %	0.35
	LHV (MJ/kg)	17.9
Ultimate Analysis (Dry Basis)	Carbon %	50.4
	Hydrogen %	5.6
	Oxygen %	43.4
	Nitrogen %	0.12
	Sulphur %	0.017
	Chlorine %	0.019

According to the analysis above, Birchwood has a lower amount of moisture in comparison to some of the other wood types [33–35], which is an important aspect for gasification. Apart from the moisture content, Birchwood has shown a lower amount of ash, Cl, and S, while higher values of calorific value and volatile matter [34] has been indicated. It is obvious that Birchwood has a higher calorific value because of the higher amount of carbon content [33]. In addition, when a biomass consists of less ash, it can lead to a higher conversion process which has lesser slag [14]. Hence, Birchwood has some positive characteristics in the gasification perspective. According to some of the literature, K, Si, and Ca can have a small effect in determining gasification characteristics. However, in this work, those are not considered as they are not measured in the ultimate analysis [36].

3. Methods

In this section, first, the experimental setup and the experimental process is described. Then, a brief description of the equations which use for the calculations is presented. It is then followed by the numerical study and simulation process.

3.1. Experimental Study

For the experiment, initially, about 6 kg of biomass is fed into the gasifier along with 1 kg of charcoal. Charcoal is added to the system to help the ignition process. The simulations are also carried out following the same procedure. The average size of the Birchwood chip which was used for the experiments is approximately 11.5 mm × 11.5 mm. A schematic diagram of the system is shown in Figure 1.

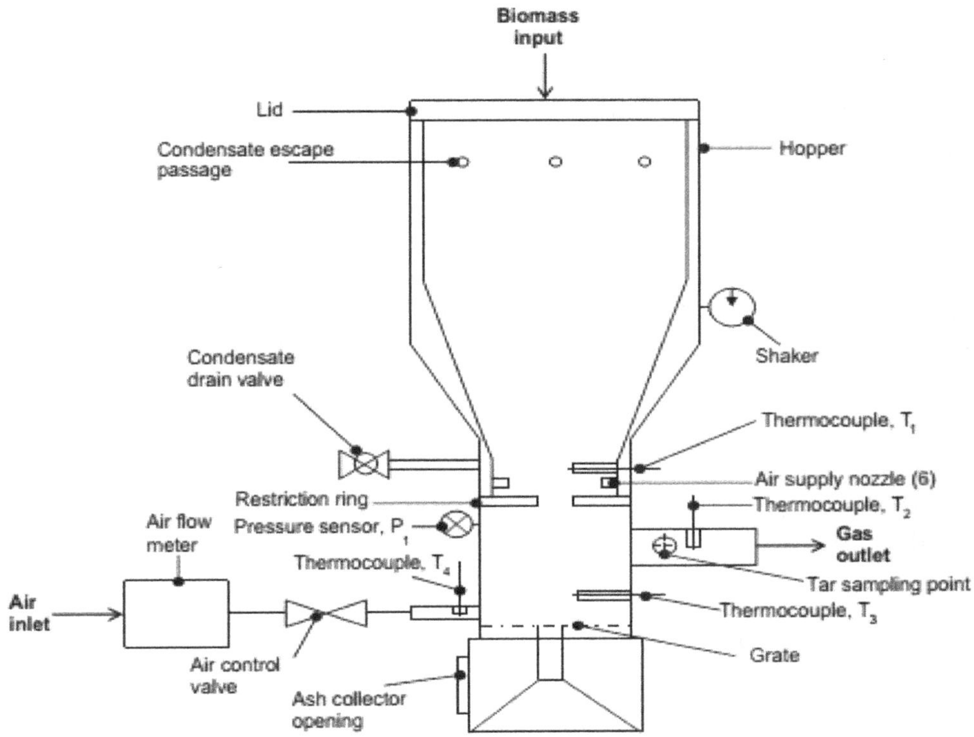

Figure 1. Schematic diagram of the gasifier system [34].

The hopper is a conically shaped stainless-steel structure with a volume of 0.13 m^3 and double-walled thick walls. It has a radius of 0.5 m. Biomass is fed through this hopper, and it is connected to an electrical shaker which can provide 15 shakes per minute (shakes/min). This shaking process helps to reduce or avoid some common problems, such as channeling and bridging. To drain out the condensed moisture, a drain valve is also attached to the bottom of the hopper.

The primary gasification reactor has a diameter of 0.26 m. It is 0.6 m in height. The air inlet system consists of six same size nozzles with equal spacing among each of a 5 mm diameter, and they are located from about 0.4 m from the bottom of the reactor. At the bottom of the reactor part, a reciprocally oscillating grate is connected and takes out the ash to the lower part of the ash collector. The grate is oscillating 30 s/min. To measure the weight loss during the gasification process, a mass scale locates at the bottom of the gasifier system. In addition, two pressure sensors and five K-type thermocouples are connected to the gasifier assembly to monitor and measure the parameters.

All the necessary experimental data is measured via a Lab view program which is specially designed to obtain data from the gasifier system. Further, to measure the composition of the syngas, a gas sample is sent to a gas analyzer, and throughout the experiment, changes in the composition of the syngas output is produced by the gas analyzer.

3.2. Theoretical and Numerical Study

In this study, the effect of biomass particle size and ER on the behavior of gasification parameters is examined using Birchwood as the feedstock. Thus, the variance of producer gas composition, producer gas yield, LHV of the syngas, and CGE is mainly assessed on two different values of biomass particle sizes.

Although downdraft gasifier demonstrates better performances at the ER ratio of 0.25, acceptable performances showed even close to the ER value of 0.43 [33,37]. Therefore, ER value range from 0.2 to 0.5 is selected in this study to investigate the best performance. ER is calculated using the following expression Equation (1) [8,15].

$$\text{ER} = \frac{\text{Actual air fuel ratio}}{\text{Stoichiometric air fuel ratio}} \tag{1}$$

here, the actual air-fuel ratio is calculated from the actual values measured from the experiment, while the stoichiometric air-fuel ratio is determined using Equation (2).

$$\text{CH}_a\text{O}_b + x(\text{O}_2 + 3.76\text{N}_2) \rightarrow y\text{H}_2\text{O} + z\text{CO}_2 + 3.76x\text{N}_2 \tag{2}$$

where a and b can be determined using the results of the ultimate analysis of the biomass (Birchwood). Here, y and z represent the stoichiometric coefficients of H_2O and CO_2, respectively, while x represent the stoichiometric coefficient of O_2 and 3.25 times of the N_2 amount. Further, the other parameters are calculated using the following Equation (3) to Equation (5) [33,38–40].

$$LHV_{syngas} = ((25.7 \times \text{H}_2) + 30 \times \text{CO} + (85.4 \times \text{CH}_4)) \times \frac{4.2}{1000} \tag{3}$$

Since the Hydro-Carbons (HC) higher than CH_4 are not considered for this study, they have not been used to calculate LHV_{syngas}.

$$\text{Producer gas yield } (V_{syngas}) = \frac{(Q_{air} \times 79)}{\text{N}_2 \times m_b} \tag{4}$$

$$\text{Cold gas efficiency (CGE)} = \frac{LHV_{syngas} \times V_{syngas}}{LHV_{biomass}} \times 100 \tag{5}$$

where LHV_{syngas} represents the dry base lower heating value of the syngas in MJ/Nm^3, V_{syngas} represents the syngas yield Nm^3/kg feed stock, Q_{air} represents the air input in Nm^3/h, m_b represents the biomass input in kg/h, $LHV_{biomass}$ represents the lower heating value of the feedstock in MJ/kg, and N_2 represents nitrogen mass fraction of the output gas.

In the simulation model, a spherical particle is used for modeling. Hence, spherical particles with a diameter of 11.5 mm (available wood chip size in the lab) and 9.18 mm are utilized for the simulations (a spherical particle which has the same surface area as an actual wood chip, with a diameter of 9.18 mm consider here).

In the presented work, the STAR CCM+ software package is used for the modeling and simulation. A 3D geometric model of the actual gasifier is designed to create the computational model. For the gasifier model, 0.01m mesh is used, while in the air inlets and air-fuel mixing region, a 0.001 m mesh has been used. Figure 2 shows the customized meshfor the gasifier (0.01 m) and for air inlets (0.001 m).

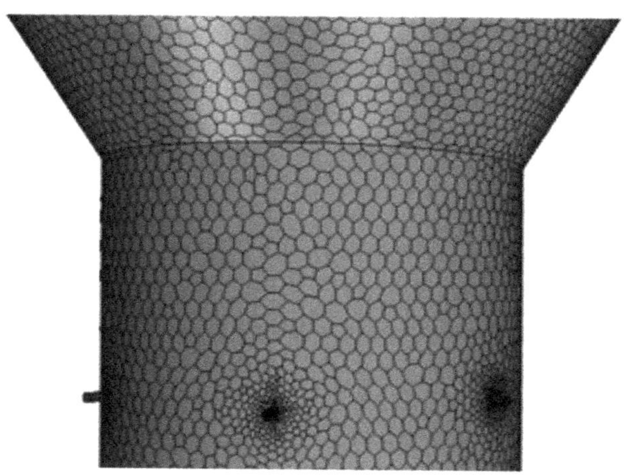

Figure 2. Meshed gasifier geometry and customized mesh at air inlets.

A segregated flow model is implemented to solve the transport equations. In addition, the Lagrangian particle modeling method is used to model the particle behavior (solid phase).

3.2.1. Assumptions

The underlying assumptions of the modeling process are summarized as follows:

- Steady flow is considered inside the gasifier.
- The flow inside the domain is considered as incompressible and turbulent.
- Spherical particles are used.
- Evenly distributed particle regime is utilized.
- The No-slip condition is imposed on inside wall surfaces.

3.2.2. Eulerian-Lagrangian Method

Reynolds averaged Navier-Stokes (RANS) equations are solved using a Eulerian-Lagrangian reference frame to solve the numerical scheme. The gas phase is considered as a reacting gas phase, and Lagrangian particle models are used to address the solid (particle) phase. Standard k-ε model is used for turbulence modeling. In addition, both finite rates of chemical reactions and eddy dissipation rates are used in the computational model. Moreover, for the modeling work, the conservation equations for transport of energy, momentum, and mass are used (Appendix A). Also, the default sub models available for moisture evaporation, devolatilization and char oxidation are employed (Appendix A). In the Eulerian-Eulerian approach, gas phase, as well as the solid phase, is considered as continuum [41]. In the Lagrangian approach, usually a large number of particles are tracked transiently. The method is started by solving the transient momentum equation for each particle. In order to calculate the trajectory of a particle, Newton's second law is used and written in a Lagrangian reference frame [41–44]. Momentum balance for particles is described in Equation (6),

$$m_p \frac{dV_p}{dt} = F_s + F_b \tag{6}$$

where V_p represents the velocity of the particle, m_p is the mass of the particle, and t is time. F_s and F_b represent the surface force and body force, respectively. When these forces are observed in depth, each of them is a composition of few forces available in the system. The following Equations (7) and (8) describe the two forces.

$$F_s = F_d + F_p + F_{vm} \tag{7}$$

$$F_b = F_g + F_u \tag{8}$$

where F_d, F_p, and F_{vm} represent the drag force on the particle, pressure force as a gradient in the static pressure, and the extra virtual mass added while the acceleration in the phase, respectively. F_g and F_u represent the gravitational force and the user defined force added to the system, respectively. In addition, the drag force can be defined as follows in Equation (9).

$$F_d = \frac{1}{2} C_D \rho A_p |V_s| V_s \tag{9}$$

where C_D is the drag coefficient, ρ is the density, A_p represents the surface area of the particle, and V_s is the slip velocity of the particle.

Using the Nusselt number (Nu_p), the heat transfer coefficient can be found for the Lagrangian particle, which is the energy model for the Lagrangian particle. This is shown in Equation (10),

$$h_p = \frac{Nu_p}{D_p} \cdot k \tag{10}$$

where k has represented the thermal conductivity of the phase and D_p is the particle diameter. Nusselt number (Nu_p) is expressed in the following Equation (11), where it is calculated by the Ranz-Marshall correlation and presented below in Equation (12).

$$Nu_p = 2\left(1 + 0.3Re_p^{\frac{1}{2}} Pr^{\frac{1}{3}}\right) \quad (11)$$

where Re represents the Reynolds number, while the expression Pr expresses the Prandtl number [45,46].

$$N_u = h\frac{d_p}{k} = a + cRe^m Pr^n \quad (12)$$

where h is the heat coefficient, d_p is the particle diameter, k represents the thermal conductivity of the gas, Re is the Reynolds number, and Pr represents the prandtl number. The letters a, c, m, and n are numerical constants which are determined by the flow field. According to the Ranz-Marshall correlation, at a fluid velocity (V) equal to zero, the heat transfer can only be affected by conduction. therefore, the values of a, c, m, and n become 2, 0.6, 0.5, and 0.33, respectively [47].

3.2.3. Reactions Modeling

In this work, the following approach is used to simulate the gasification process. This method has been used by [8,48]. In this study, it is assumed that the moisture is evaporated completely in the drying phase [49].

$$\text{Biomass (wet)} \rightarrow \text{Biomass (dry)} + H_2O\ \text{(steam)} \quad (R1)$$

In the pyrolysis phase, the Biomass (dry) is converted into char, tar, ash, and volatile matter [49,50].

$$\text{Biomass (dry)} \rightarrow \text{volatiles} + \text{char} + \alpha\text{tar} + \text{ash} \quad (R2)$$

where, α is the stoichiometric coefficient of tar.

Here, the volatile matter is described as follows. It is assumed that the volatile matter consists of only CO, CO_2, H_2, and CH_4.

$$\text{Volatile} \rightarrow \alpha_1 CO + \alpha_2 CO_2 + \alpha_3 H_2 + \alpha_4 CH_4 \quad (R3)$$

where, $\alpha_1, \alpha_{2,}\alpha_{3,}$, and α_4 are the stoichiometric coefficients of CO, CO_2, H_2, and CH_4 respectively.

The following method is used to calculate the mass distribution in volatile matter and mass fractions in dry biomass. For the calculations, the data available in Table 1 is used. These values can be used to combine with the ultimate analysis. In this method, a set of linear equations is used to solve the C, O, and H components balance in the system [8,48].

$$\frac{12}{28}x + \frac{12}{44}y + \frac{12}{16}z = Y_{C,vol} = 0.449 \quad (13)$$

$$\frac{16}{28}x + \frac{32}{44}y = Y_{O,vol} = 0.488 \quad (14)$$

$$\frac{4}{16}z + w = Y_{H,vol} = 0.063 \quad (15)$$

here, x, y, z, and w are unknown mass fractions of CO, CO_2, CH_4, and H_2, respectively in volatile species after devolatilization. $Y_{C,vol}$, $Y_{O,vol}$, and $Y_{H,vol}$ are, respectively, the mass fractions of C, O, and H in biomass volatile.

In the above equations, the coefficients are the ratios of molar mass of each atom to the molar mass of each species. (As an example, in Equation (15), $\frac{4}{16} = \frac{M_H}{M_{CH_4}}$). Here, with the molar masses, the numerical values are rounded up for the simplification of the calculations. For finding the mass fractions $Y_{C,vol}$, $Y_{O,vol}$, and $Y_{H,vol}$, the following method is used along with the data from the ultimate analysis of the biomass.

$$\text{Volume fraction} = Y_{C\ total} - Y_{C\ fixed} + Y_O + Y_H = 0.89 \tag{16}$$

$$Y_{C,vol} = \frac{Y_{C\ total} - Y_{C\ fixed}}{\text{Volume fraction}} = 0.449 \tag{17}$$

$$Y_{O,vol} = \frac{Y_O}{\text{Volume fraction}} = 0.488 \tag{18}$$

$$Y_{H,vol} = \frac{Y_H}{\text{Volume fraction}} = 0.063 \tag{19}$$

According to the mass conservation,

total mass before devolatilization − mass of fixed carbon = total mass after devolatilization

Thus, according to the above criteria,

$$x + y + z + w = Y_{C,vol} + Y_{O,vol} + Y_{H,vol} = 1 \tag{20}$$

The ratio between CO and CO_2 is temperature dependent. In this study, the ratio has been determined as 2.43 according to [8]. For the calculations, the bed temperature is used as 907 K, which was obtained during the experiments. Below Equation (21) is used to determine the connection between CO and CO_2 at 907 K.

$$\frac{x}{y} = 2400e^{-\left(\frac{6234}{T}\right)} \tag{21}$$

$$\frac{x}{y} = 2.43 \tag{22}$$

The linear equations from Equation (9) to Equation (15) are solved using the MATLAB program, and the following values are obtained from the program.

$$x = 0.5605$$

$$y = 0.2306$$

$$z = 0.1945$$

$$w = 0.0144$$

According to the above results, 1 kg of volatile matter produce devolatilized products according to the following expression.

$$1\text{kg}_{vol} \rightarrow 0.5605\text{kg}_{CO} + 0.2306\text{kg}_{CO_2} + 0.1945\text{kg}_{CH_4} + 0.0144\text{kg}_{H_2} \tag{23}$$

Using the following molar mass values, the above Equation (23) can be used as a molar based equation.

$$M_{CO_2} = 44.008\ \text{kg/mol},\ M_{CO} = 28.009\ \text{kg/mol},\ M_{CH_4} = 16.042\ \text{kg/mol},\ M_{H_2} = 2.016\ \text{kg/mol}$$

$$1\text{kg}_{vol} \rightarrow 0.02CO + 0.00524CO_2 + 0.0072H_2 + 0.0122CH_4 \tag{24}$$

Therefore, according to the following (R4), the devolatilization of the volatiles is modeled.

$$0.0375CH_{1.69}O_{0.81} \rightarrow 0.02CO + 0.00524CO_2 + 0.0072H_2 + 0.0122CH_4 \tag{R4}$$

Based on the above devolatilization reaction (R4), the gasification process is modeled. In addition, for the modeling work, the following homogeneous and heterogeneous reactions are considered [8,28,51].

Water-gas shift reaction:

$$CO_{(g)} + H_2O_{(g)} \leftrightarrow CO_{2(g)} + H_{2(g)};\ \Delta H = +41.98\ \text{kJ/mol} \tag{R5}$$

Methanation reaction:

$$CO_{(g)} + 3H_{2(g)} \leftrightarrow CH_{4(g)} + H_2O_{(g)};\ \Delta H = -227\ \text{kJ/mol} \tag{R6}$$

Dry reforming reaction:

$$CH_{4(g)} + CO_{2(g)} \leftrightarrow 2CO_{(g)} + 2H_{2(g)};\ \Delta H = -793\ \text{kJ/mol} \tag{R7}$$

Steam reforming reaction:

$$CH_{4(g)} + H_2O_{(g)} \leftrightarrow CO_{(g)} + 3H_{2(g)};\ \Delta H = +206\ \text{kJ/mol} \tag{R8}$$

mbustion reaction:

$$C_{(s)} + O_{2(g)} \rightarrow CO_{2(g)};\Delta H = -408.8\ \text{kJ/mol} \tag{R9}$$

Boudouard reaction:

$$C_{(s)} + CO_{2(g)} \rightarrow 2CO_{(g)};\ \Delta H = +172\ \text{kJ/mol} \tag{R10}$$

Water-gas reaction:

$$C_{(s)} + H_2O_{(g)} \rightarrow CO_{(g)} + H_{2(g)};\ \Delta H = +131\ \text{kJ/mol} \tag{R11}$$

Hydrogasification:

$$C_{(s)} + 2H_{2(g)} \rightarrow CH_{4(g)};\ \Delta H = +75\ \text{kJ/mol} \tag{R12}$$

Carbon partial reaction:

$$2C_{(g)} + O_{2(g)} \leftrightarrow 2CO_{(g)};\ \Delta H = -246.4\ \text{kJ/mol} \tag{R13}$$

4. Results and Discussions

In this section, the computational model validation is conducted by comparing the available experimental data and numerical simulation results. Moreover, the comparison of the two sizes of biomass particle sizes is carried out using the gasification characteristics.

4.1. Development of Mass Fractions of Considered Gases in the Syngas

A specially designed lab view program was used to measure all the gas component mass fractions throughout the experiment. From the obtained data, the variation of the mass fractions of the gas components during the experiment is presented below in Figure 3. For the tests, a biomass woodchip with an average size of 11.5 mm $\times$ 11.5 mm is used, which is available in the laboratory.

In the simulation model, for two iterations, one data point has been recorded. 600 iterations are used in the simulations, and 300 data points are recorded throughout the simulation. Using the obtained data points gas component mass fractions are graphed in Figure 4. For the simulations, two types of wood particle sizes are used. One of them is 11.5 mm in diameter, while the other is 9.18 mm in diameter. Below, Figure 4 was obtained using 11.5 mm diameter wood particle.

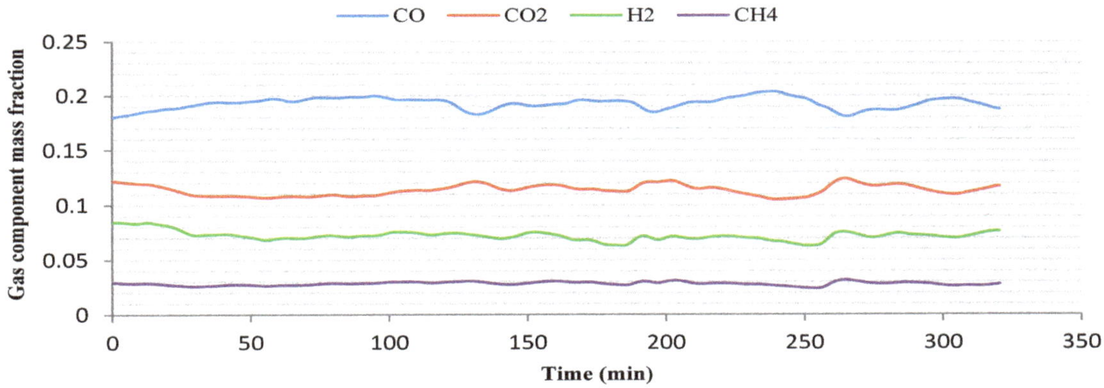

Figure 3. Mass fraction variation with time in the experimental study.

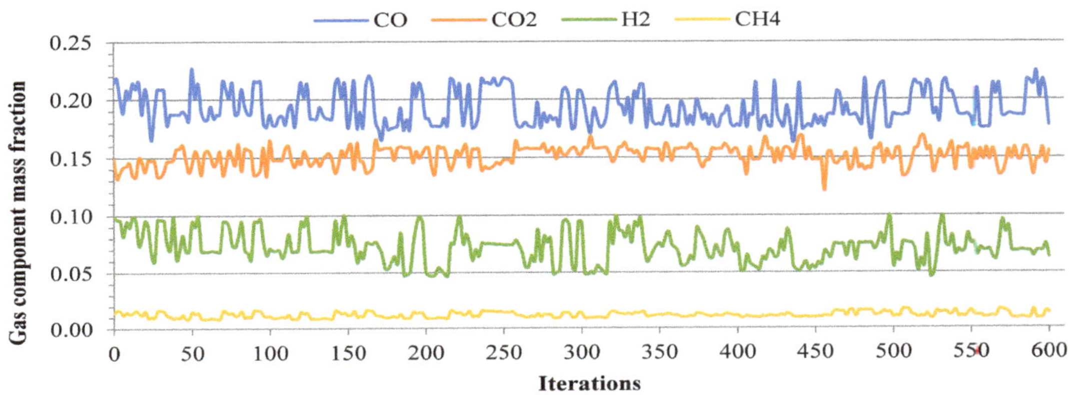

Figure 4. Mass fraction variation with iteration in the simulation model.

By comparing Figures 3 and 4, it shows the fluctuation in gas components mass friction in the simulation values, whereas in the experiment, it shows smooth variation. It can be due to the fact that the interval between two data points are bigger. However, an average value is used to plot graphs.

The data obtained from the model is also compared with some of the data collected from the literature (Study 1 [52], and Study 2 [53]), as well as with the data achieved during the experiment of the present work. Figure 5 shows the comparison of the predicted data along with the experimental data. According to the graph, predicted data shows good agreement with the literature and experimental data.

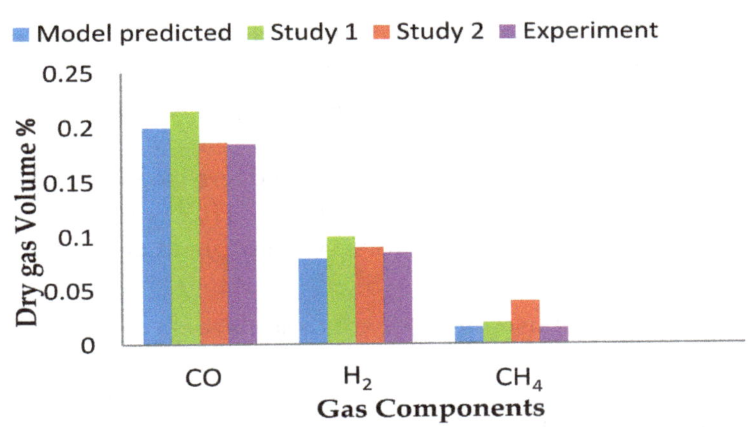

Figure 5. Composition of the syngas predicted by model compared with experimental results.

4.2. Effect of ER and Biomass Particle Size on Quality of Producer Gas

The variation of gas components in producer gas in both experiments and simulation models is illustrated in the following Figure 6. For the simulations, a biomass particle with a diameter of 11.5 mm and a biomass particle with a diameter of 9.18 mm were used. (The change of gas component contours in various ER values are illustrated in Appendix A (Figures A1–A3)).

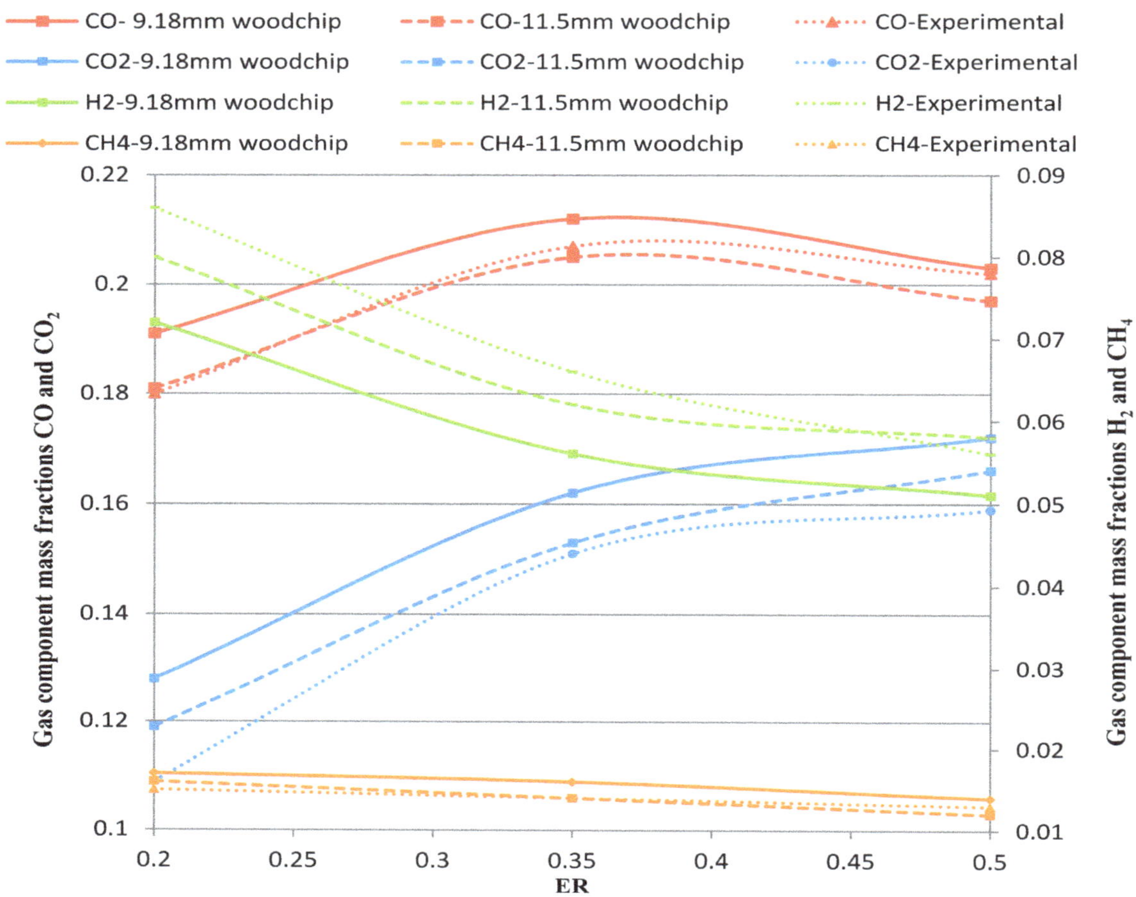

Figure 6. Variation of gas component mass fractions as a function of ER.

According to Figure 6, mass fractions of CO and CO_2 are increased when the ER is increased up to a certain level (ER = 0.2 to 0.35). Then, ER contributes to a reduction of the mass fraction of CO, while further increasing CO_2. This can be a result of the combined effect of Combustion reaction (R9), Boudouard reaction (R10), and Carbon partial reaction (R13). Therefore, in lower ER values, (R10) tends to produce more CO to the system at the expense of CO_2, while in high ER values, (R9) and (R13) convert more CO into CO_2 [2].

Rgearding CH_4 and H_2, increasing ER is effected on a decrease in both gas species. Nevertheless, H_2 is varied in a higher mass fraction range than of CH_4. This is mainly due to the effects of dry reforming (R7) and steam reforming (R8) reactions, which produce more H_2 to the system at the expense of CH_4.

Since a variable air inflow rate is used in the experiments, an average value of the air inflow rate is used for the calculation of ER values. However, a fixed air inflow rate is used in the simulations. Therefore, the impact of variable air inflow may have caused the deviation in the graphs of the experimental and simulation data. In the simulations, spherical shape wood particles are used, while wood chips with different shapes are used for the experiments. The difference of behavior of spherical wood particle and divergement shaped wood chips can also create the inequalities in the graphs. In addition, during the experiments, channeling and bridging can also have taken place [7]. The effect of that phenomenon may have also affected the experimental results.

According to Figure 6, all the considered gas component mass fractions, except H_2, are increased for the 9.18 mm wood particle over 11.5 mm wood particle. This pattern can be found in some of the literature [54,55]. When the particle size is getting smaller, the area to volume ratio of the wood particle is increased. This can result in an increased contact with the reactants [54–56]. Hence, the reaction rates can be increased and gasification reactions can also be improved. Further, the enhancement of hydrogasification (R12) is crucial for the behavior of the gas component mass fractions. When the biomass particle is getting smaller, it can enhance the reactivity and produce more CH_4 with R12.

4.3. Effect of Equivalences Ratio (ER) and Biomass Particle Size on Syngas Yield, Lower Heating Value (LHV), and Cold Gas Efficiency (CGE)

Figure 7 shows the variation of syngas yield, LHV and CGE of the syngas as a function of ER. According to Figure 7, the behavior of syngas yield could be an effect of higher volatilization [28].

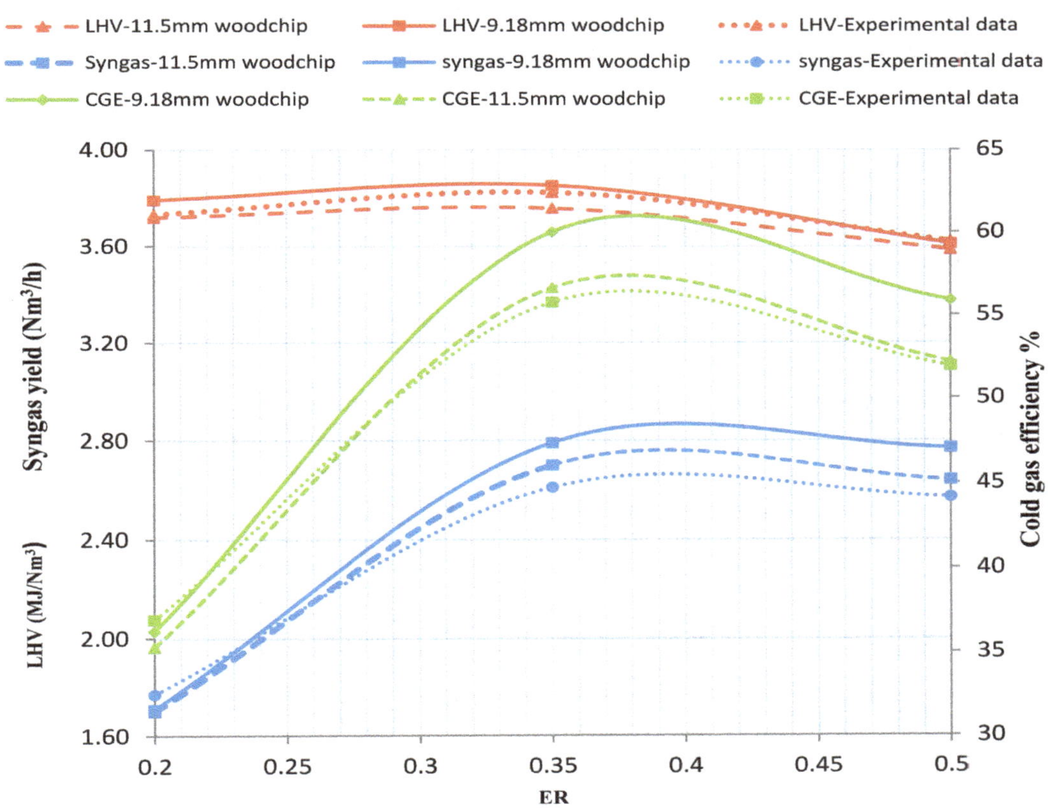

Figure 7. Variation of Syngas yield lower heating value (LHV) and cold gas efficiency (CGE) of syngas as a function of Equivalences Ratio (ER).

During the simulations, both biomass particles are evaluated in the same ER range. However, the gas yields with two separate sizes of biomass particles have shown a difference. This difference of syngas yield should be due to a different conversion. For the context of LHV of syngas, increasing ER has shown a positive effect within the range from 0.2 to 0.35. It is evident that, as the gas species mass fractions increase, LHV of syngas increases as well.

According to Figure 7, the smaller particle size has resulted in a significant increment of the LHV of the gas, although it has followed the same variation along the ER range as the bigger biomass particle. The increment of LHV can be mainly due to the increment of CH_4 with the smaller biomass particle. As mentioned before, LHV is a direct function of combustible gas species, and CH_4 has the highest contribution for the LHV among the gas species.

For CGE, a positive effect of increasing ER is observed from 0.2 to 0.35 of ER. Within this ER range, CGE could have been improved by increasing airflow. Increasing airflow (increasing ER) can contribute to creating a high composition of combustible gas species in syngas by enhancing the reaction chemistry of homogeneous and heterogeneous reactions. Therefore, as the gas species are increased up to 0.35 of ER, so does the CGE. As the gas species have shown a decrement in the mass fractions in the ER range from 0.35 to 0.5, CGE has also shown a decrement in its value. With more O_2 input to the system, the system can move towards combustion rather than gasification. This can be a result of the reduction of CGE after the ER value of 0.35. According to the graph in Figure 7, with the 9.18 mm biomass particle, CGE has shown an average maximum value of 60% at ER value of 0.35, compared to an average maximum value of 56.63 % with 11.5 mm diameter biomass particle.

For both particle sizes, same ER range is considered. Hence, an increase of LHV, syngas yield, and combustible gas spices in syngas with smaller biomass particle could be the reason for the increase of the CGE.

4.4. Temperature Profile

Temperature profile inside the gasifier in the simulation model is illustrated in Figure 8. In Figure 9, the temperature values recorded in the experiments and the simulation model are demonstrated in a graph. Figure 8 is mainly used to help understand the actual placement of the thermocouples. In the simulation model, seven temperature measurements are taken along the mid-axis, which is shown in Figure 9. However, in the experimental setup, only four thermocouples are available. Hence, to develop the curve, those four temperature measurement values are used. According to Figure 9, the maximum temperature is shown close to the air inlets in both experimental and simulation setups. In the experimental setup, the thermo couples are placed on the gasifier wall. Hence, the actual temperature can be slightly higher than what it is shown here.

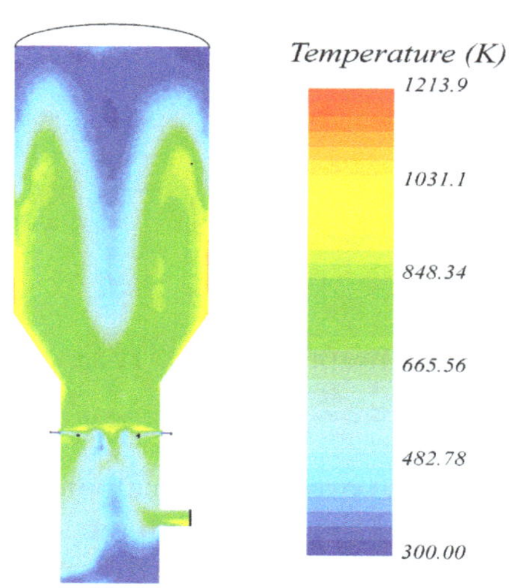

Figure 8. Temperature profile inside the gasifier.

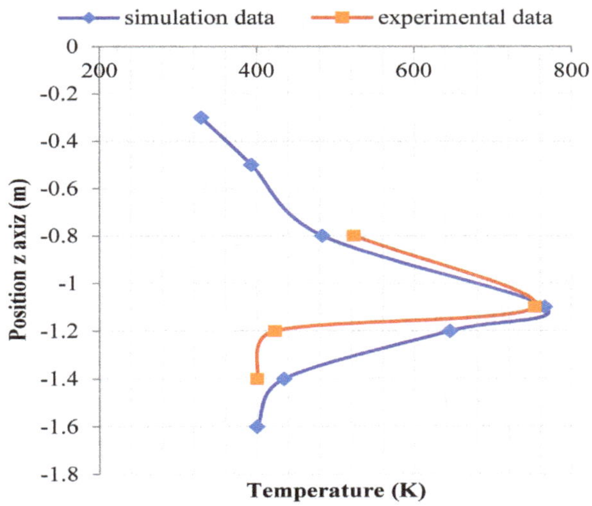

Figure 9. Temperature variation along the mid-axis of the gasifier simulation values vs. experimental values.

5. Conclusions

Based on the comparative study, it is concluded that most of the measured and calculated parameters (including syngas yield, LHV of syngas, and CGE) have shown an increasing trend as ER is increased up to 0.35, and when biomass particle size is reduced. Furthermore, the average value of CGE, for instance, reached a maximum value of 60% with the 9.18 mm diameter wood particle, and with a corresponding ER about 0.35. Moreover, 11.5 mm diameter wood particle has shown an average maximum value of 56.6% at the same ER value, which has been both an improvement and a significant output of this work. Reaching an average value of 2.79 Nm^3/h with a corresponding ER of 0.35, the average value of Syngas yield has shown an improvement of 0.09 Nm^3/h with the 9.18 mm diameter wood particle over 11.5 mm wood particle, which has been an important aspect in energy harvesting from the syngas. Therefore, as the result shows, an ER value close to 0.35 shows the best results in CGE, syngas yield, and LHV of the syngas. Furthermore, smaller biomass particles show improvements in the results. Therefore, according to our study, in order to obtain better outputs, ER should be kept close to 0.35. In the future, more simulations and experiments could be done with finer biomass particles. Further, more user defined functions could be used to improve the gasification process and to represent more realistic biomass particle shapes.

Acknowledgments: Authors would like to thank Department of Engineering and Sciences, UiA, for supporting to do the experiment at Bio-energy lab. In addition, we would appreciate the contribution and guidance given by Hans Jorgen Morch, CEO of CFD Marine AS, especially in obtaining the license for the STAR CCM+ software.

Author Contributions: Rukshan Jayathilake and Souman Rudra conceived and designed the experiments; Rukshan Jayathilake performed the experiments; Rukshan Jayathilake and Souman Rudra analyzed the data; Rukshan Jayathilake and Souman Rudra wrote the paper.

Appendix A

Appendix A.1. Conservation Equations

In order to develop the numerical model, conservation equations have been used, along with some other important equations. The equation for mass conservation is presented below in Equation (A1) [39],

$$\frac{\partial \rho}{\partial t} + \frac{\partial \rho v_i}{\partial x_i} = 0 \tag{A1}$$

where ρ is the density of the fluid, is the velocity tensor in the compact form in Einstein notation, t represents the time, and x_i represents the special first order tensor.

The momentum conservation equation is:

$$\rho\frac{\partial u_i}{\partial t} + \rho u_j\frac{\partial u_i}{\partial x_j} = \rho g_i - \nabla P + \frac{\partial}{\partial x}\left[\mu\left(\frac{\partial v_i}{\partial x} + \frac{\partial v_j}{\partial x}\right)\right] \tag{A2}$$

where ρ is the density of the fluid and g_i is the gravity force. P represents the pressure, while the dynamic viscosity is represented by μ.

The energy equation is presented below in Equation A3 [57],

$$\rho\left(\frac{\partial h}{\partial t} + u_i\frac{\partial h}{\partial x_i}\right) = \frac{\partial P}{\partial t} + u_i\frac{\partial P}{\partial x_i} + \frac{\partial}{\partial x_j}(k\nabla T) + \tau'_{ij}\frac{\partial u_i}{\partial x_j} \tag{A3}$$

where ρ represents the density of the fluid P represents the pressure, k represents the thermal conductivity, and τ'_{ij} represents the viscous stress tensor.

Appendix A.2. Submodels

Appendix A.2.1. Two Ways Coupling

When the Lagrangian particles are relatively smaller, a one-way coupling model is used, since the heat transfer and drag force are the only things which affect through continues phase. However, once the particle size is much larger compared to the volume cells, then a more advance model has to be introduced. Once the two-way coupling is used, an active particle or the Lagrangian phase can exchange mass momentum and energy with fluid phase [58–60].

Appendix A.2.2. Coal Combustion Model

Coal combustion model is a set of sub models consist of moisture evaporation, two step devolatilization, and char oxidation models. Thus, in the following part, each of these models will be discussed.

Appendix A.2.3. Moisture Evaporation Model

The evaporation model has been built on a small assumption. It assumes the moisture has covered the whole particle with a thin film. According to that, the moisture has to be evaporated before the volatiles are released because moisture is at the outermost layer. In this model, there are no properties associated. Here, the Ranz Marshall correlation has been used to calculate the relevant Nusselt and Sherwood numbers. In order to formulate this model, the Quasi-Steady single component droplet evaporation model has been applied to a water droplet [61].

Appendix A.2.4. Devolatilization Model

In order to simulate the volatile release from the coal particles, a devolatilization model has to be used. In STAR CCM+, there are two devolatilization models that can be found: the single step devolatilization model and two step devolatilization models [46]. In Equation (A4), the devolatilization model for n steps has been shown.

$$\text{rawcoal} \overset{k_{pn}}{\rightarrow} VM_{pa(g)} + (1 - VM_{pa})(Char)_p \tag{A4}$$

In the above equation, $(Char)_p$ represents the amount of char in the particle, while VM_{pa} represents the volatile matter in the proximate analysis. The kinetic rate for the above equation is shown in the below Equation (A5).

$$r_{vpn} = k_{pn} V M_{pa} \gamma_{cp} \tag{A5}$$

here, k_{pn} represents the reaction rate constant of the devolatilization reaction, and γ_{cp} is the coal mass fraction from the proximate analysis. Thus, the reaction rate constant k_{pn} is expressed in the following Equation (A6).

$$k_{pn} = A_{pn} \exp\left(-\frac{E_{pn}}{RT_p}\right) \tag{A6}$$

E_{pn} represents the activation energy of the nth reaction for the particle and A_{pn} is the pre-exponential factor for the reaction. T_p is the particle temperature, while R is the universal gas constant. For the modeling work a single step devolatilization model is used. Hence, the pre-exponent factor $A = 6 \times 10^{13}\ \mathrm{s}^{-1}$ and activation energy of $E = 2.5 \times 10^{8}\mathrm{J/kg\ mol}$ is used for the study, which are generally used as the pre-exponent and activation energy values for wood [62].

Appendix A.3. Figures

In the following Figure A1, change in CO contours against various ER values are shown.

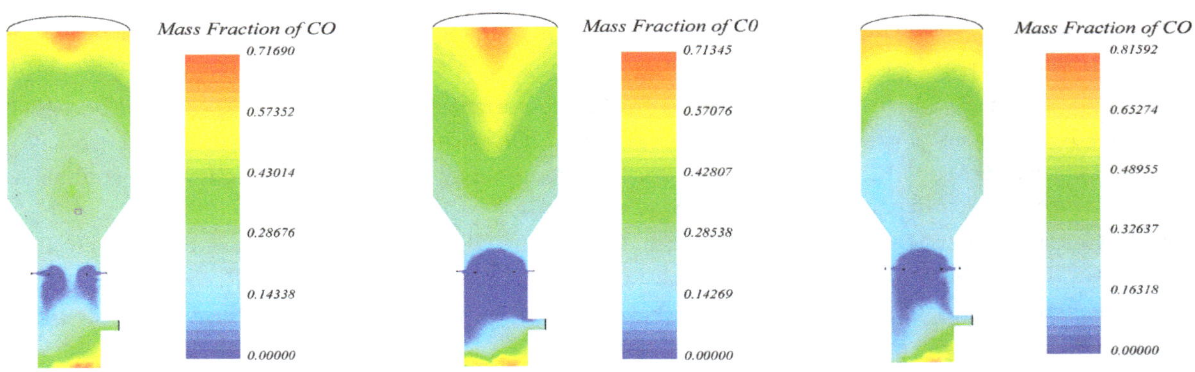

(a) (b) (c)

Figure A1. CO contours with ER values of (**a**) 0.2; (**b**) 0.35; (**c**) 0.5 from left to right.

In the following Figure A2, change in H_2 contours against various ER values are shown.

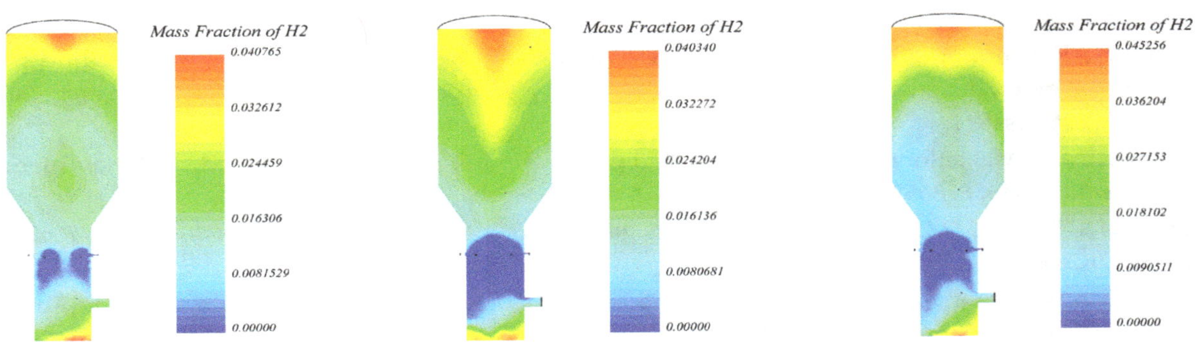

(a) (b) (c)

Figure A2. H_2 contours with ER values of (**a**) 0.2; (**b**) 0.35; (**c**) 0.5 from left to right.

In the following Figure A3, change in CO_2 contours against various ER values are shown.

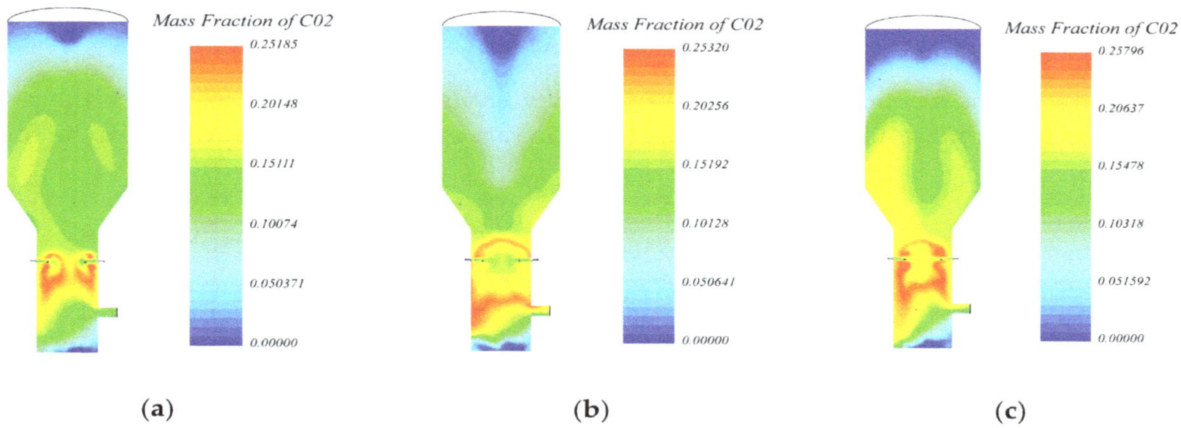

Figure A3. CO_2 contours with ER values of (**a**) 0.2; (**b**) 0.35; (**c**) 0.5 from left to right.

References

1. Caliandro, P.; Tock, L.; Ensinas, A.V.; Marechal, F. Thermo-economic optimization of a solid oxide fuel cell-gas turbine system fuelled with gasified lignocellulosic biomass. *Energy Convers. Manag.* **2014**, *85*, 764–773. [CrossRef]
2. Sarker, S.; Bimbela, F.; Sánchez, J.L.; Nielsen, H.K. Characterization and pilot scale fluidized bed gasification of herbaceous biomass: a case study on alfalfa pellets. *Energy Convers. Manag.* **2015**, *91*, 451–458. [CrossRef]
3. Saxena, R.C.; Adhikari, D.K.; Goyal, H.B. Biomass-based energy fuel through biochemical routes: A review. *Renew. Sustain. Energy Rev.* **2009**, *13*, 167–178. [CrossRef]
4. Neri, E.; Cespi, D.; Setti, L.; Gombi, E.; Bernardi, E.; Vassura, I.; Passarini, F. Biomass Residues to Renewable Energy: A Life Cycle Perspective Applied at a Local Scale. *Energies* **2016**, *9*, 922. [CrossRef]
5. Cotana, F.; Cavalaglio, G.; Coccia, V.; Petrozzi, A. Energy opportunities from lignocellulosic biomass for a biorefinery case study. *Energies* **2016**, *9*, 748. [CrossRef]
6. Cavalaglio, G.; Gelosia, M.; D'Antonio, S.; Nicolini, A.; Pisello, A.L.; Barbanera, M.; Cotana, F. Lignocellulosic ethanol production from the recovery of stranded driftwood residues. *Energies* **2016**, *9*, 634. [CrossRef]
7. Alauddin, Z.A.B.Z.; Lahijani, P.; Mohammadi, M.; Mohamed, A.R. Gasification of lignocellulosic biomass in fluidized beds for renewable energy development: A review. *Renew. Sustain. Energy Rev.* **2010**, *14*, 2852–2862. [CrossRef]
8. Basu, P. *Biomass Gasification and Pyrolysis: Practical Design and Theory*; Academic Press: Burlington, MA, USA, 2010.
9. Rudra, S.; Rosendahl, L.; Blarke, M.B. Process analysis of a biomass-based quad-generation plant for combined power, heat, cooling, and synthetic natural gas production. *Energy Convers. Manag.* **2015**, *106*, 1276–1285. [CrossRef]
10. Zaccariello, L.; Mastellone, M.L. Fluidized-bed gasification of plastic waste, wood, and their blends with coal. *Energies* **2015**, *8*, 8052–8068. [CrossRef]
11. Baruah, D.; Baruah, D.C. Modeling of biomass gasification: A review. *Renew. Sustain. Energy Rev.* **2014**, *39*, 806–815. [CrossRef]
12. James, R.A.M.; Yuan, W.; Boyette, M.D. The effect of biomass physical properties on top-lit updraft gasification of woodchips. *Energies* **2016**, *9*, 283. [CrossRef]
13. Shi, H.; Si, W.; Li, X. The concept, design and performance of a novel rotary kiln type air-staged biomass gasifier. *Energies* **2016**, *9*, 67. [CrossRef]
14. Rajvanshi, A.K.; Goswami, D.Y. *Biomass Gasification in Alternative Energy in Agriculture Vol. II*; CRC Press: Boca Raton, FL, USA, 1986.
15. Brown, R.C.; Stevens, C. *Thermochemical Processing of Biomass: Conversion into Fuels, Chemicals and Power*; Wiley series in Renewable Resources: Chichester, UK, 2011.

16. Sheth, P.N.; Babu, B.V. Experimental studies on producer gas generation from wood waste in a downdraft biomass gasifier. *Bioresour. Technol.* **2009**, *100*, 3127–3133. [CrossRef] [PubMed]
17. Sharma, A.K. Equilibrium and kinetic modeling of char reduction reactions in a downdraft biomass gasifier: A comparison. *Sol. Energy* **2008**, *82*, 918–928. [CrossRef]
18. Wang, Y.; Yan, L. CFD studies on biomass thermochemical conversion. *Int. J Mol. Sci* **2008**, *9*, 1108–1130. [CrossRef] [PubMed]
19. Gerber, S.; Behrendt, F.; Oevermann, M. An Eulerian modeling approach of wood gasification in a bubbling fluidized bed reactor using char as bed material. *Fuel* **2010**, *89*, 2903–2917. [CrossRef]
20. Ku, X.; Li, T.; Løvås, T. CFD-DEM simulation of biomass gasification with steam in a fluidized bed reactor. *Chem. Eng. Sci.* **2015**, *122*, 270–283. [CrossRef]
21. Biglari, M.; Liu, H.; Elkamel, A.; Lohi, A. Application of scaling-law and CFD modeling to hydrodynamics of circulating biomass fluidized bed gasifier. *Energies* **2016**, *9*, 504. [CrossRef]
22. Patel, K.D.; Shah, N.K.; Patel, R.N. CFD Analysis of spatial distribution of various parameters in downdraft gasifier. *Procedia Eng.* **2013**, *51*, 764–769. [CrossRef]
23. Gerun, L.; Paraschiv, M.; Vijeu, R.; Bellettre, J.; Tazerout, M.; Gøbel, B.; Henriksen, U. Numerical investigation of the partial oxidation in a two-stage downdraft gasifier. *Fuel* **2008**, *87*, 1383. [CrossRef]
24. Ali, D.A.; Gadalla, M.A.; Abdelaziz, O.Y.; Hulteberg, C.P.; Ashour, F.H. Co-gasification of coal and biomass wastes in an entrained flow gasifier: Modelling, simulation and integration opportunities. *J. Nat. Gas Sci. Eng.* **2017**, *37*, 126–137. [CrossRef]
25. Slezak, A.; Kuhlman, J.M.; Shadle, L.J.; Spenik, J.; Shi, S. CFD simulation of entrained-flow coal gasification: Coal particle density/sizefraction effects. *Powder Technol.* **2010**, *203*, 98–108. [CrossRef]
26. Rogel, A.; Aguillon, J. The 2D Eulerian approach of entrained flow and temperature in a biomass stratified downdraft gasifier. *Am. J. Appl. Sci.* **2006**, *3*, 2068–2075. [CrossRef]
27. Wu, Y.; Zhang, Q.; Yang, W.; Blasiak, W. Two-dimensional computational fluid dynamics simulation of biomass gasification in a downdraft fixed-bed gasifier with highly preheated air and steam. *Energy Fuels* **2013**, *27*, 3274–3282. [CrossRef]
28. Janajreh, I.; Al Shrah, M. Numerical and experimental investigation of downdraft gasification of wood chips. *Energy Convers. Manag.* **2013**, *65*, 783–792. [CrossRef]
29. Ismail, T.M.; El-Salam, M.A.; Monteiro, E.; Rouboa, A. Eulerian-Eulerian CFD model on fluidized bed gasifier using coffee husks as fuel. *Appl. Therm. Eng.* **2016**, *106*, 1391–1402. [CrossRef]
30. Monteiro, E.; Ismail, T.M.; Ramos, A.; El-Salam, M.A.; Brito, P.S.D.; Rouboa, A. Assessment of the miscanthus gasification in a semi-industrial gasifier using a CFD model. *Appl. Therm. Eng.* **2017**, *123*, 448–457. [CrossRef]
31. Silva, V.; Monteiro, E.; Couto, N.; Brito, P.; Rouboa, A. Analysis of syngas quality from portuguese biomasses: An experimental and numerical study. *Energy Fuels* **2014**, *28*, 5766–5777. [CrossRef]
32. Couto, N.D.; Silva, V.B.; Monteiro, E.; Rouboa, A. Assessment of municipal solid wastes gasification in a semi-industrial gasifier using syngas quality indices. *Energy* **2015**, *93*, 864–873. [CrossRef]
33. Sarker, S.; Nielsen, H.K. Preliminary fixed-bed downdraft gasification of birch woodchips. *Int. J. Environ. Sci.* **2015**, *12*, 2119. [CrossRef]
34. Sarker, S.; Nielsen, H.K. Assessing the gasification potential of five woodchips species by employing a lab-scale fixed-bed downdraft reactor. *Energy Convers. Manag.* **2015**, *103*, 801–813. [CrossRef]
35. Atnaw, S.M.; Sulaiman, S.A.; Yusup, S. Influence of fuel moisture content and reactor temperature on the calorific value of syngas resulted from gasification of oil palm fronds. *Sci. World J.* **2014**, *2014*, 121908. [CrossRef] [PubMed]
36. Bouraoui, Z.; Jeguirim, M.; Guizani, C.; Limousy, L.; Dupont, C.; Gadiou, R. Thermogravimetric study on the influence of structural, textural and chemical properties of biomass chars on CO_2 gasification reactivity. *Energy* **2015**, *88*, 703–710. [CrossRef]
37. Zainal, Z.A.; Ali, R.; Lean, C.H.; Seetharamu, K.N. Prediction of performance of a downdraft gasifier using equilibrium modeling for different biomass materials. *Energy Convers. Manag.* **2001**, *42*, 1499–1515. [CrossRef]
38. Sarkar, M.; Kumar, A.; Tumuluru, J.S.; Patil, K.N.; Bellmer, D.D. Gasification performance of switchgrass pretreated with torrefaction and densification. *Appl. Energy* **2014**, *127*, 194–201. [CrossRef]
39. Syed, S.; Janajreh, I.; Ghenai, C. Thermodynamics equilibrium analysis within the entrained flow gasifier environment. *Int. J. Therm. Environ. Eng.* **2012**, *4*, 47–54. [CrossRef]

40. Renganathan, T.; Yadav, M.V.; Pushpavanam, S.; Voolapalli, R.K.; Cho, Y.S. CO_2 utilization for gasification of carbonaceous feedstocks: A thermodynamic analysis. *Chem. Eng. Sci.* **2012**, *83*, 159. [CrossRef]
41. Chiesa, M.; Mathiesen, V.; Melheim, J.A.; Halvorsen, B. Numerical simulation of particulate flow by the Eulerian-Lagrangian and the Eulerian-Eulerian approach with application to a fluidized bed. *Comput. Chem. Eng.* **2005**, *29*, 291–304. [CrossRef]
42. Zhang, Z.; Chen, Q. Comparison of the Eulerian and Lagrangian methods for predicting particle transport in enclosed spaces. *Atmos. Environ.* **2007**, *41*, 5236–5248. [CrossRef]
43. White, F.M. *Viscous Fluid Flow*; McGraw-Hill Higher Education: Boston, MA, USA, 2006.
44. Yu, X.; Hassan, M.; Ocone, R.; Makkawi, Y. A CFD study of biomass pyrolysis in a downer reactor equipped with a novel gas-solid separator-II thermochemical performance and products. *Fuel Process. Technol.* **2015**, *133*, 51–63. [CrossRef]
45. Aissa, A.; Abdelouahab, M.; Noureddine, A.; Elganaoui, M.; Pateyron, B. Ranz and Marshall correlations limits on heat flow between a sphere and its surrounding gas at high temperature. *Therm. Sci.* **2015**, *19*, 1521–1528. [CrossRef]
46. Luan, Y.T.; Chyou, Y.P.; Wang, T. Numerical analysis of gasification performance via finite-rate model in a cross-type two-stage gasifier. *Int. J. Heat Mass Transf.* **2013**, *57*, 558–566. [CrossRef]
47. Sharma, A.; Pareek, V.; Wang, S.; Zhang, Z.; Yang, H.; Zhang, D. A phenomenological model of the mechanisms of lignocellulosic biomass pyrolysis processes. *Comput. Chem. Eng.* **2014**, *60*, 231. [CrossRef]
48. Molcan, P.; Caillat, S. Modelling approach to woodchips combustion in spreader stoker boilers. In Proceedings of the 9th European Conference on Industrial Furnaces and Boilers, Estoril, Portugal, 26–29 April 2011.
49. Wu, Y.; Yang, W.; Blasiak, W. Energy and exergy analysis of high temperature agent gasification of biomass. *Energies* **2014**, *7*, 2107–2122. [CrossRef]
50. Allesina, G. Modeling of coupling gasification and anaerobic digestion processes for maize bioenergy conversion. *Biomass Bioenergy* **2015**, *81*, 444–451. [CrossRef]
51. Allesina, G.; Pedrazzi, S.; Tartarini, P. Modeling and investigation of the channeling phenomenon in downdraft stratified gasifers. *Bioresour. Technol.* **2013**, *146*, 704–712. [CrossRef] [PubMed]
52. Giltrap, D.L.; McKibbin, R.; Barnes, G.R.G. A steady state model of gas-char reactions in a downdraft biomass gasifier. *Sol. Energy* **2003**, *74*, 85–91. [CrossRef]
53. González-Vázquez, M.P.; García, R.; Pevida, C.; Rubiera, F. Optimization of a bubbling fluidized bed plant for low-temperature gasification of biomass. *Energies* **2017**, *10*, 306. [CrossRef]
54. Zhang, C. Numerical modeling of coal gasification in an entrained-flow gasifier. *ASME Int. Mech. Eng. Congr. Expo.* **2012**, *6*, 1193–1203. [CrossRef]
55. Lv, P.; Chang, J.; Wang, T.; Fu, Y.; Chen, Y.; Zhu, J. Hydrogen-rich gas production from biomass catalytic gasification. *Int. J. Hydrogen Energy* **2009**, *34*, 1260–1264. [CrossRef]
56. Hernández, J.J.; Aranda-Almansa, G.; Bula, A. Gasification of biomass wastes in an entrained flow gasifier: Effect of the particle size and the residence time. *Fuel Process. Technol.* **2010**, *91*, 681–692. [CrossRef]
57. Cornejo, P.; Farias, O. Mathematical modeling of coal gasification in a fluidized bed reactor using an Eulerian Granular description. *Int. J. Chem. React. Eng.* **2011**, *9*. [CrossRef]
58. Discrete Element method (DEM) in STAR CCM+. Available online: https://mdx2.plm.automation.siemens.com/presentation/discrete-element-method-dem-star-ccm (accessed on 10 August 2017).
59. Alletto, M.; Breuer, M. One-way, two-way and four-way coupled LES predictions of a particle-laden turbulent flow at high mass loading downstream of a confined bluff body. *Int. J. Multiph. Flow* **2012**, *45*, 70–90. [CrossRef]
60. Benra, F.K.; Dohmen, H.J.; Pei, J.; Schuster, S.; Wan, B. A comparison of one-way and two-way coupling methods for numerical analysis of fluid-structure interactions. *J. Appl. Math.* **2011**, *2011*, 853560. [CrossRef]
61. Kent, J.C. Quasi-steady diffusion-controlled droplet evaporation and condensation. *Appl. Sci. Res.* **1973**, *28*, 315–360. [CrossRef]
62. Belosevic, S. Modeling approaches to predict biomass co-firing with pulverized coal. *Open Thermodyn. J.* **2010**, *4*, 50–70. [CrossRef]

13

Numerical Study on the Effect of Tunnel Aspect Ratio on Evacuation with Unsteady Heat Release Rate Due to Fire in the Case of Two Vehicles

Younggi Park, Youngman Lee, Junyoung Na and Hong Sun Ryou *

School of Mechanical Engineering, Chung-Ang University,82, Heukseok-ro, Dongjak-gu, Seoul 06974, Korea; pyg0511@cau.ac.kr (Y.P.); ymlee@alllitelife.com (Y.L.); junyoung628@naver.com (J.N.)

* Correspondence: cfdmec@cau.ac.kr.

Abstract: In this study, the characteristics of fires in case of two vehicles in a tunnel are analyzed by Computational Fluid Dynamics analysis for varying tunnel aspect ratios. Unsteady heat release rates over time are set as the input conditions of fire sources considering real phenomena. Unsteady heat release rate values are obtained from experiments. As a result, the smoke velocities above the fire source appear faster in the case of tunnels with a large aspect ratio because the higher the height of the tunnel, the faster the smoke velocity caused by buoyancy forces. The smoke velocity in the longitudinal direction increases quickly. However, the temperature distribution in the vicinity of the ceiling is low when the tunnel aspect ratio is large because the height of the tunnel is not directly affected by the flames. Also, the higher the height of the tunnel, the lower the visibility distance due to the heat and smoke coming down along the wall surface. However, in the tunnels represented in this study, it is considered that the visibility of evacuees is sufficiently secured.

Keywords: aspect ratio; evacuation; fire propagation; tunnel vehicle fire; unsteady heat release rate

1. Introduction

Tunnels have been continuously constructed to overcome topographical barriers in mountainous terrains and urban areas, as well as to reduce waste of time. Generally, fires in tunnels have a complex flow structure because it is a physical phenomenon affected by the tunnel geometry, ventilation system, and fan location, as well as chemical reactions and heat and mass transfer. Therefore, it is one of the most challenging research topics. When there are fires in tunnels, there are often critical casualties because smoke spreads rapidly along the longitudinal direction of the tunnel. According to the National Fire Protection Association (NFPA) fire statistics report, more than 70% of all tunnel fire casualties are caused by suffocation due to smoke and toxic gas inhalation [1]. Therefore, lots of studies regarding fires in tunnels have been undertaken using experimental and numerical approaches.

In the case of experimental studies, real-scale or reduced-scale tunnel experiments with pool fires have normally been conducted to figure out the heat release rate (HRR), temperature, smoke velocity and so forth. Some studies have conducted experiments with real vehicles, not in tunnels but rather in large scale experimental rooms. By changing the initial ignition location of the fire Katsuhiro et al. figured out that the temperature distribution and maximum heat release rate reached 3 MW. They found that as soon as a window breaks, the fire suddenly diffused due to the inflow of exterior air [2]. Most experimental studies were conducted considering single vehicle situations [3,4], but in reality, vehicle accidents normally occur as collisions between two vehicles, however, there are only a few such studies. Many researchers have conducted pool fire experiments and the pool size was calculated using the scaling law [5], and based on the type of vehicles considering maximum Heat Release Rate (HRR). However, lots of reduced-scale experimental studies have been conducted, even if there are different temperature and HRR distributions between vehicles and pool fires due to material properties. Lee et al. [6] conducted small-scale experimental studies to investigate the change of fire characteristics such as ceiling jet flow and critical velocity for varying tunnel aspect ratios. They applied new dimensionless velocity and heat release rates to suggest modified critical velocities.

Numerical methods have been used many studies because the results of experimental studies can be compared with numerical studies and fire phenomena in tunnel fires possibly verified under various conditions [7–9]. In particular, the aspect ratio of the tunnel is the one of the most interesting factors in tunnel smoke propagation studies. The heat flow generated in a fire is different for varying tunnel aspect ratios, and this directly affects the evacuation result. In tunnels with the same hydraulic diameter, the critical velocity to prevent the smoke spreading for evacuees is different for various aspect ratios. Therefore, investigating the fire characteristics according to different aspect ratios is an important parameter when designing smoke control systems [10]. In previously studies, model tunnels have been constructed in which one can change the tunnel aspect ratio and determine the effect on mass loss rate, maximum smoke temperature and temperature distribution. It was found that aspect ratio is an important factor that affects the smoke temperature distribution. As can be seen from related papers regarding the effect of aspect ratio with numerical study, aspect ratio is a decisive parameter for designing of smoke control systems. Figure 1 presents a schematic diagram of the basic concepts about varying aspects ratio on tunnels. Therefore, it is very important to investigate the fire characteristics in case of two vehicle fires in real situations. Thus, in this study, we numerically investigated the effect of tunnel aspect ratios on unsteady heat release rates.

Figure 1. Schematic diagram of the basic concepts about varying aspect ratios of tunnels.

2. Numerical Method

The values of heat release rate obtained from a large scale cone calorimeter experimental apparatus are set as the fire input conditions for the numerical analysis. Park et al. have presented detailed information about this experimental study, currently under review for the journal *Fire Technology* [11]. Figure 2 presents whether the heat release rate shows the same result values when applied under the conditions of a fire source. In addition, unlike the pool fire experiments, in real vehicle experiments, there are some sections where the heat release rate decreases or stabilizes for the various points at which the fire source is propagated. Section 1 indicates that the fire spreads from the passenger seat to the driver's seat and rear seat. In addition, Section 2 is the time when the fire propagated from the vehicle where the initial fire occurred to the vehicle next to it, and Section 3 represents the time to propagate from indoors to outside vehicle components such as fuel tank, engine room and bumpers. Section 4 was a period for comparing Sections 1–3 that represent other trends. This is because, unlike the usual pool fire experiment, the heat release rates does not increase continuously but are steady or decrease. In this study, temperature, smoke velocities and visibility and so forth changes caused by the aspect ratio at the point of reaching the maximum heat release rate in four sections are analyzed using Fire Dynamic Simulator (FDS) numerical analysis program [12].

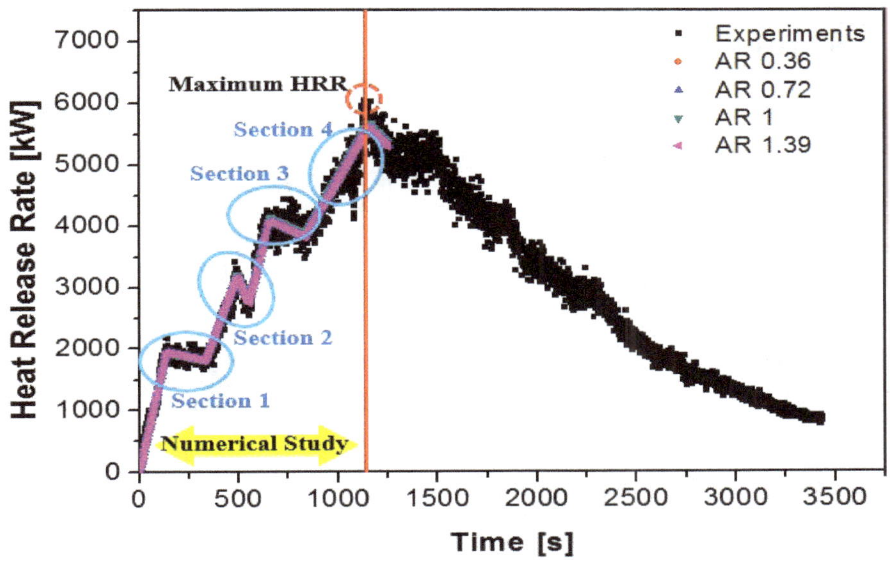

Figure 2. Heat release rate (HRR) results from experiments and numerical studies.

Figure 3a presents the computational domain of a tunnel with 100 m of longitudinal direction without any ventilation system. Vehicle sizes are 1.8 m (W) × 4.7 m (L) × 1.4 (H) and distance between vehicles is 0.6 m. Figure 3b represents the location of thermocouples and analysis planes. It depends on tunnel aspect ratio, however, the thermocouples are attached 0.2 m below the ceiling and 0.6 m above the vehicles. The interval between each thermocouple is 0.6 m. Figure 3c presents four difference tunnel aspect ratios but the total cross section area is almost constant. The aspect ratio is defined as the height divided by the width. Adiabatic conditions are set on the tunnel wall, and pressure outlet conditions at the tunnel exits. Figure 3d represents the three various flow directions occurring in a tunnel. The smoke velocity above fire source, velocities against the ceiling, and ceiling jet flows are affected by the tunnel's aspect ratio caused by heat gases and temperatures inside the tunnel. These flows are represented as flow 1, flow 2 and flow 3, respectively. The number of grids was approximately 770,000 and applied to differentiate the number of regions to improve the accuracy and efficiency of analysis. The size of the grid was 0.1, 0.2 and 0.4 m from the above of fire source to the exit of the tunnel, respectively. To distinguish the optimum grid size, the size of the minimum grid required for the flow modeling is determined by the characteristic flame diameter taking into account the heat release rate of the fire source [13]. Also, in order to improve the accuracy and efficiency of the analysis, the grid size is applied differently considering the flow characteristics:

$$D^* = \left(\frac{\dot{Q}}{\rho_\infty C_p T_\infty \sqrt{g}} \right)^{2/5} \tag{1}$$

where ρ_∞ [kg/m^3], C_p [kJ/(kg·K)], T_∞ [°C] represent the density, specific heat and temperature for ambient air (20 °C, 1 atm), respectively. For simulations involving buoyant plumes, the flow field is given by non-dimensional expression $D^*/\delta x$, where D^* is the characteristic fire diameter and δx is the nominal size of a mesh cell. The quantity,$\dot{Q}$, is the total heat release rate of the fire. The quantity $D^*/\delta x$ can be thought of as the number of computational cells spanning the characteristic diameter of the fire. Generally, a reasonable numerical results can be obtained at a grid size of $0.05 < \overline{\Delta}/D* < 0.10$ [14] where, $\overline{\Delta}$ represents the average grid size. In this study, the average grid size is 0.35 m. Therefore, an average grid size of approximately 770,000 is considered when considering characteristic flame diameters.

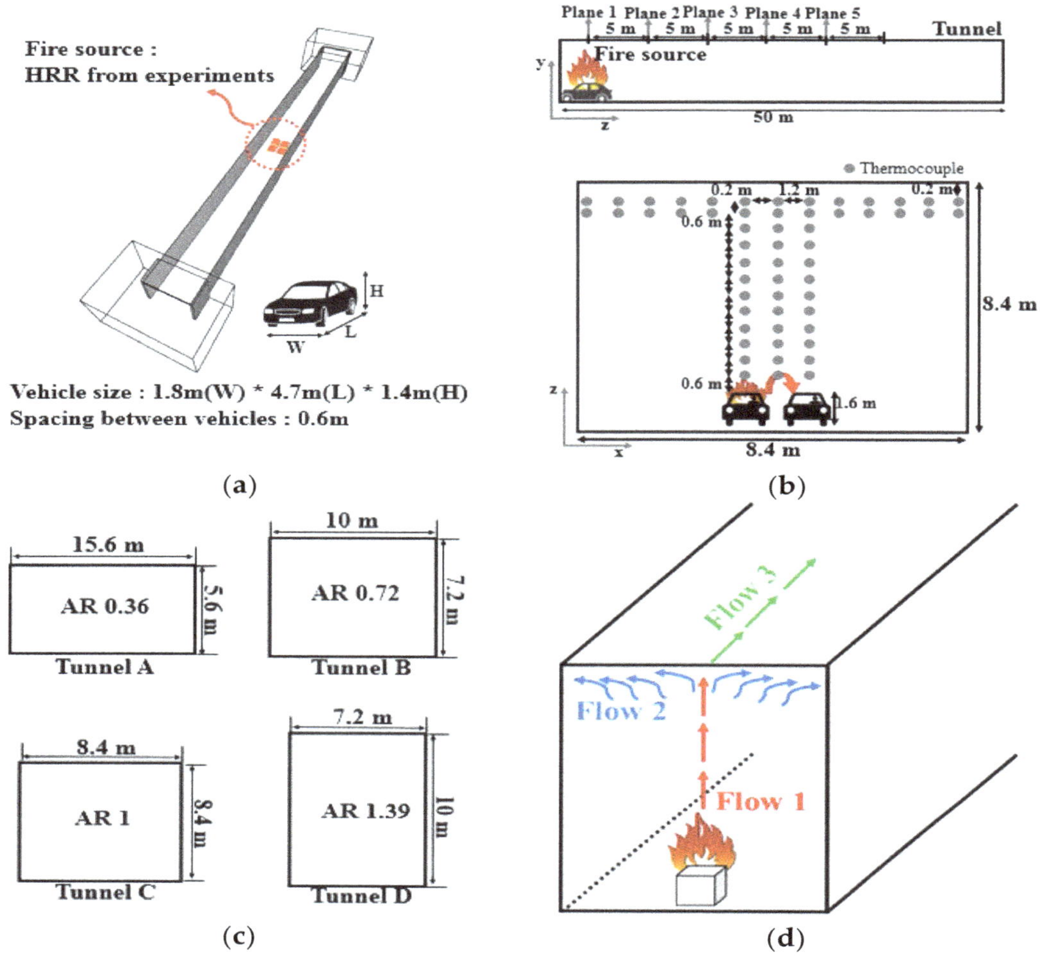

Figure 3. Numerical study: (**a**) computational domain for tunnel, (**b**) location of thermocouples and analysis planes, (**c**) cross-section of tunnel for varying aspect ratio and (**d**) main flow in case of tunnel fire.

3. Results and Discussion

3.1. Effect of Aspect Ratio on Tunnel

Figure 4 presents the temperature and smoke velocity contours for varying tunnel aspect ratios at 1500 s where it concerns the maximum heat release rate. When the aspect ratio of the tunnel is small, the temperature distribution above the fire source and temperature distribution in the horizontal direction are high. The lower the height of the tunnel, the more affected by the flame. Conversely, in case of smoke velocity, the higher the tunnel height, the more the momentum forces on the fire source increase, so that the smoke velocity appears faster. Also, the higher the height of the tunnel, the stronger the buoyancy force applied. Figure 5 presents the values on the change of temperature and smoke velocities for varying aspect ratios measured 0.2 m below the ceiling at 1500 s underpinning the effects of tunnel height on smoke velocities. To compensate the severe fluctuations, the value on 1500 s was calculated as an average value before the 10 seconds and after the 10 seconds at the specific time. Because of the different tunnel heights, the results shown are obtained by normalizing the tunnel height.

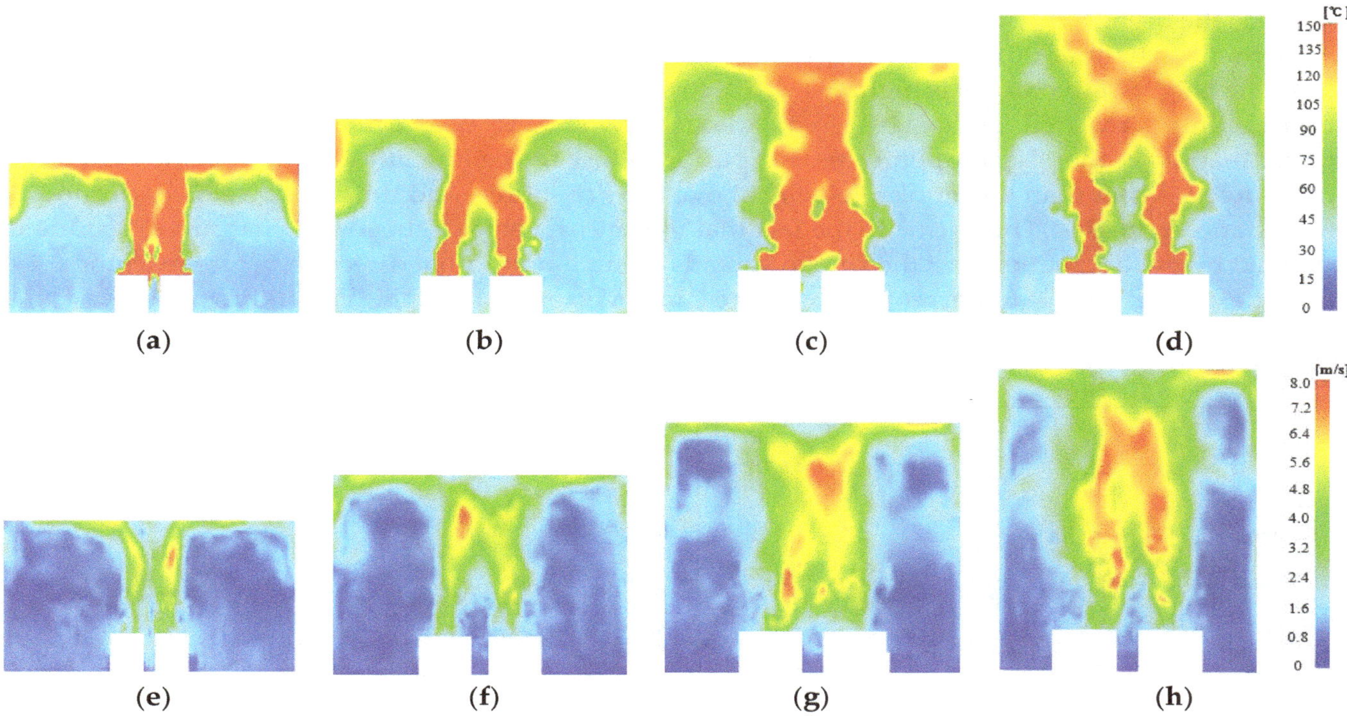

Figure 4. Temperature and smoke velocity contours at 1500 s; (**a**–**d**) represents the temperature distribution where the tunnel aspect ratio is 0.36, 0.72, 1 and 1.39, respectively, (**e**–**h**) represents the smoke velocities.

The value of 0 on the Y axis represents the temperature just above the fire source and the value of 1 represents at a point 0.2 m beneath from the ceiling. When the maximum heat release rate occurs, the temperature above the fire source is from about 120 to 200°C, and the temperature at the ceiling is from about 150 to 450°C. It varied by almost two- and three-fold depending on the location and tunnel aspect ratio. The smoke velocity is the fastest near the center of the tunnel, with a speed difference about 1.4 times higher. Furthermore, the smoke velocity is the fastest and has a tendency to decrease at the middle of the tunnel. This is due to buoyancy forces, which accelerate the smoke velocity over time, but near the tunnel wall, the smoke velocity is lowered due to no-slip conditions, friction and the normalized tunnel height to analyze the tendency of the tunnel aspect ratio.

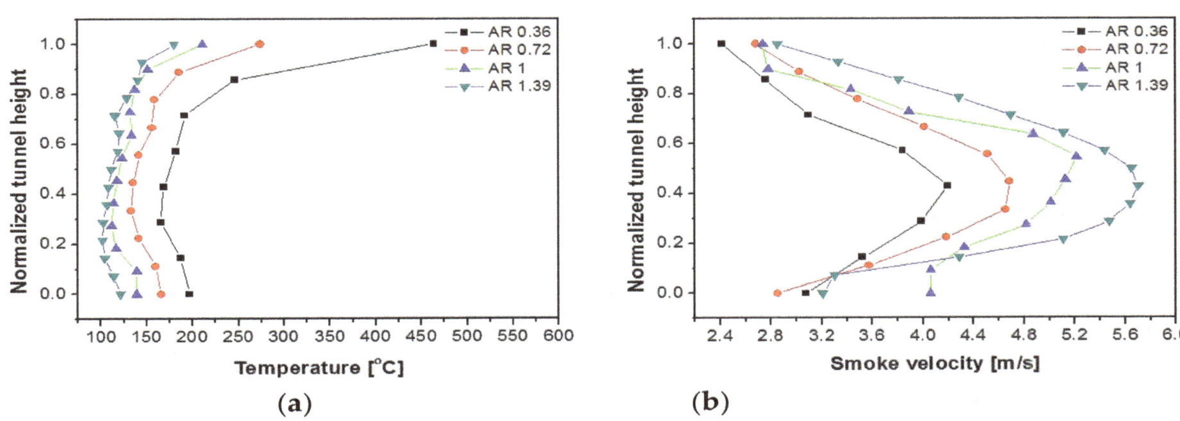

Figure 5. Temperature and smoke velocity distributions where it is located below 0.2 m of ceiling for varying aspect ratio at 1500 s. (**a**) Temperature distribution, (**b**) Smoke velocities.

Figure 6a presents the temperature changes 0.2 m below the tunnel ceiling for varying aspect ratios and distances. In the case of the tunnel with the smallest aspect ratio, the temperature above the fire source is the highest. In addition, the slope of the temperature distribution along the tunnel distances show that the temperature drop rate greatly changes when the aspect ratio is low. This is because when the height of the tunnel is high, the width of the tunnel is relatively small and the heat loss to the tunnel ceiling is reduced. Figure 6b presents the smoke velocities at the same point with various aspect ratios. It appears that the temperature distribution below the ceiling at the point 5 m away from the fire source is faster than the smoke velocity above the fire source, and the smoke velocity declines as the distance increases. This is due to the fact that the smoke velocity above the fire source is suddenly changed by the fluctuation and the smoke spreads along longitudinal direction before reaching 0.2 m from the ceiling. This is underpinned in Figure 7 by a vector in the longitudinal direction.

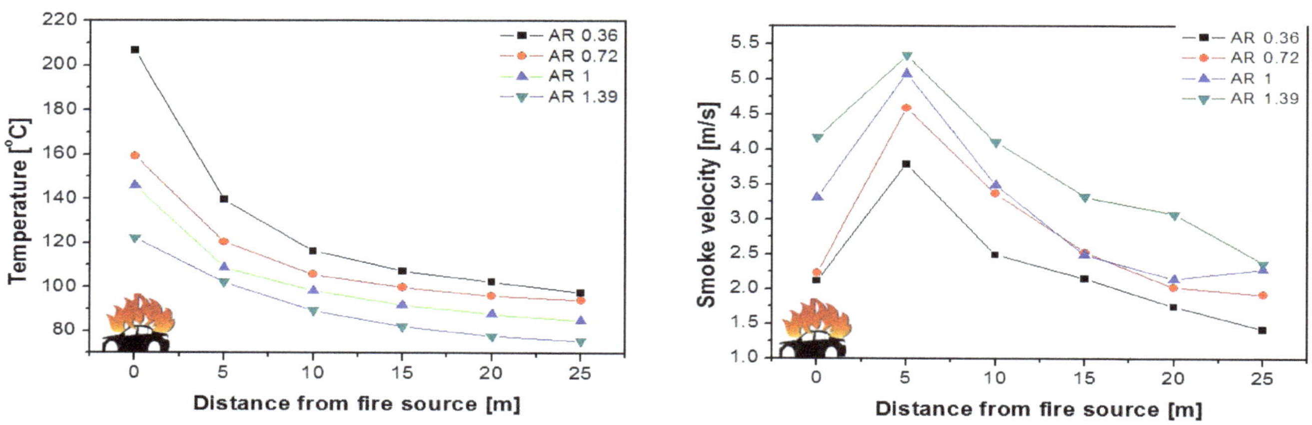

(a) (b)

Figure 6. Temperatures and smoke velocities below 0.2 m of the tunnel ceiling for varying aspect ratio and distances from fire source at 1500 s. (**a**) Temperature distribution, (**b**) Smoke velocity distribution.

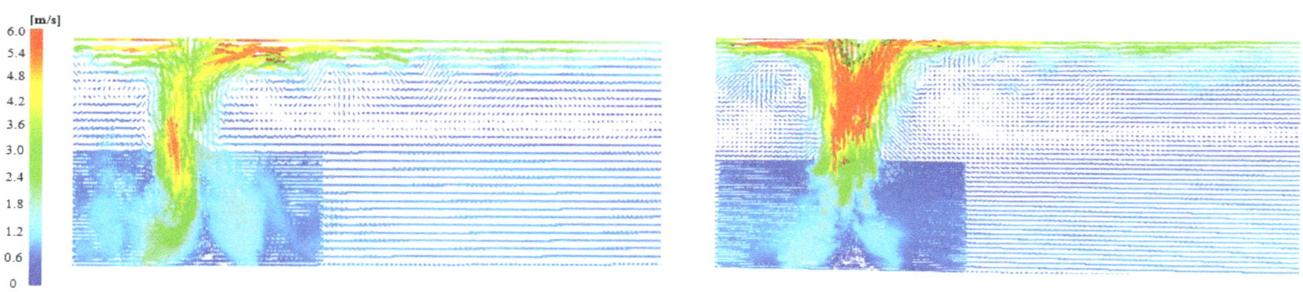

(a) (b)

Figure 7. Velocity vector on longitudinal direction at 1500 s, (**a**,**b**) represent the tunnel aspect ratio of 0.36 and 1.39, respectively.

Also, in the case of the smoke velocity distribution according to distance, when the tunnel is high, buoyancy forces are have a great influence and it seems that the momentum force of the ceiling jet flow spreading in the longitudinal direction acts stronger. Therefore, when the height of the tunnel is high, the exit time to the tunnel exit is also fastest and this is represented in Figure 8. The time for the initial smoke to exit the tunnel was different as 90 s, 86 s, 79 s and 75 s, respectively, for the various aspect ratios. The smaller the aspect ratio, the longer it took for smoke to be emitted. The lower the height of the tunnel, the higher the temperature is because the flame directly reaches the ceiling, and the smoke temperature spreads in the longitudinal and lateral directions. However, the smoke velocity at which the smoke is spreading increased as the height is increased by the buoyancy forces.

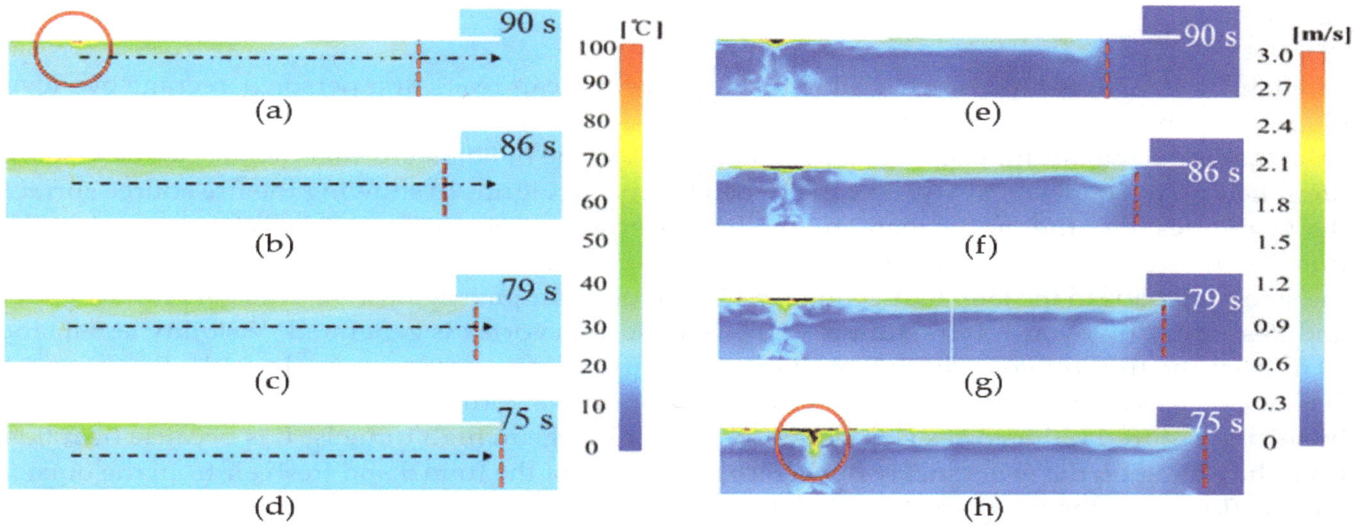

Figure 8. Temperature and smoke velocity distributions according to the initial time when the smoke exits the tunnel, (**a–d**) represents the various aspect ratio of 0.36, 0.72, 1 and 1.39, respectively. (**e–h**) represents the smoke velocity for varying aspect ratio of 0.36, 0.72, 1 and 1.39, respectively.

The temperature of the upper part of the tunnel increases continuously as time goes by. However, the temperature presents almost constant values because external cold air continuously flows into the lower part of the tunnel. Therefore, in order to secure the evacuation route of the evacuees, a flow is introduced in the opposite direction of the ceiling jet flow when the fire occurs. As the ceiling jet flow changes according to the aspect ratio, Lee et al. [9] suggested the formula considering aspect ratio as follows:

$$V_c = 0.73{AR_{tu}}^{0.2}\sqrt{g\overline{H}}\left(\frac{Q}{\rho_0 c_{p.0} T_0 \sqrt{AR_{tu} g \overline{H}^5}}\right)^{1/3} \quad (2)$$

where, AR represents the tunnel aspect ratio and ρ_∞, C_p, T_∞ represent the density, specific heat and temperature for the ambient air (20 °C, 1 atm) and $\overline{H}$ is the hydraulic diameter of the tunnel. Based on the formula presented by Lee, the critical velocities for varying aspect ratios are simply 1.74, 1, 0.77 and 0.59, respectively. However, from an energy efficiency point of view, it is essential to calculate the appropriate air flow rate over time. Therefore, the critical velocity considering the heat release over time is shown in Figure 9. The critical velocity should be introduced considering the specific shape of the tunnel.

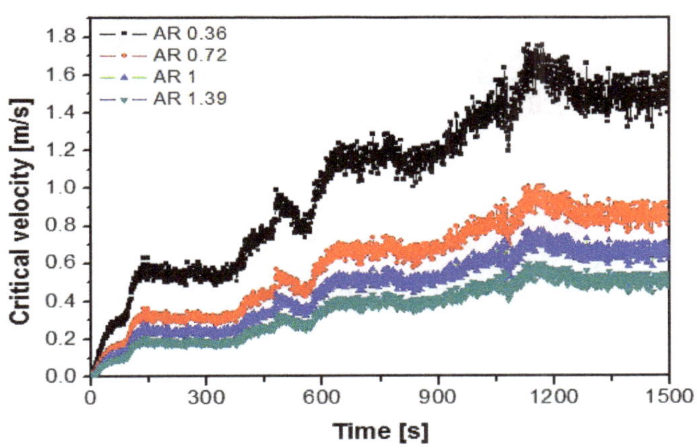

Figure 9. Critical velocity considering the aspect ratio of a tunnel with unsteady heat release rate.

3.2. Visibility

Figure 10 shows how much visible distance the fire can reach at a person's average eye level of 1.7 m. It shows the contour at the point of maximum heat release rate based on the FDS setting value having a maximum visible distance of 30 m. A value of 30 means that people can be visually identified up to 30 meters away. As can be seen from A in Figure 10 showing the fire source surface, the visibilities near the tunnel wall are low in a narrow tunnel width. On the other hand, in wide tunnels, the visual distance is relatively clear. This is because as the tunnel width becomes narrower, the smoke spreads in the lateral direction, and the smoke bumping against the wall generates a vortex and descends downward. If the tunnel width is wide, less vortex is generated. Visibility at 5 m and 10 m from the fire source is similar, but visibility is lower for a narrow tunnel width. Furthermore, velocity vectors are represented in Figure 11 to underpin the vortex for varying aspect ratios. As can be seen in Figure 11, the vortex size can be found to be large in high tunnels. It is because heat flow from the ceiling on the wall quickly flows into the bottom of the tunnel and fresh air from the outside quickly flows into the top of the tunnel, resulting in a larger vortex and it results in reduced visibility for evacuees. However, although the visibility varies depending on the aspect ratio of the tunnel, the visibility at a point where the average height of a person is 1.7 m seems to be clear.

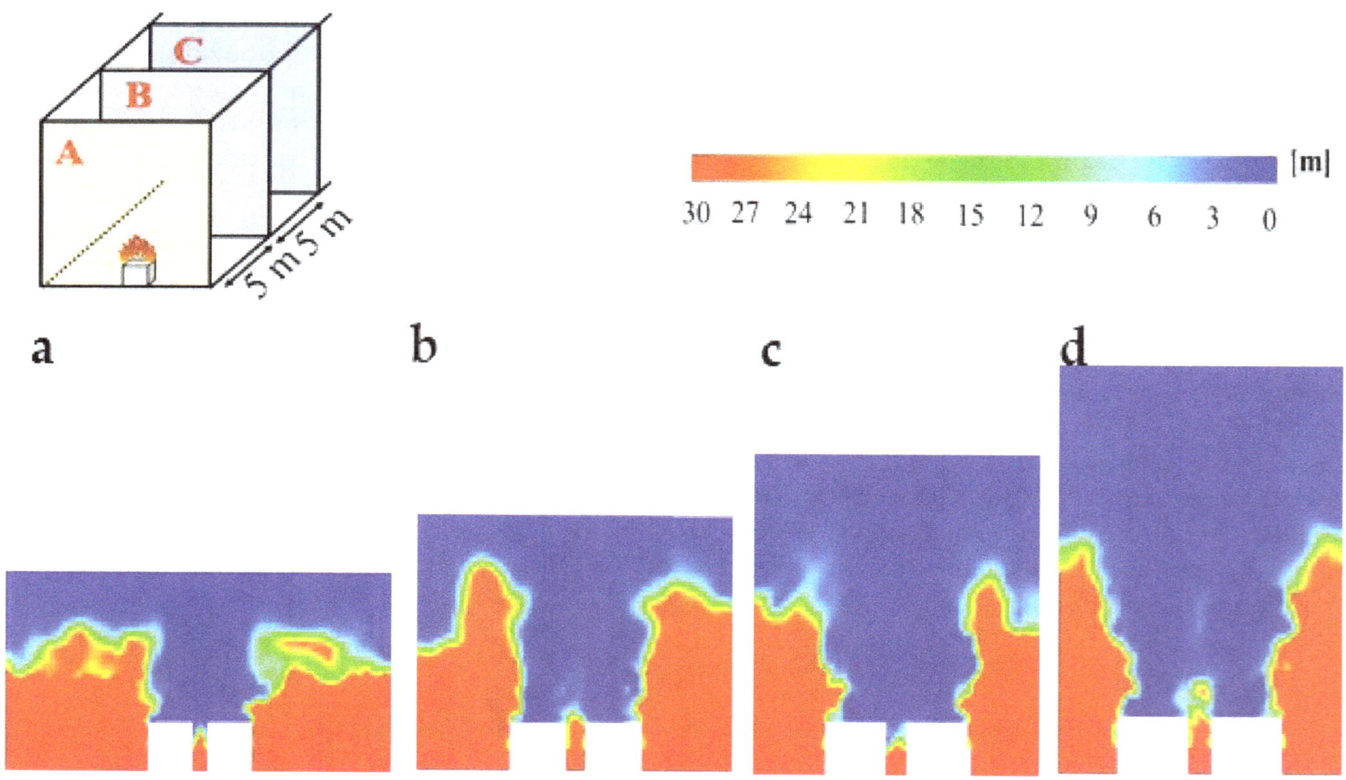

(A)

Figure 10. *Cont.*

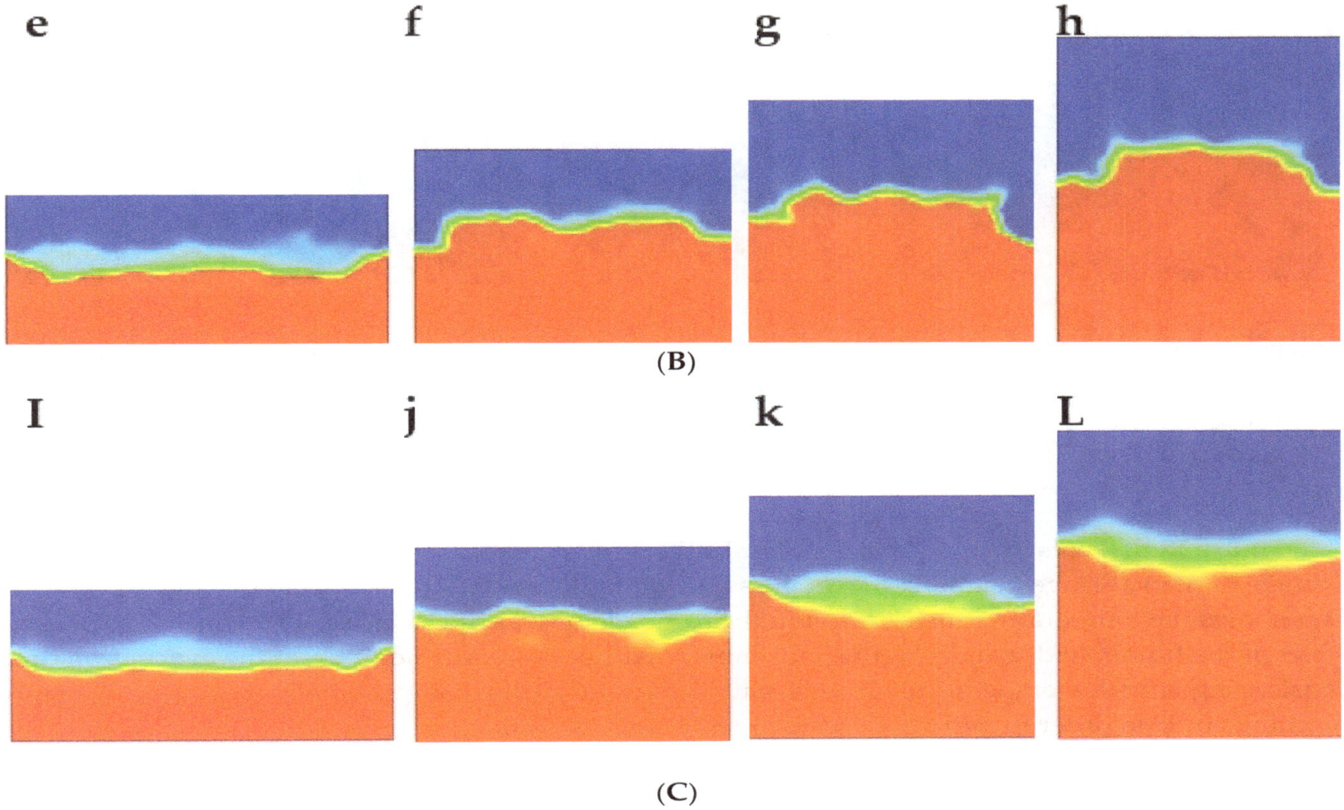

Figure 10. Visibility depending on distances, (**A**) represents on Fire plane, (**B**,**C**) represent the planes of 5 m and 10 m away from the fire sources, respectively. The aspect ratios of a, e, and I are 0.36, b, f, and j are aspect ratios of 0.72, c, g, and k are aspect ratios of 1, and d, b, and L are aspect ratios of 1.39.

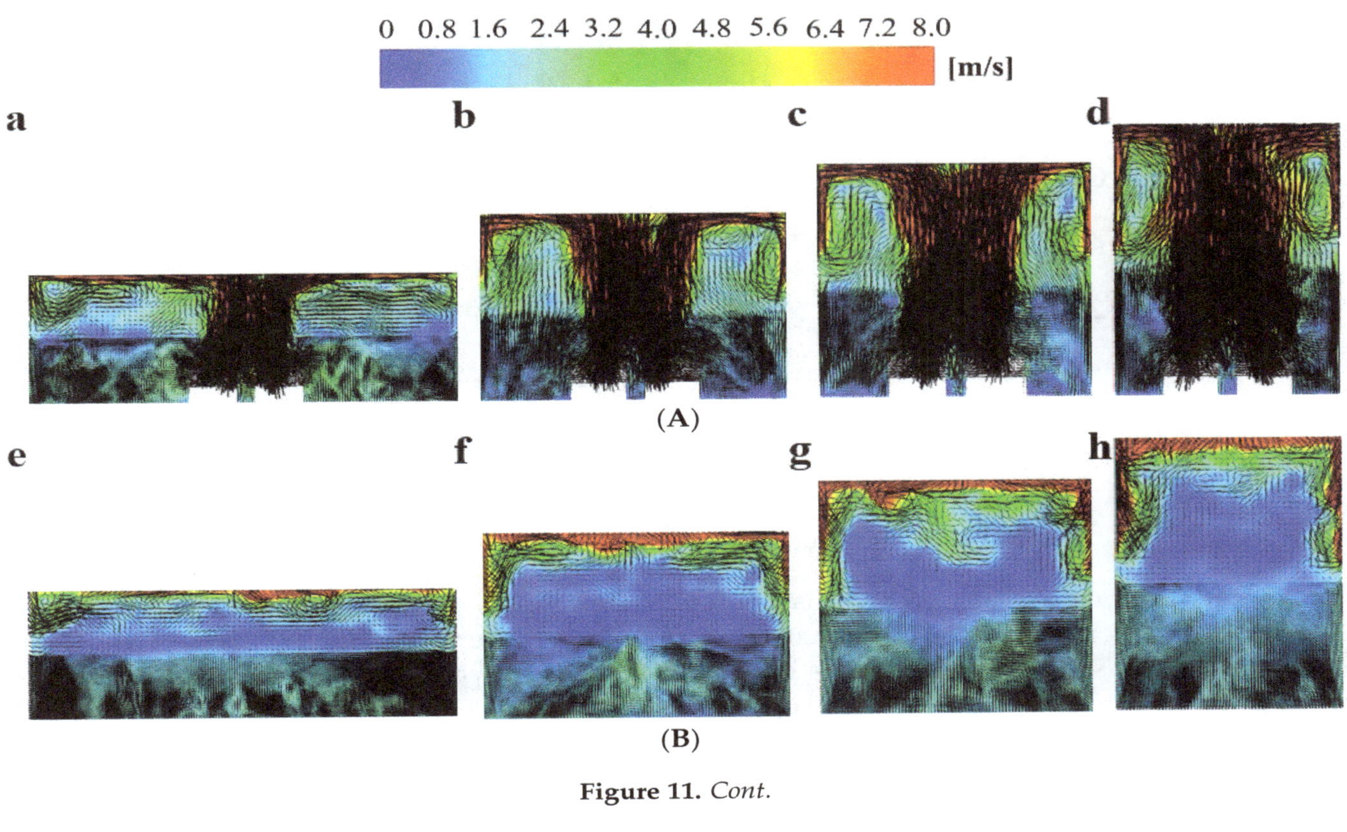

Figure 11. *Cont.*

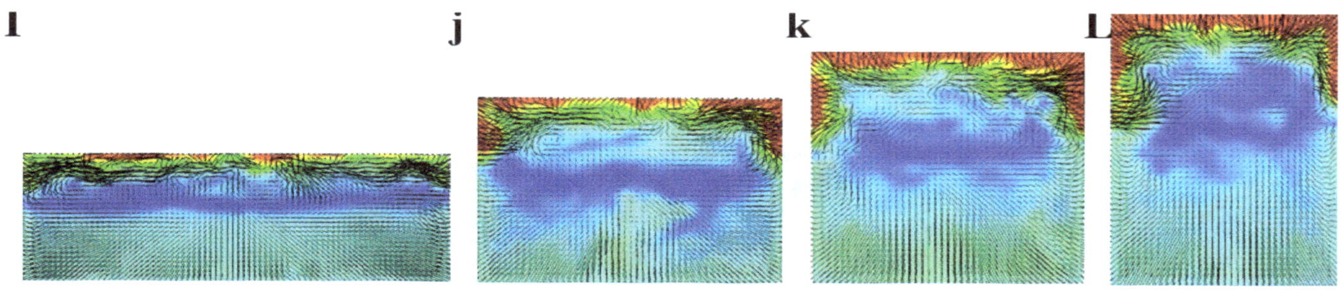

(C)

Figure 11. Velocity vector to figure out vortex depending on distances and aspect ratio, (**A**) represents on Fire plane, (**B**,**C**) represent the planes of 5 m and 10 m away from the fire sources, respectively. The aspect ratios of a, e, and I are 0.36, b, f, and j are aspect ratios of 0.72, c, g, and k are aspect ratios of 1, and d, b, and L are aspect ratios of 1.39.

Figure 12 presents the temperature distributions over time at the height of 1.7 m to consider an average human average height. It shows the same temperature change irrespective of the aspect ratio. This shows the temperature distribution in the same vicinity as the location of the fire source, so the effect of the tunnel width and length is not shown. Although it is difficult to set the evacuation time considering only the temperature, the evacuation must be within completed 450 s (7 min 30 s) before reaching 60 °C in order to reduce deaths due to fire in a tunnel.

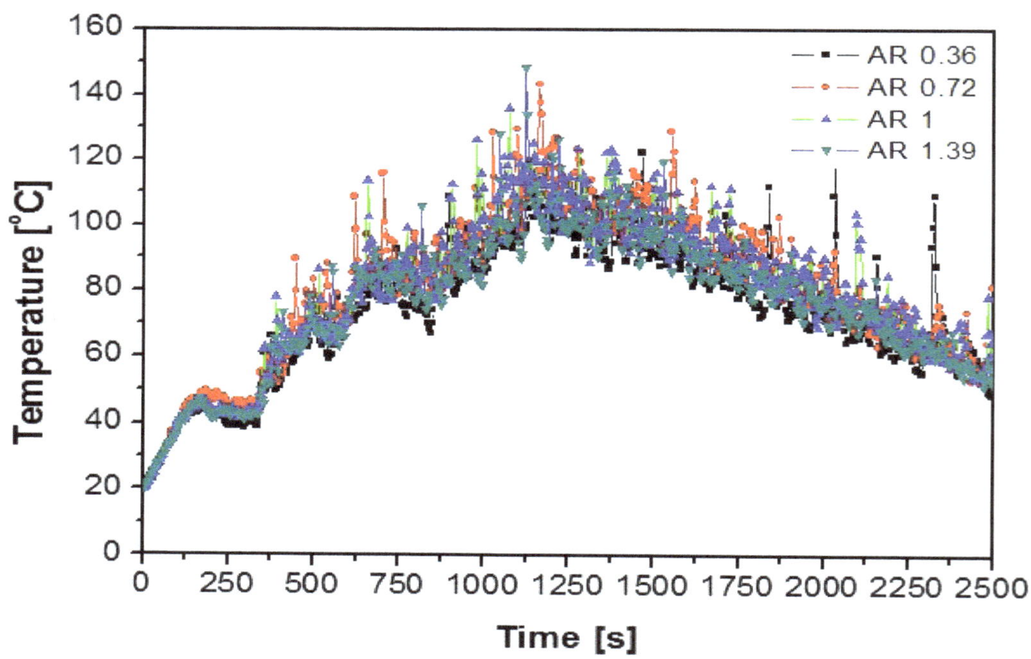

Figure 12. Temperature changes over time at the height of 1.7 m.

Figure 13 shows the temperature distribution at the specific section for the various aspect ratios in Sections 1–4, 0.2 m below the tunnel ceiling. Sections 1–3 are the sections where the heat release rate is stable or decreases, whereas Section 4 represents the section where the heat release rate increases radically in order to compare with other sections. As presented in Figure 13, the temperature increases and decreases are relatively stable. However, the temperature distribution in the region where the heat release rate increases rapidly differs from that in other regions.

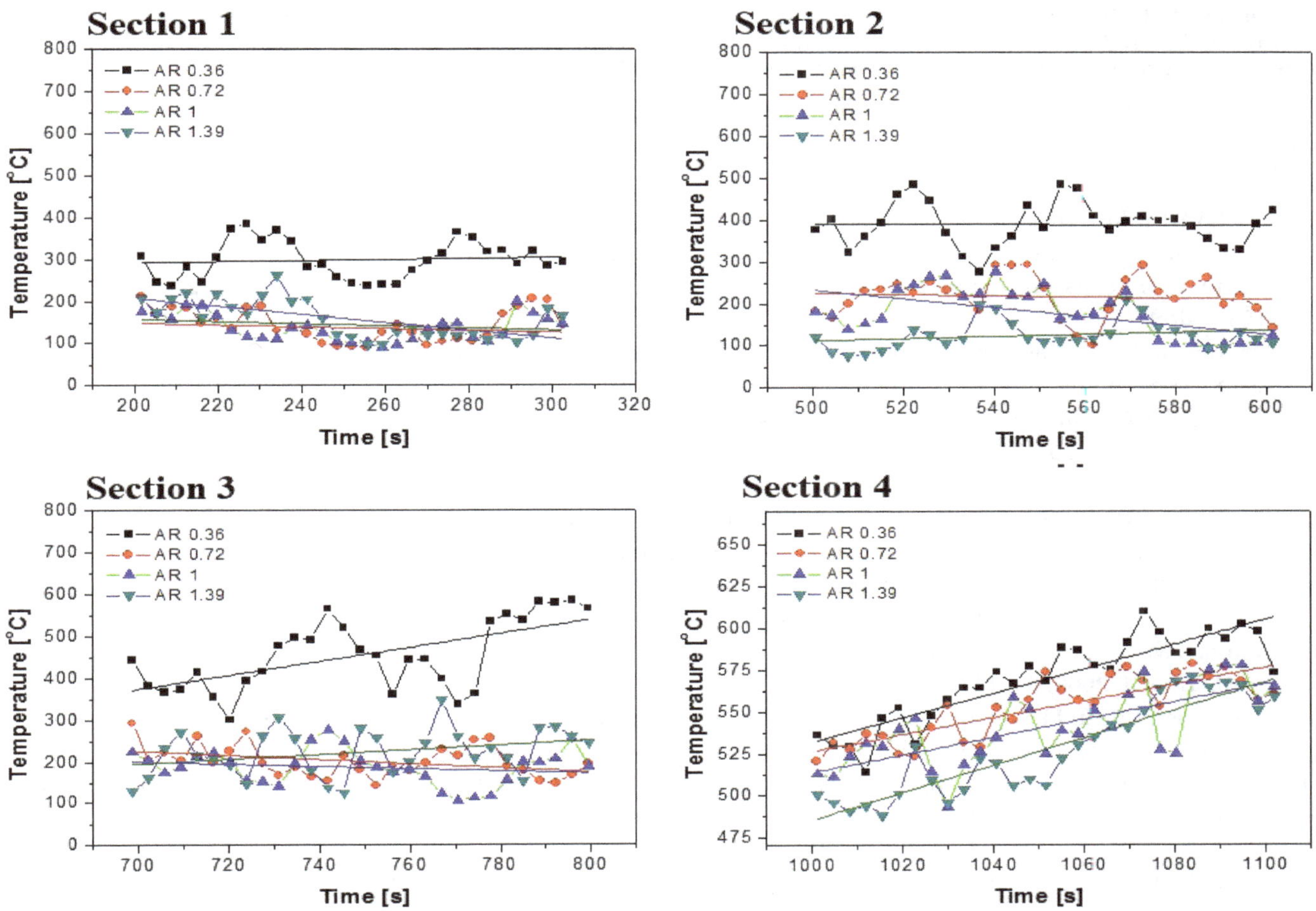

Figure 13. Temperature distribution at the specific times for the various aspect ratio at the point of where below 0.2 m of ceiling, (**Section 1**) represents the temperature values from 200 s to 300 s, (**Section 2**) represents the temperature values from 500 s to 600 s, (**Section 3**) represents the temperature values from 700 s to 800 s and (**Section 4**) represents the temperature values from 1000 s to 1100 s to compared to other sections.

4. Conclusions

The aim of this study was to investigate the fire characteristics of the tunnel vehicle fires for varying the aspect ratio of the tunnel cross-section using FDS. For this purpose, the heat release rates obtained from experimental study in case of two vehicles were measured. The following conclusions were obtained:

1) In a tunnel with a large aspect ratio, the smoke velocity above the fire source was generated rapidly. The buoyancy force and momentum force which affected the smoke velocity in the direction of longitudinal is greater in the case of a small aspect ratio tunnel.

2) As the aspect ratio becomes smaller, the temperature distribution in the cross-section area of the tunnel was higher because the flame directly affects the below of the ceiling but the flame was relatively small when the tunnel aspect ratio was large.

3) When the tunnel width was narrow, the vortex generated by the wall was generated strongly, and the smoke falls along the wall surface, so it is difficult to secure visibility.

4) The temperature changes in the section where the heat release rate was constant are insufficient, but the temperature change in the section where the heat release rate increases was apparent.

Author Contributions: Conceptualization, methodology, investigation: Y.P. and H.S.R.; software, validation, writing—original draft preparation: Y.P.; formal analysis, investigation: J.N. and Y.L.; writing—review and editing, supervision, project administration, funding acquisition: H.S.R.

Acknowledgments: This research was supported by the Chung-Ang University research grant in 2018.

References

1. *NFPA 92A: Recommended Practice for Smoke control Systems*; Natl Fire Protection Assn: Quincy, MA, USA, 1988.
2. Katsuhiro, O.; Norimichi, W.; Yasuaki, H.; Yasuaki, H.; Tadaomi, C.; Ryoji, M.; Hitoshi, M.; Satoshi, O.; Hideki, S.; Yohsuke, T.; et al. Burning behavior of sedan passenger cars. *Fire Saf. J.* **2009**, *44*, 301–310. [CrossRef]
3. Mangs, J.; Keski-Rahkonen, O. Characterization of the Fire Behaviour of a Burning Passenger Car. Part1: Car Fire Experiments. *Fire Saf. J.* **1994**, *23*, 17–35. [CrossRef]
4. Katsuhiro, O.; Takuma, O.; Hiroki, M.; Masakatsu, H.; Norimichi, W. Burning behavior of minivan passenger cars. *Fire Saf. J.* **2013**, *62*, 272–280. [CrossRef]
5. Quintiere, J.G. Scaling Applications in Fire Research. *Fire Saf. J.* **1989**, *15*, 3–29. [CrossRef]
6. Lee, S.R.; Ryou, H.S. An Experimental Study of the Effect of the Aspect Ratio on the Critical Velocity in Longitudinal Ventilation Tunnel Fires. *Fire Sci. J.* **2005**, *23*, 119–138. [CrossRef]
7. Jain, S.; Kumar, S.; Kumar, S.; Sharma, T.P. Numerical simulation of fire in a tunnel: Comparative study of CFAST and CFX predictions. *Tunn. Undergr. Space Technol.* **2008**, *23*, 160–170. [CrossRef]
8. Banjac, M.; Nikolic, B. Numerical Study of Smoke Flow Control in Tunnel Fires Using Ventilation Systems. *FME Trans J.* **2008**, *36*, 145–150.
9. Lee, S.R.; Ryou, H.S. A numerical study on smoke movement in longitudinal ventilation tunnel fires for different aspect ratio. *Build. Environ. J.* **2006**, *41*, 719–725. [CrossRef]
10. Ji, J.; Bi, Y.; Venkatasubbaiah, K.; Li, K. Influence of aspect ratio of tunnel on smoke temperature distribution under ceiling in near field of fire source. *Appl. Therm. Eng.* **2016**, *106*, 1094–1102. [CrossRef]
11. Park, Y.; Kim, J.; Ryou, H.S. Experimental Study on the Fire-Spreading Characteristics and Heat Release Rates of Burning Vehicles using a Large-Scale Cone Calorimeter. *Fire Technol. J.* **2018**. submitted.
12. McGrattan, K.; Hostikka, S.; McDermott, R.; Floyd, J.; Weinschenk, C.; Overholt, K. Fire Dynamics Simulator Technical Reference Guide Volume 1: Mathematical Model. *NIST Publ.* **2017**. [CrossRef]
13. Bounagui, A.; Benichou, N.; McCarteny, C.; Kashef, A. Optimizing the grid size used in CFD sumulations to evaluate fire safety in houses. In Proceedings of the 3rd NRC Symposium on Computational Fluid Dynamics, High performance Computing Virtual Reality, Ottawa, ON, Canada, 4 December 2003.
14. McGrattan, K.B.; Jason, F.; Forney, G.P.; Baum, H.R.; Hoskikka, S. Improved radiation and combustion routines for a large eddy simulation fire model. *Fire Saf. Sci.* **2003**, *7*, 827–883. [CrossRef]

14

An Investigation of the Restitution Coefficient Impact on Simulating Sand-Char Mixing in a Bubbling Fluidized Bed

Xinjun Zhao, Qitai Eri * and Qiang Wang

School of Energy and Power Engineering, Beihang University, Beijing 100191, China; kunpengzhao@buaa.edu.cn (X.Z.); qwang518@buaa.edu.cn (Q.W.)

* Correspondence: eriqitai@buaa.edu.cn.

Academic Editor: Bjørn H. Hjertager

Abstract: In the present work, the effect of the restitution coefficient on the numerical results for a binary mixture system of sand particles and char particles in a bubbling fluidized bed with a huge difference between the particles in terms of density and volume fraction has been studied based on two-fluid model along with the kinetic theory of granular flow. Results show that the effect of restitution coefficient on the flow characteristics varies in different regions of the bed, which is more evident for the top region of the bed. The restitution coefficient can be categorized into two classes. The restitution coefficients of 0.7 and 0.8 can be included into one class, whereas the restitution coefficient of 0.9 and 0.95 can be included into another class. Moreover, four vortices can be found in the time-averaged flow pattern distribution, which is very different from the result obtained for the binary system with the similar values between particles in density and volume fraction.

Keywords: restitution coefficient; segregation; flow pattern; bubbling fluidized bed; binary particles; sand; char

---

1. Introduction

Bubbling fluidized beds are commonly used in industrial processes [1], such as combustion and gasification, due to their good mixing ability and heat transfer characteristic between the gas and solid phases. In the process of biomass gasification, the first step is pyrolysis. The gas component products after pyrolysis are non-condensable gases and tars, and char is left as a solid residue [2]. Due to the irregular shape and low density of char particles, it is difficult to attain a stable fluidization status. Generally, inert particles such as sand particles are added to the fluidized bed to improve the fluidization status and the heat transfer effect. During the process of fluidization, char particles will accumulate toward the top of the bed, whereas the inert particles will sink towards the bottom of the bed. Hence, segregation is a widespread phenomenon in a binary system of particles in a bubbling fluidized bed [3,4].

The complicated flow characteristics in the bubbling fluidized bed have a strong effect on the reaction process. Therefore understanding the flow behavior is important to design a fluidized bed and optimize the operation conditions. Although the most accurate method is still based on the experimental data, the application of this method has limitations, in terms of the longer time required and high costs. With advances in computational fluid dynamics (CFD), simulation studies have become a useful method to analyze the flow characteristics in bubbling fluidized beds [5,6]. The most widely used approach for simulating dense gas-solid flow is the Eulerian–Eulerian model (two fluid model) along with the kinetic theory of granular flow [7,8].

The main parameters affecting the simulation results include the drag model, particle collision characteristics, and solid phase wall boundary conditions. The particle collision characteristics are defined based on the restitution coefficient, and the wall boundary conditions are defined in at least three ways, which are the traditional no-slip boundary conditions and two partial slip conditions [9,10]. The effects of these parameters on the simulation results have been studied by some researchers. Chao et al. [10] and Bai et al. [11] focused on the effect of the drag model on the mixing and segregation behavior of biomass mixtures in a fluidized bed. Tagliaferri et al. [12] and Mostafazadeh et al. [13] have studied the influence of the restitution coefficient on the flow dynamics of a binary solid mixture. Zhong et al. [14] have investigated the influence of wall boundary conditions on the concentration and velocity distribution of particles. In these studies, the density or volume fraction of binary particles was usually similar. However, in actual biomass gasification, the char has lower density than the sand. Meanwhile, the char generated from the pyrolysis will continue to react with the gases, hence, the volume fraction of the char in the solid phase mixtures is very low [15]. When the simulation conditions do not agree with actual situation, the reasonability and accuracy of simulation results is questionable.

Many studies [16–18] have studied the suitability of different drag models. Generally, the Gidaspow drag model is suitable for simulation in a dense bubbling fluidized bed. Loha et al. [6] reported that the model predictions were sensitive to the specularity coefficient and the simulation results with a higher value of specularity coefficient were in good agreement with the experimental results. According to a comparison of the differences between the specularity coefficients of one wall boundary conditions and traditional no-slip wall boundary conditions, Zhong et al. [19] found that the no-slip wall boundary condition was more suitable for simulating the dynamic segregation process of binary particles. In this study, based on a Gidaspow drag model and no-slip wall boundary conditions, a simulation is carried out to investigate the effect of the restitution coefficient on the segregation and flow characteristics of a binary particles system in a bubbling fluidized bed with a huge difference between the particles in terms of density and volume fraction. The particle system consists of char particles and sand particles with bulk density of 120 and 1590 kg/m^3 and volume fraction of 11% and 89%. According to the comparison of results between computation and experiment, a suitable scope of the restitution coefficient is determined, and meanwhile, the effect of different particles characteristics in the binary mixture on the fluid dynamic behavior is also compared.

2. Model Description

In the two fluid model (TFM), the gas and solid phases are considered as inter-penetrating continua, hence the governing equations for the solid phase are similar to those for the gas phase. For the case of cold fluidization with no chemical reactions, the conservation equations for mass and momentum are represented as follows:

The continuity equation for the gas phase is expressed as follows:

$$\frac{\partial(\alpha_g \rho_g)}{\partial t} + \nabla \cdot (\alpha_g \rho_g \vec{u}_g) = 0 \tag{1}$$

The continuity equation for the sth solid phase is written as:

$$\frac{\partial(\alpha_s \rho_s)}{\partial t} + \nabla \cdot (\alpha_s \rho_s \vec{u}_s) = 0 \tag{2}$$

The momentum balance equation for the gas phase is as follows:

$$\frac{\partial(\alpha_g \rho_g \vec{u}_g)}{\partial t} + \nabla \cdot (\alpha_g \rho_g \vec{u}_g \vec{u}_g) = -\alpha_g \nabla p + \nabla \cdot \overline{\overline{\tau}}_g + \alpha_g \rho_g \vec{g} + \sum_{m=1}^{N} K_{mg}(\vec{u}_m - \vec{u}_g) \tag{3}$$

The momentum balance equation for the sth solid phase can be written as:

$$\frac{\partial(\alpha_s\rho_s\vec{u}_s)}{\partial t} + \nabla\cdot(\alpha_s\rho_s\vec{u}_s\vec{u}_s) = -\alpha_s\nabla p - \nabla p_s + \nabla\cdot\overline{\overline{\tau}}_s + \alpha_s\rho_s\vec{g} + \sum_{n=1}^{N} K_{ns}(\vec{u}_n - \vec{u}_s) \tag{4}$$

The total volume fraction of the phases is equal to 1 and it is expressed in the following way:

$$\alpha_g + \sum\alpha_s = 1 \tag{5}$$

where t is the time. The subscript "g" refers to the gas phase and the subscript "s" refers to the sth solid phase. $\vec{g}$, $\vec{u}$, and $\overline{\overline{\tau}}$ are the acceleration due to the gravity, velocity vector, and stress-strain tensor, respectively. p, α, and ρ are the pressure, volume fraction, and density, respectively. K is the momentum exchange coefficient between phases.

The solid pressure is evaluated using the expression proposed by Lun et al. [20], and it is composed of a kinetic term and a second term due to particle collisions:

$$p_s = \alpha_s\rho_s\Theta_s + 2\rho_s(1+e_{ss})\alpha_s^2 g_{0,ss}\Theta_s \tag{6}$$

where $g_{0,ss}$ and Θ_s are the radial distribution function and granular temperature of the sth solid phase, respectively. e_{ss} is the restitution coefficient for particle collision, which quantifies the elasticity of particle collision.

3. Simulation Setup

In this study, the experimental setup by Park et al. [21] forms the basis for the simulation. The geometry of the rectangular fluidized bed with 200 mm width and 50 mm depth is schematically shown in Figure 1. In the experiments, sand particles and char particles are used. The properties of the particles in this study are summarized in Table 1.

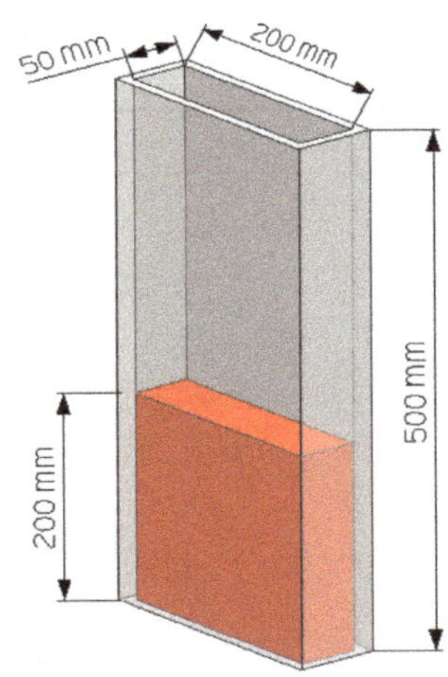

Figure 1. Schematic representation of the fluidized bed.

Table 1. Properties of particles.

Particles	Mean Diameter (μm)	Void Fraction	Bulk Density (kg/m^3)
Sand	387	0.333	1590
Char	957	0.693	120

The 2D simulation is carried out using the TFM method, and the computational domain is accordingly set to be 200 mm width and 500 mm height. The mesh geometry and coordinate system is schematically shown in Figure 2. The Z direction represents the axial direction or height direction, whereas X direction represents the radial direction or lateral direction.

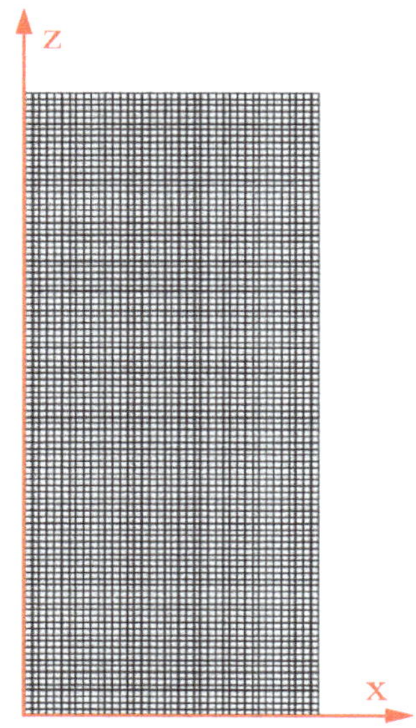

Figure 2. Mesh geometry and coordinate system.

In the simulation, ambient air is used as the fluidizing medium. Meanwhile, the uniform gas velocity is specified at the bottom of the bed and the atmospheric pressure boundary condition is used at outlet of the bed. The detailed parameters including the material properties and operating conditions for simulation are summarized in Table 2.

Table 2. Summary of the simulation parameters.

Parameter	Value or Model
Bed height (m)	0.5
Bed width (m)	0.2
Minimum fluidization velocity (m/s)	0.14
Superficial gas velocity (m/s)	0.19
Total particle weight (g)	Sand: 2000 Char: 40
Particle volume fraction	Sand: 89% Char: 11%
Drag coefficient	Gidaspow
Granular viscosity	Syamlal-O'Brien
Granular bulk viscosity	Lun et al.
Restitution coefficient	e_{ss} = 0.7, 0.8, 0.9, 0.95
Wall boundary condition	No-slip

At the beginning of the computation, the sand particles and char particles are perfectly mixed at the indicated ratio. A time step of 5×10^{-4} s was employed in the simulation using Fluent 15.0 software. When TFM is used for the simulation, the grid size should be much smaller than the physical dimensions of the geometry. Meanwhile, it should also be bigger than the particle diameter, which will ensure that the solid phase can be treated as a continuous flow. Therefore, the grid size has a drastic effect on the flow behavior [22]. A coarse grid could lead to an overprediction of the solid expansion height of the bed [23]. Sande et al. [17] found that the numerical results agree well with the experimental results when the grid size is about 10 times the particle diameter and there is no improvement in capturing homogeneous expansion when the mesh is further refined. Hence, a grid size of 5 mm was chosen in this simulation.

4. Results and Discussion

4.1. Stationary Condition

In the fluidization of the binary mixture, the particles that sink at the bottom of the bed are known as jetsam, while those that accumulate at the top of the bed are known as flotsam. In this study, the sand particles are the jetsam particles, and the char particles are the flotsam particles.

For the case with e_{ss} = 0.9, the time evolution of particle volume fraction in different layers along the height direction is monitored to obtain the statistical steady-state for solution, as shown in Figure 3. As time increases, the flotsam moves upwards while the jetsam moves downwards. Hence, with increase in time, the jetsam volume fraction increases while the flotsam volume fraction decreases at the height of 0.053 m, and the jetsam volume fraction decreases while the flotsam volume fraction increases at the height of 0.153 m. At the beginning of fluidization, the volume fractions vary rapidly, whereas the values change slowly after 30 s, corresponding to complete fluidization. Hence, the time-averaged variables are computed between 30 s and 50 s. Moreover, the result shows that the simulation time corresponding to the stationary state should be increased with increase in restitution coefficient. For the cases with e_{ss} = 0.7 and 0.8, the simulation time of 50 s is long enough. However, for e_{ss} = 0.95, the simulation has not converged to a stationary state until the time exceeds 65 s. Hence, for the case with e_{ss} = 0.95, the time-averaged variables are computed between 65 s and 85 s. When the restitution coefficient increases, it means that there is lesser dissipation of kinetic energy of particle, due to a more significant elastic particle-particle collision. This may be the reason why a longer simulation time is required before the mixing pattern reaches a stationary state when the restitution coefficient increases.

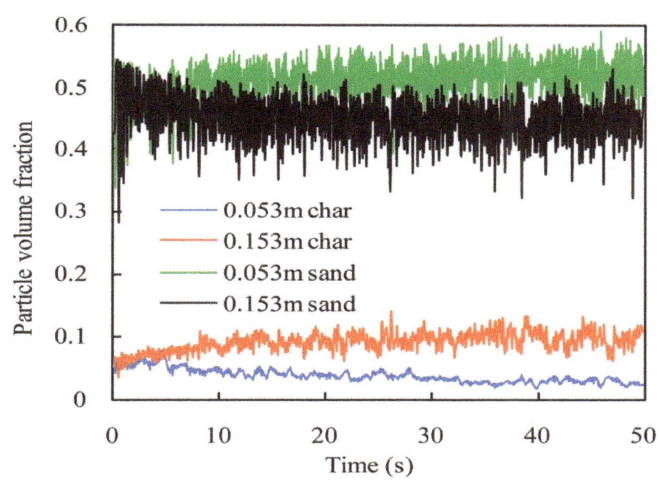

Figure 3. Time profiles of the particle volume fraction in different layers along the height direction.

4.2. Particle Velocity and Flow Pattern

Figure 4 shows the lateral profiles of time-averaged axial velocity of the flotsam at the bed height of 0.153 m. In this figure, the velocity profiles of the flotsam for e_{ss} = 0.7, 0.8, 0.9, and 0.95 are similar. This clearly shows that the axial velocity of the flotsam is directed upward in the center of the bed and the axial velocity at the wall is nearly equal to zero, which represents the characteristic of the no-slip wall boundary condition.

In order to reflect the movement relevance between the jetsam and flotsam, in the case of e_{ss} = 0.9, the predicted flow patterns for the flotsam and jetsam are shown in Figures 5 and 6, respectively. The flow patterns for the flotsam and jetsam are quite similar. Two vortices show at the bottom of the bed and two vortices close to the top of the bed. At the bottom of the bed, the particles generally rise towards the wall, whereas they fall down towards the central region. However, at the top of the bed, the movement direction of particles is reversed. The flow pattern obtained in this study is very different from the result obtained by other researchers [24]. In those studies, only two vortices can be found throughout the entire region of the dense bed layer. However, in this study, four vortices are obtained in the simulation. Moreover, the two vortices at the bottom of the bed are relatively close to the centerline of the bed, whereas the two vortices at the top of the bed are generated near the wall. It can also be seen that the distance between the two vortices at the top of the bed is farther than the value for the vortices at the bottom of the bed. This phenomenon might be due to the relationship between the binary particles in density and volume fraction. When the density and volume fraction between the binary particles changes not much [14,24], the segregation of particles does not play a leading role in fluidization process. Hence, in the top and bottom regions of the bed, uniform patterns can be found. However, in this study, a binary particles system with a huge difference between particles in density and volume fraction is used, which causes a much more notable segregation. Hence, the flow pattern of particles at the top region of the bed is different from the result for the bottom region.

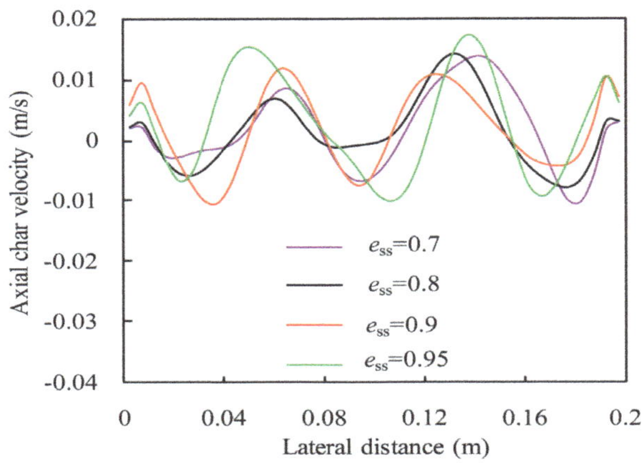

Figure 4. Lateral profiles of the time-averaged axial velocity of the flotsam at Z = 0.153 m.

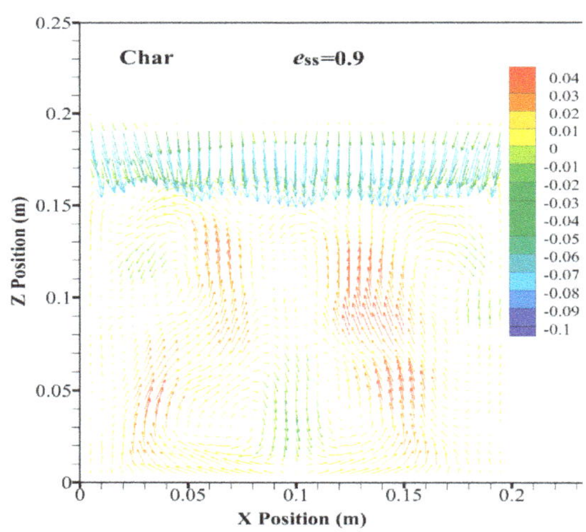

Figure 5. Time-averaged particle velocity distribution of the flotsam for e_{ss} = 0.9 (the color legend represents the axial velocity of the particles).

The small bubbles are generally generated near the wall and the larger bubbles form near the centerline region of the bed as small bubbles move upwards [25]. Based on the analysis of positive axial velocity of particles, from Figures 5 and 6, it clearly shows that the motion of particles obtained in the simulation has the same characteristics.

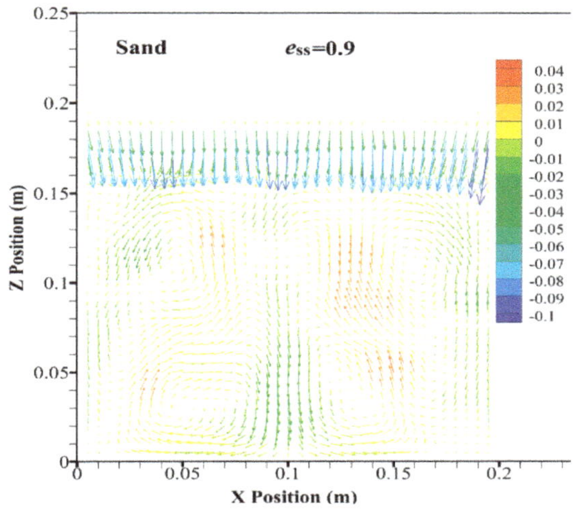

Figure 6. Time-averaged particle velocity distribution of the jetsam for e_{SS} = 0.9 (the color legend represents the axial velocity of the particles).

Due to the similar flow patterns for the flotsam and jetsam, as shown in Figures 5 and 6, in the cases with same restitution coefficient, the lateral profile of the time-averaged axial velocity of the jetsam at Z = 0.153 m is similar to the profile for the flotsam. Some studies [26] have shown that the central region of the profile for jetsam changed more gently and the minimum value appeared near the wall when the layer height was at the top of the bed, which was above the jetsam-rich layer. The shape of the profile mentioned in these studies is similar with the shape of the curve obtained in the case of e_{SS} = 0.95 in this study, as shown in Figure 4.

In the cases of e_{SS} = 0.7, 0.8, and 0.95, the flow patterns of the flotsam are shown in Figures 7–9, respectively. The vortex structures for different restitution coefficients are similar. However, due to the lesser dissipation of granular energy, with the increase in restitution coefficient, the vortex intensity increases. At the bottom of the bed, due to the effect of the inlet gas, the restitution coefficient does not much affect the vortex intensity. However, for the two vortices at the top of the bed, with increase in the height of the bed, the dissipation of granular energy is accumulated. Hence, the effect of restitution coefficient on the flow pattern of particles at the top of the bed is more evident. It can be seen that the vortex characteristic of the two vortices at the top of the bed, as shown in Figure 9, is more obvious when the restitution coefficient increases from 0.9 to 0.95.

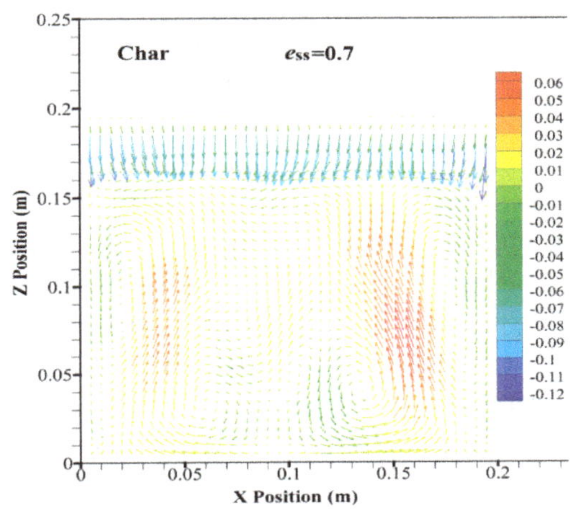

Figure 7. Time-averaged particle velocity distribution of the flotsam for e_{SS} = 0.7 (the color legend represents the axial velocity of the particles).

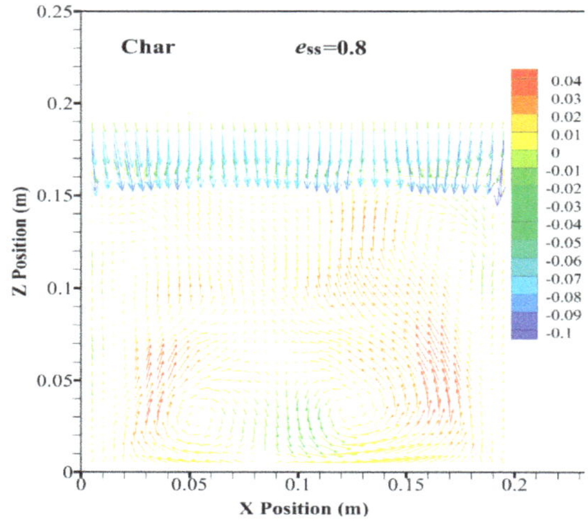

Figure 8. Time-averaged particle velocity distribution of the flotsam for e_{ss} = 0.8 (the color legend represents the axial velocity of the particles).

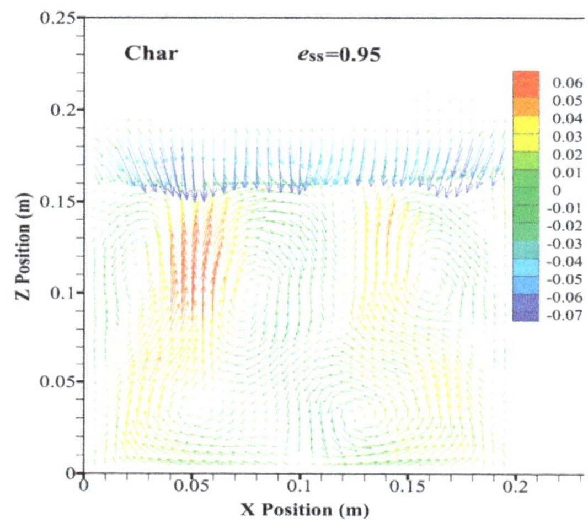

Figure 9. Time-averaged particle velocity distribution of the flotsam for e_{ss} = 0.95 (the color legend represents the axial velocity of the particles).

4.3. Granular Temperature Distribution

Granular temperature is a concept in the field of the kinetic theory of granular flow, and it is used for assessing fluctuating energy of the particles suspended in the gas flow. The lateral profiles of the time-averaged granular temperature of the flotsam at a height of 0.153 m are shown in Figure 10. When the restitution coefficient is no more than 0.9, the granular temperature of the flotsam closer to the wall is high and decreases towards the central region of the bed. Meanwhile, the curve becomes almost flat at the center of the bed. However, for the case of e_{ss} = 0.95, at the center of the bed, the curve is no longer flat. Two peaks of granular temperature can be found, which is related to the big values of the axial velocity, as shown in Figures 4 and 9. With restitution coefficient increases, the values of the granular temperature increase, since there is less dissipation of the randomly fluctuating kinetic energy of particles. For a restitution coefficient below than 0.9, the granular temperature increases slightly. However, the granular temperature increases dramatically when the restitution coefficient increases from 0.9 to 0.95, especially for the center region of the bed.

Figure 11 presents the axial profiles of the time-averaged granular temperature of the flotsam. In the region of the gas-solid interface, the granular temperature is very high, due to the collapse of bubbles and granular splash [27]. The value of granular temperature increases as the restitution coefficient increases. When the restitution coefficient increases from 0.7 to 0.8, the granular temperature along the axial direction increases slightly. However, with increase in restitution coefficient from 0.8 to 0.9, the granular temperature at the top of the bed increases obviously. Due to the similar flow patterns for the flotsam and jetsam, it can be seen than the effects of the restitution coefficient with different values on the granular temperature are also different.

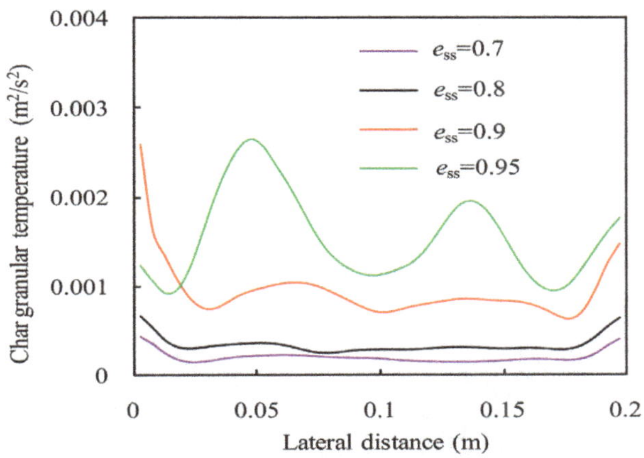

Figure 10. Lateral profiles of the time-averaged granular temperature of the flotsam at Z = 0.153 m.

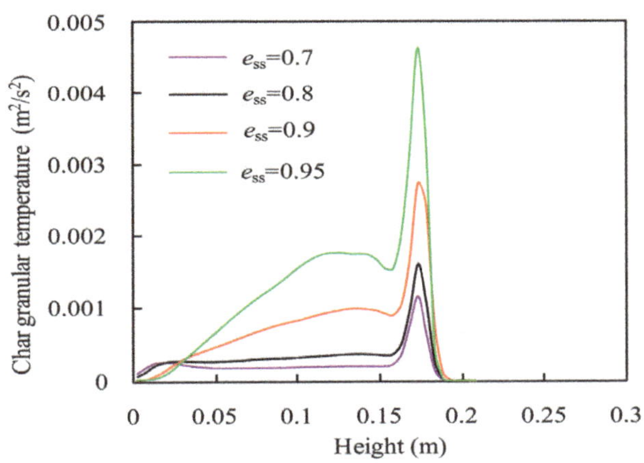

Figure 11. Axial profiles of the time-averaged granular temperature of the flotsam.

4.4. Distribution of Particle Volume Fraction

The effects of restitution coefficient on the instantaneous volume fraction of char and sand particles are shown in Figures 12 and 13, respectively. Closer to the gas-solid interface, the volume fraction of sand particles is relatively low, which is just the opposite of the value of the char particles. This shows that a strong segregation phenomenon exists in the fluidization process of a binary particle system. With increase in the restitution coefficient, more char particles accumulate at the top of the bed, which shows that the tendency of segregation is more remarkable.

More studies [26,28] are concerned with the distribution characteristics of jetsam, however, in the gasification process, the flotsam will react with gas. Hence, the study on distribution characteristics of flotsam is much more significant. Lateral distributions of the time-averaged volume fraction of the flotsam at a height of 0.153 m are plotted in Figure 14. For e_{ss} = 0.7, 0.8, and 0.9, the effect of the restitution coefficient on the volume fraction of the flotsam is not obvious, hence, the profiles are very similar. However, when the restitution coefficient increases to 0.95, the values of volume fraction increase dramatically.

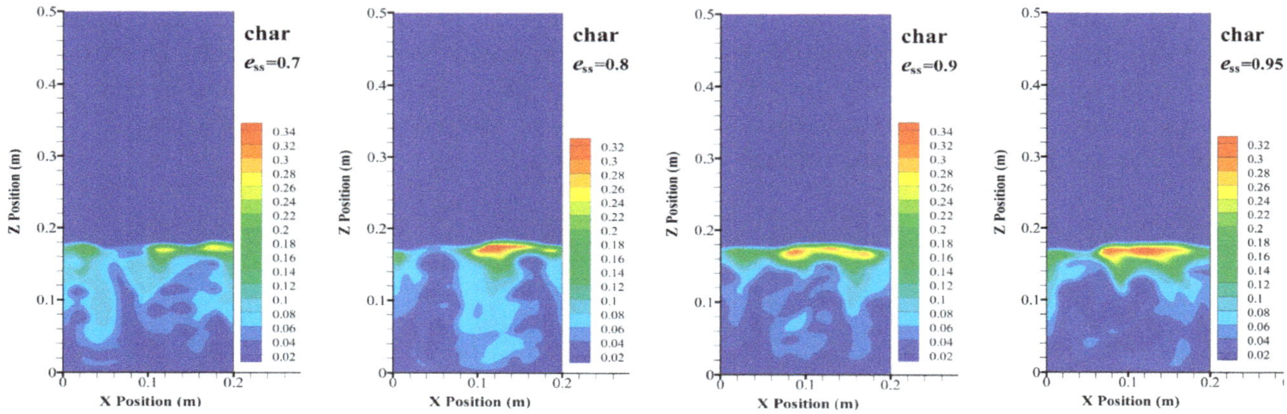

Figure 12. Distributions of instantaneous volume fraction of char particles.

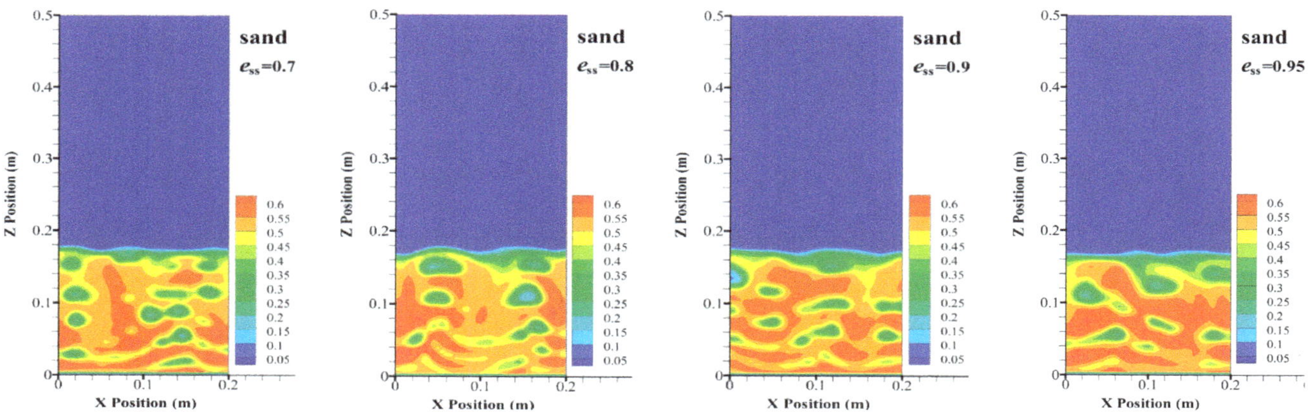

Figure 13. Distributions of instantaneous volume fraction of sand particles.

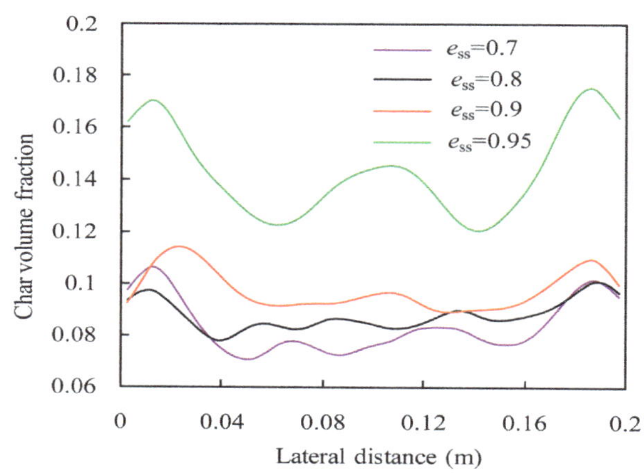

Figure 14. Lateral profiles of the time-averaged particle volume fraction of the flotsam at Z = 0.153 m.

The effect of the restitution coefficient on the distribution of char mass fraction in the axial direction is shown in Figure 15. The char mass fraction is defined as the ratio of the char mass corresponding to the specific height range to total char mass in the bed. With increase in restitution coefficient, the char mass fraction in the upper part of the bed gradually increases, whereas the char mass fraction in the lower part of the bed decreases. In the case of e_{ss} = 0.95, the predicted profile well fits the experimental result. In the binary particles system with less difference between particles in density or volume fraction [12], they found that there was a small difference between the simulation results and the experimental data only for low restitution coefficient such as 0.8. However, for the binary particles system with huge difference in density and volume fraction between particles, the predicted degree of segregation compared with experimental result may be too low, when a relatively small value of restitution coefficient is adopted. From Figure 15, it can be seen that the mass fraction profile of char particles along the height direction for $e_{ss} = 0.95$ is quite close to the experimental result.

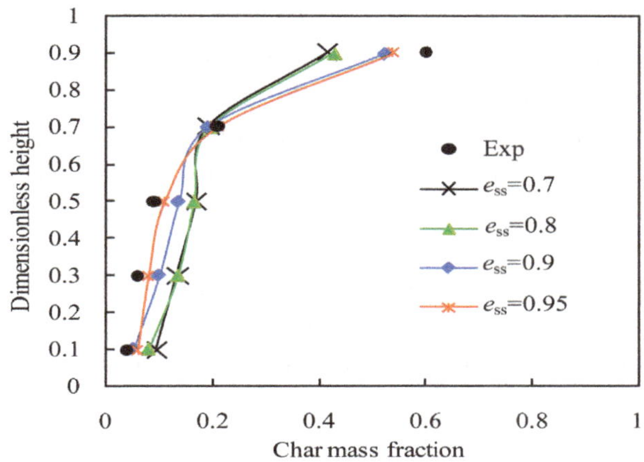

Figure 15. Effect of restitution coefficient on the axial mass fraction profile of the flotsam.

Figure 16 presents the axial profiles of the time-averaged particle volume fraction of the jetsam. Initially, the particle volume fraction along the bed height shows a slight increase, due to the effect of the inlet gas, and then gradually decreases. Closer to the gas-solid interface, the particle volume fraction decreases more sharply. This shows that the splash zone formed by the collapse of bubbles is a small area, which is consistent with previous studies [29,30]. The result from Figure 16 shows that the restitution coefficient had little effect on the axial profile of the jetsam volume fraction.

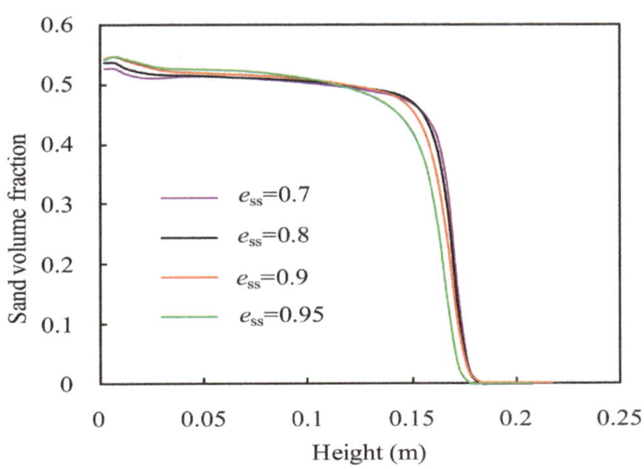

Figure 16. Axial profiles of the time-averaged particle volume fraction of the jetsam.

There are some studies about the comparison of results between 2D simulation and 3D simulation in the rectangular fluidized bed for monoparticle flow. For a pseudo-2D fluidized bed [31], the result has shown that no major differences are observed between 2D and 3D simulations in predicting the mean pressure drop and bed expansion. Just for bubble diameter and rise velocity, the 3D simulations are better agreement with experiments than the corresponding 2D simulations, whereas, for bubble aspect ratio, the 2D simulation has a better agreement with the experimental data. Xie et al. [32] also used an Eulerian-Eulerian model to investigate the differences between 2D and 3D simulations of a rectangular fluidized bed. They found that for a bubbling fluidized bed a satisfactory qualitative agreement between 2D and 3D simulations is observed. Hence, for monoparticle flow, though 2D simulations have certain limitations, they can provide reasonable results compared to experimental observations. Meanwhile, due to the requirement for lower computational resources, 2D simulation is widely used. However, for a binary mixture system in the rectangular fluidized bed, there are few related studies. Geng et al. [33] investigated the difference between the results of 2D and 3D simulations for a binary mixture system in a pseudo-2D rectangular bubbling fluidized bed. The results showed that the flotsam (coal particles) is nearly constant along the height direction, which totally deviates from the experimental observation. This means that 2D simulation is not suitable for modeling a pseudo-2D fluidized bed, which is entirely different from the conclusion for monoparticle flow. They also found that when the thickness of the rectangular bed was larger than a critical value (20 mm in this study), 2D simulation can provide a reasonable result. To sum up, the difference between 2D simulation and 3D simulation might change with the composition of the particle system. For a binary particle system, the relationship between the critical value in the thickness direction of the rectangular bed and the dimension parameters of the bed or the flow parameters may be important, and how to determine the relationship may be a key task in a follow-up study.

5. Conclusions

In the present work, the effect of the restitution coefficient on the numerical results for a binary particles system in a bubbling fluidized bed with the huge difference between the particles in terms of density and volume fraction has been studied based on two-fluid model along with the kinetic theory of granular flow. The effect of the restitution coefficient on the particle velocity, particle flow patterns, and mass fraction distribution varies in the different regions of the bed. At the bottom of the bed, the restitution coefficient does not affect the flow characteristic of particles significantly. However, in the top region of the bed, due to the cumulative effect of the dissipation of granular energy, the restitution coefficient has an obvious influence on the flow characteristic of the particles.

With an increase in the restitution coefficient, the degree of segregation increases. However, it does not change linearly with the restitution coefficient. Considering the effect of the restitution coefficient on the degree of segregation and flow pattern of particles in the top region of the bed, the restitution coefficient can be categorized into two classes: restitution coefficients of 0.7 and 0.8 can be included in one class, whereas the restitution coefficients of 0.9 and 0.95 can be included in another class.

For a binary particles system with the huge density and volume fraction difference between the particles, two vortices at the bottom of the bed and two vortices at the top of the bed are observed in the flow pattern distribution. The time-averaged flow pattern of particles in this study is very different from the result obtained for the system with similar values between particles in density and volume fraction, in which only two vortices can be found in the entire region of the dense bed layer.

Author Contributions: Xinjun Zhao did the simulation and wrote the manuscript. Qitai Eri and Qiang Wang supervised the investigation and checked the model.

References

1. Fan, X.; Yang, Z.; Parker, D.J. Impact of solid sizes on flow structure and particle motions in bubbling fluidization. *Powder Technol.* **2011**, *206*, 132–138. [CrossRef]
2. Mohd Salleh, M.A.; Kisiki, N.H.; Yusuf, H.M.; Ab Karim Ghani, W.A.W. Gasification of Biochar from Empty Fruit Bunch in a Fluidized Bed Reactor. *Energies* **2010**, *3*, 1344–1352. [CrossRef]
3. Liu, H.; Zhao, Y.; Ding, J.; Gidaspow, D.; Wei, L. Investigation of mixing/segregation of mixture particles in gas-solid fluidized beds. *Chem. Eng. Sci.* **2007**, *62*, 301–317.

4. Coroneo, M.; Mazzei, L.; Lettieri, P.; Paglianti, A.; Montante, G. CFD prediction of segregating fluidized bidisperse mixtures of particles differing in size and density in gas-solid fluidized beds. *Chem. Eng. Sci.* **2011**, *66*, 2317–2327. [CrossRef]
5. Sun, J.; Wang, J.; Yang, Y. CFD investigation of particle fluctuation characteristics of bidisperse mixture in a gas-solid fluidized bed. *Chem. Eng. Sci.* **2012**, *82*, 285–298. [CrossRef]
6. Loha, C.; Chattopadhyay, H.; Chatterjee, P.K. Euler-Euler CFD modeling of fluidized bed: Influence of specularity coefficient on hydrodynamic behavior. *Particuology* **2013**, *11*, 673–680. [CrossRef]
7. Bakshi, A.; Altantzis, C.; Bates, R.B.; Ghoniem, A.F. Eulerian–Eulerian simulation of dense solid–gas cylindrical fluidized beds: Impact of wall boundary condition and drag model on fluidization. *Powder Technol.* **2015**, *277*, 47–62. [CrossRef]
8. Wang, Y.; Zou, Z.; Li, H.; Zhu, Q. A new drag model for TFM simulation of gas-solid bubbling fluidized beds with Geldart-B particles. *Particuology* **2014**, *15*, 151–159. [CrossRef]
9. Johnson, P.C.; Jackson, R. Frictional-collisional constitutive relations for granular materials with application to plane shearing. *J. Fluid Mech.* **1987**, *176*, 67–93. [CrossRef]
10. Chao, Z.; Wang, Y.; Jakobsen, J.P.; Fernandino, M.; Jakobsen, H.A. Derivation and validation of a binary multi-fluid Eulerian model for fluidized beds. *Chem. Eng. Sci.* **2011**, *66*, 3605–3616. [CrossRef]
11. Bai, W.; Keller, N.K.G.; Heindel, T.J.; Fox, R.O. Numerical study of mixing and segregation in a biomass fluidized bed. *Powder Technol.* **2013**, *237*, 355–366. [CrossRef]
12. Tagliaferri, C.; Mazzei, L.; Lettieri, P.; Marzocchella, A.; Olivieri, G.; Salatino, P. CFD simulation of bubbling fluidized bidisperse mixtures: Effect of integration methods and restitution coefficient. *Chem. Eng. Sci.* **2013**, *102*, 324–334. [CrossRef]
13. Mostafazadeh, M.; Rahimzadeh, H.; Hamzei, M. Numerical analysis of the mixing process in a gas-solid fluidized bed reactor. *Powder Technol.* **2013**, *239*, 422–433. [CrossRef]
14. Zhong, H.; Gao, J.; Xu, C.; Lan, X. CFD modeling the hydrodynamics of binary particle mixtures in bubbling fluidized beds: Effect of wall boundary condition. *Powder Technol.* **2012**, *230*, 232–240. [CrossRef]
15. Xue, Q.; Heindel, T.J.; Fox, R.O. A CFD model for biomass fast pyrolysis in fluidized-bed reactors. *Chem. Eng. Sci.* **2011**, *66*, 2440–2452. [CrossRef]
16. Askarishahi, M.; Salehi, M.-S.; Molaei Dehkordi, A. Numerical investigation on the solid flow pattern in bubbling gas–solid fluidized beds: Effects of particle size and time averaging. *Powder Technol.* **2014**, *264*, 466–476. [CrossRef]
17. Sande, P.C.; Ray, S. Mesh size effect on CFD simulation of gas-fluidized Geldart A particles. *Powder Technol.* **2014**, *264*, 43–53. [CrossRef]
18. Wang, Y.; Chao, Z.; Jakobsen, H.A. A sensitivity study of the two-fluid model closure parameters (β, e) Determing the main gas-solid flow pattern characteristics. *Ind. Eng. Chem. Res.* **2010**, *49*, 3433–3441. [CrossRef]
19. Zhong, H.; Lan, X.; Gao, J.; Zheng, Y.; Zhang, Z. The difference between specularity coefficient of 1 and no-slip solid phase wall boundary conditions in CFD simulation of gas-solid fluidized beds. *Powder Technol.* **2015**, *286*, 740–743. [CrossRef]
20. Lun, C.C.K.; Savage, S.B.; Jeffrey, D.J.; Chepurniy, N. Kinetic theories for granular flow: Inelastic particles in Couette flow and slightly inelastic particles in a general flow field. *J. Fluid Mech.* **1984**, *140*, 223–256. [CrossRef]
21. Park, H.C.; Choi, H.S. The segregation characteristics of char in a fluidized bed with varying column shapes. *Powder Technol.* **2013**, *246*, 561–571. [CrossRef]
22. Gelderbloom, S.J.; Gidaspow, D.; Lyczkowski, R.W. CFD Simulations of Bubbling/Collapsing Fluidized Beds for Three Geldart Groups. *AIChE J.* **2003**, *49*, 844–858. [CrossRef]
23. Chen, J.; Yu, G.; Dai, B.; Liu, D.; Zhao, L. CFD simulation of a bubbling fluidized bed gasifier using a bubble-based drag model. *Energy Fuels* **2014**, *28*, 6351–6360. [CrossRef]
24. Chao, Z.; Wang, Y.; Jakobsen, J.P.; Fernandino, M.; Jakobsen, H.A. Multi-fluid modeling of density segregation in a dense binary fluidized bed. *Particuology* **2012**, *10*, 62–71. [CrossRef]
25. Verma, V.; Deen, N.G.; Padding, J.T.; Kuipers, J.A.M. Two-fluid modeling of three-dimensional cylindrical gas–solid fluidized beds using the kinetic theory of granular flow. *Chem. Eng. Sci.* **2013**, *102*, 227–245. [CrossRef]

26. Zhong, H.; Lan, X.; Gao, J.; Xu, C. Effect of particle frictional sliding during collisions on modeling the hydrodynamics of binary particle mixtures in bubbling fluidized beds. *Powder Technol.* **2014**, *254*, 36–43. [CrossRef]
27. Yang, S.; Luo, K.; Fang, M.; Fan, J. LES–DEM investigation of the solid transportation mechanism in a 3-D bubbling fluidized bed. Part II: Solid dispersion and circulation properties. *Powder Technol.* **2014**, *256*, 395–403. [CrossRef]
28. Di Renzo, A.; Di Maio, F.P.; Girimonte, R.; Vivacqua, V. Segregation direction reversal of gas-fluidized biomass/inert mixtures—Experiments based on particle segregation model predictions. *Chem. Eng. J.* **2015**, *262*, 727–736. [CrossRef]
29. Li, T.; Grace, J.; Bi, X. Study of wall boundary condition in numerical simulations of bubbling fluidized beds. *Powder Technol.* **2010**, *203*, 447–457. [CrossRef]
30. Zhao, Y.; Lu, B.; Zhong, Y. Influence of collisional parameters for rough particles on simulation of a gas-fluidized bed using a two-fluid model. *Int. J. Multiph. Flow* **2015**, *71*, 1–13. [CrossRef]
31. Asegehegn, T.W.; Schreiber, M.; Krautz, H.J. Influence of two- and three-dimensional simulations on bubble behavior in gas-solid fluidized beds with and without immersed horizontal tubes. *Powder Technol.* **2012**, *219*, 9–19. [CrossRef]
32. Xie, N.; Battaglia, F.; Pannala, S. Effects of using two-versus three-dimensional computational modeling of fluidized beds. *Powder Technol.* **2008**, *182*, 1–13. [CrossRef]
33. Geng, S.; Jia, Z.; Zhan, J.; Liu, X.; Xu, G. CFD modeling the hydrodynamics of binary particle mixture in pseudo-2D bubbling fluidized bed: Effect of model parameters. *Powder Technol.* **2016**, *302*, 384–395. [CrossRef]

Permissions

All chapters in this book were first published in MDPI; hereby published with permission under the Creative Commons Attribution License or equivalent. Every chapter published in this book has been scrutinized by our experts. Their significance has been extensively debated. The topics covered herein carry significant findings which will fuel the growth of the discipline. They may even be implemented as practical applications or may be referred to as a beginning point for another development.

The contributors of this book come from diverse backgrounds, making this book a truly international effort. This book will bring forth new frontiers with its revolutionizing research information and detailed analysis of the nascent developments around the world.

We would like to thank all the contributing authors for lending their expertise to make the book truly unique. They have played a crucial role in the development of this book. Without their invaluable contributions this book wouldn't have been possible. They have made vital efforts to compile up to date information on the varied aspects of this subject to make this book a valuable addition to the collection of many professionals and students.

This book was conceptualized with the vision of imparting up-to-date information and advanced data in this field. To ensure the same, a matchless editorial board was set up. Every individual on the board went through rigorous rounds of assessment to prove their worth. After which they invested a large part of their time researching and compiling the most relevant data for our readers.

The editorial board has been involved in producing this book since its inception. They have spent rigorous hours researching and exploring the diverse topics which have resulted in the successful publishing of this book. They have passed on their knowledge of decades through this book. To expedite this challenging task, the publisher supported the team at every step. A small team of assistant editors was also appointed to further simplify the editing procedure and attain best results for the readers.

Apart from the editorial board, the designing team has also invested a significant amount of their time in understanding the subject and creating the most relevant covers. They scrutinized every image to scout for the most suitable representation of the subject and create an appropriate cover for the book.

The publishing team has been an ardent support to the editorial, designing and production team. Their endless efforts to recruit the best for this project, has resulted in the accomplishment of this book. They are a veteran in the field of academics and their pool of knowledge is as vast as their experience in printing. Their expertise and guidance has proved useful at every step. Their uncompromising quality standards have made this book an exceptional effort. Their encouragement from time to time has been an inspiration for everyone.

The publisher and the editorial board hope that this book will prove to be a valuable piece of knowledge for researchers, students, practitioners and scholars across the globe.

List of Contributors

Michalina Kurkus-Gruszecka and Piotr Krawczyk
Institute of Heat Engineering, Faculty of Power and Aeronautical Engineering, Warsaw University of Technology, Nowowiejska 21/25, 00-665 Warsaw, Poland

Chao Yu, Xiangyao Xue, Kui Shi, Mingzhen Shao and Yang Liu
Changchun Institute of Optics, Fine Mechanics and Physics, Chinese Academy of Sciences, Changchun 130033, China

Vidya Chandran
Department of Mechanical Engineering, Karunya Institute of Technology and Sciences, Coimbatore, Tamil Nadu 600018, India

Sekar M.
Department of Mechanical Engineering, AAA College of Engineering and Technology, Sivakasi, Tamil Nadu 600018, India

Sheeja Janardhanan
Department of Mechanical Engineering, SCMS School of Engineering and Technology, Ernakulam, Kerala 673307, India

Varun Menon
Department of Computer Science and Engineering, SCMS School of Engineering and Technology, Ernakulam, Kerala 673307, India

Yixiang Liao
Gesellschaft für Anlagen-und Reaktorsicherheit, Boltzmannstraße 14, 85748 Garching bei München, Germany
Helmholtz-Zentrum Dresden-Rossendorf, Institute of Fluid Dynamics, Bautzner Landstraße 400, 01328 Dresden, Germany

Dirk Lucas
Helmholtz-Zentrum Dresden-Rossendorf, Institute of Fluid Dynamics, Bautzner Landstraße 400, 01328 Dresden, Germany

Mingyuan Zhang
State Key Laboratory of Ocean Engineering, Shanghai Jiao Tong University, Shanghai 200240, China

Guocheng Zhao, Longfei Xiao and Tao Peng
State Key Laboratory of Ocean Engineering, Shanghai Jiao Tong University, Shanghai 200240, China
Collaborative Innovation Center for Advanced Ship and Deep-Sea Exploration, Shanghai 200240, China

S. M. Sayeed-Bin-Asad, Tord Staffan Lundström and Anders Gustav Andersson
Division of Fluid and Experimental Mechanics, Luleå University of Technology, SE-971 87 Luleå, Sweden

Zhongchao Zhao, Yong Zhang, Yanrui Zhang, Yimeng Zhou and Hao Hu
School of Energy and Power, Jiangsu University of Science and Technology, Zhenjiang 212000, China

Lijuan Qian and Fubing Bao
College of Mechanical and Electrical Engineering, China Jiliang University, Hangzhou 310018, China

Jianzhong Lin
College of Mechanical and Electrical Engineering, China Jiliang University, Hangzhou 310018, China
State Key Laboratory of Fluid Power and Mechatronic Systems, Zhejiang University, Hangzhou 310027, China

Di Zhu, Ruofu Xiao and Fujun Wang
College of Water Resources and Civil Engineering, China Agricultural University, Beijing 100083, China
Beijing Engineering Research Center of Safety and Energy Saving Technology for Water Supply Network System, China Agricultural University, Beijing 100083, China

Ran Tao
Beijing Engineering Research Center of Safety and Energy Saving Technology for Water Supply Network System, China Agricultural University, Beijing 100083, China
Department of Energy and Power Engineering, Tsinghua University, Beijing 100083, China

Christian Suárez and José Antonio Salva
AICIA, Andalusian Association for Research & Industrial Cooperation, Camino de los Descubrimientos s/n, Edf. Escuela Superior de Ingenieros de Sevilla, 41092 Seville, Spain

Alfredo Iranzo, Elvira Tapia and José Guerra
Thermal Engineering Group, Energy Engineering Department, School of Engineering, University of Seville, Camino de los Descubrimientos s/n, 41092 Seville, Spain

Gonzalo Barea
Hispacold, C/Pino Alepo 1, Polígono Industrial El Pino, 41016 Seville, Spain

Wenjiao Qi, Qinghua Deng, Yu Jiang, Qi Yuan and Zhenping Feng
Shaanxi Engineering Laboratory of Turbomachinery and Power Equipment, Institute of Turbomachinery, School of Energy and Power Engineering, Xi'an Jiaotong University, Xi'an 710049, China

Rukshan Jayathilake and Souman Rudra
Department of Engineering and Sciences, University of Agder, Grimstad 4879, Norway

Younggi Park, Youngman Lee, Junyoung Na and Hong Sun Ryou
School of Mechanical Engineering, Chung-Ang University, 82, Heukseok-ro, Dongjak-gu, Seoul 06974, Korea

Xinjun Zhao, Qitai Eri and Qiang Wang
School of Energy and Power Engineering, Beihang University, Beijing 100191, China

Index